# PLANET FORMATION
Theory, Observations, and Experiments

It is just over ten years since the first planet outside our Solar System was detected. Since then, much work has been done to try to understand how extrasolar planets may form. This volume addresses fundamental questions concerning the formation of planetary systems in general, and of our Solar System in particular. Drawing from recent advances in observational, experimental, and theoretical research, it summarizes our current understanding of the planet formation processes, and addresses major open questions and research issues. Chapters are written by leading experts in the field of planet formation and extrasolar planet studies. The book is based on a meeting held in 2004 at Ringberg Castle in Bavaria, where experts gathered together to present and exchange their ideas and findings. It is a comprehensive resource for graduate students and researchers, and is written to be accessible to newcomers to the field.

The Cambridge Astrobiology series aims to facilitate the communication of recent advances in astrobiology, and to foster the development of scientists conversant in the wide array of disciplines needed to carry astrobiology forward. Books in the series are at a level suitable for graduate students and researchers, and are written to be accessible to scientists working outside the specific area covered by the book.

HUBERT KLAHR and WOLFGANG BRANDNER are both at the Max-Planck-Institute for Astronomy in Heidelberg. Hubert Klahr is Head of the Theory Group for Planet and Star Formation, and Wolfgang Brandner is a staff researcher and Head of the Adaptive Optics Lab.

T0210653

Cambridge Astrobiology

**Series Editors**

Bruce Jakosky, Alan Boss, Frances Westall, Daniel Prieur, and Charles Cockell

*Books in the series:*

1. Planet Formation: Theory, Observations, and Experiments
   Edited by Hubert Klahr and Wolfgang Brandner
   ISBN-10   0-521-86015-6
   ISBN-13   978-0-521-86015-4

# PLANET FORMATION

## Theory, Observations, and Experiments

*Edited by*

### HUBERT KLAHR

*Max-Planck-Institut für Astronomie, Heidelberg, Germany*

### WOLFGANG BRANDNER

*Max-Planck-Institut für Astronomie, Heidelberg, Germany*

CAMBRIDGE
UNIVERSITY PRESS

CAMBRIDGE UNIVERSITY PRESS
Cambridge, New York, Melbourne, Madrid, Cape Town, Singapore,
São Paulo, Delhi, Dubai, Tokyo, Mexico City

Cambridge University Press
The Edinburgh Building, Cambridge CB2 8RU, UK

Published in the United States of America by Cambridge University Press, New York

www.cambridge.org
Information on this title: www.cambridge.org/9780521180740

First published 2006
First paperback edition 2010

*A catalogue record for this publication is available from the British Library*

ISBN 978-0-521-86015-4 Hardback
ISBN 978-0-521-18074-0 Paperback

*Dedicated to the one and only true planet formation specialist*
*Slartibartfast of Magrathea and his father D. N. Adams.*

# Contents

# Preface

With the words "Twas the night before Christmas ..." does a good old story start. In December 2004, just a couple of days before Christmas, not three wise men but more than 60 wise men and women came to a castle in the Bavarian mountains. They traveled through a strong snow storm, but no-one turned back; all of them arrived. They had a noble goal in mind: to discuss the current understanding of the formation of planets. The meeting took place December 19–22, 2004 at the Ringberg castle of the Max-Planck-Society. Anyone who has had the chance to attend a meeting there knows what a friendly and stimulating atmosphere for a workshop it provides.

About a year beforehand we had called them, and now they came. The idea was to have a wonderful conference at the romantic Ringberg castle and to bring together theorists and observers, as well as meteoriticists and experimental astrophysicists. Only then, we thought, could we obtain a global picture of the ideas we have about how our planetary system came into life. We wanted to collect not only the accepted ideas, but also the speculative and competing ideas.

We were quickly convinced that this conference, unique in its composition, should generate a permanent record in the form of a proceedings book. But this book should not be just one more useless compendium of unrefereed papers, but should provide a concise and accurate picture of current planet formation theory, experiment, and observation. Based on a suggestion by Alan Boss, Cambridge University Press became interested in publishing the proceedings as part of its new astrobiology series. So we convinced some of the major league players in the planet-formation and extrasolar-planet business not only to come and give presentations but also to write overview chapters on their special field of expertise. These chapters were then publicly discussed at Ringberg and also refereed by some independent experts in the field.

Participants of the Ringberg workshop on planet formation December 19–22, 2004.

PLANET FORMATION addresses fundamental questions concerning the formation of planetary systems in general and of our Solar System in particular. Drawing from recent advances in observational, experimental, and theoretical research, it summarizes our current understanding of these processes and addresses major open questions and research issues. We want this book to be, for students and other newcomers to the field, a detailed summary which, if studied along with the references contained herein, will provide sufficient information to start their work in the field of solar and extrasolar planets. At the same time we want to explore the current understanding of the state of the art of this subject ten years after the detection of the first extrasolar planets around Sun-like stars. The chapters of this book have been written by the leading scientists in the field, who have made significant contributions to the subject of planets and their formation. We aim for a comprehensive and meaningful overview including observations of exoplanets and circumstellar disks, the latest findings about our own Solar System, experiments on grain growth, and finally on the competing theories on planet formation.

If we continue to attract the brightest physicists to this field of astrophysics we hope that one day we will reach our two-fold goal: first, to understand the origin of our Solar System and of our blue Mother Earth, the only place in the universe where we are sure about the existence of life, and second, to learn how exquisite or general the conditions for life are in the rest of the vast universe, based on our predictions on how many Earth-like, potentially life-harboring planets are out there.

# Acknowledgments

This book would have been impossible without the work of our contributors and referees. Thank you all! We are also indebted for the support here at the MPIA, foremost Anders Johansen for making "cvs" work. And most of all we want to thank Jacqueline Garget, our editor at Cambridge University Press, for her enthusiasm as well as patience during the production of the book.

# 1

# Historical notes on planet formation

*UCO/Lick Observatory, Santa Cruz*

## 1.1 Introduction

The history of planet formation and detection is long and complicated, and numerous books and review articles have been written about it, e.g. Boss (1998a) and Brush (1990). In this introductory review, we concentrate on only a few specific aspects of the subject, under the general assumption that the Kant–Laplace nebular hypothesis provides the correct framework for planet formation. The first recognized "theory" of planet formation was the vortex theory of Descartes, which, along with related subsequent developments, is treated in Section 1.2. Magnetic effects (Section 1.3) were of great significance in the solution of one of the major problems of the nebular hypothesis, namely, that it predicted a very rapidly rotating Sun. The early histories of the two theories of giant planet formation that are under current debate, the disk gravitational instability theory and the core accretion-gas capture theory, are discussed in Section 1.4 and Section 1.5, respectively. In the final section, 1.6, certain specific examples in the history of the search for extrasolar planets are reviewed.

## 1.2 Descartes and von Weizsäcker: vortices

Descartes (1644) gave an extensive discussion on the formation of the Earth, planets, and major satellites, the main idea of which was that they formed from a system of vortices that surrounded the primitive Sun, in three-dimensional space. His picture did not involve a disk, the rotation axes of the vortices were not all in the same direction, and the physical basis for it is elusive. In a section of his book entitled "Concerning the creation of all of the Planets" he states:

*Planet Formation: Theory, Observation, and Experiments*, ed. Hubert Klahr and Wolfgang Brandner.
Published by Cambridge University Press. © Cambridge University Press 2006.

"... *the extremely large space which now contains the vortex of the first heaven was formerly divided into fourteen or more vortices... So that since those three vortices which had at their centers those bodies that we now call the Sun, Jupiter, and Saturn were larger than the others; the stars in the centers of the four smaller vortices surrounding Jupiter descended toward Jupiter ...* "

It is not specified how the vortices got there and how the planets formed from them, and the whole treatise may be regarded as more philosophical than scientific. Furthermore, he had to be very careful in what he said to avoid disciplinary action from the Church.

Three hundred years later, C. F. von Weizsäcker (1944) wrote an influential paper in which he envisioned a turbulent disk of solar composition rotating around the Sun, which consisted of a set of stable vortices out of which planets formed by accretion of small particles. Stability required that the vortices be counter-rotating, in the co-rotating coordinate system, to the direction of the disk's rotation. The radii of the various vortex rings corresponded roughly to the Titius–Bode law of planetary spacing. However, von Weizsäcker's main contribution was to recognize that in a turbulent disk there would be angular momentum transport as a result of turbulent viscosity, with mass flowing inward to the central object and with angular momentum flowing to the outermost material, which would expand. This realization solved one of the major problems of the Kant–Laplace nebular hypothesis – that the Sun would be spinning too fast. However the physical reality of the eddy system was criticized by Kuiper (1951) on the grounds that true turbulence involved a range of eddy sizes according to the Kolmogorov spectrum and that the typical lifetime of an eddy was too short to allow planet formation in it.

Vortices were not really taken seriously for a long time after that, until Barge and Sommeria (1995) showed that small particles could easily be captured in vortices and this process would accelerate planet formation. Although the presence of long-lived vortices was not rigorously proved, they suggested that particles could accumulate in the vortices to form the cores of the giant planets in $10^5$ yr, thereby solving one of the major problems of the core accretion hypothesis. Later Klahr and Bodenheimer (2003), in a three-dimensional hydrodynamic simulation with radiation transfer, showed that under the proper conditions of baroclinic instability, vortices could form in low-mass disks. Further study of this process is strongly indicated.

### 1.3 Magnetic effects

Hoyle (1960) invoked magnetic braking to explain the slowly rotating Sun by transfer of angular momentum to the material that formed the planets. He claimed that purely hydrodynamic effects, such as viscosity, could not result in sufficient transfer of angular momentum because the frictional effect requires that the disk material must be in contact with the Sun itself, and that therefore the Sun could be slowed only to the point where it was in co-rotation with the inner disk. He envisioned a collapsing cloud that is stopped by rotational effects and forms a disk. A gap opens between the contracting Sun and the disk, and a magnetic field, spanning the gap, transfers angular momentum from the Sun to the inner edge of the disk, forcing it outward. As the Sun contracts and tends to spin up, angular momentum continues to be transferred until the inner edge of the disk is pushed out far enough so that its orbital period is comparable with the present rotation period of the Sun. The temperature has to be $\approx 1000$ K, to get magnetic coupling, and the field has to be $\approx 1$ gauss. Beyond the inner edge of the disk he does not require the magnetic field to transfer angular momentum to the outer regions of the disk; viscosity would work in that case. The terrestrial planets form from refractory material that condenses out near the inner edge of the disk and becomes decoupled from the gas.

Actually Alfvén (1954) was the one who originally invoked magnetic braking, although his idea of how the Solar System formed was not considered very plausible. His theory did not involve a disk, but rather clouds of neutral gas of different compositions which fall toward the Sun from random directions, stopping at a distance where the ionization energy equals the infall kinetic energy (the so-called "critical velocity" effect). Once ionized, the material couples to the Solar magnetic field and angular momentum is transferred to it, forcing it outward and eventually into a disk plane. The elements with the lowest ionization potentials, such as iron and silicon, stop farthest out. The cloud, composed of hydrogen, along with elements of similar ionization potential such as oxygen and nitrogen, is envisioned to stop in the region of the terrestrial planets, while a cloud composed mainly of carbon stops at distances comparable to the orbital distances of the giant planets. The theory was criticized on the grounds that it did not explain the chemical composition of the planets, but in fact the crucial aspect of it was the magnetic braking.

Today it is known that even young stars are slowly rotating, and that the interface between disk and star in a young system is very complicated, involving accretion from disk to star as well as outflow in a wind. Modern theories (Königl, 1991; Shu *et al.*, 1994) show that the basic angular momentum-loss mechanism for the central star is magnetic transfer. However relatively large fields are required, of the order of 1000 gauss.

## 1.4 Gravitational instability

We now consider early developments in the theory of planet formation. The basic condition needed for the formation of a planet by gravitational instability in a gaseous disk goes back to Jeans (1929): in a medium of uniform density $\rho$ and uniform sound speed $c_s$ a density fluctuation is unstable to collapse under self gravity if its wavelength $\lambda$ satisfies the condition

$$\lambda^2 > \frac{\pi c_s^2}{G\rho}. \tag{1.1}$$

Although the physical assumptions leading to the derivation were inconsistent, this criterion still gives the correct approximate conditions for gravitational collapse.

Kuiper (1951) suggested that giant planets could form by this mechanism; he combined the Jeans criterion with the condition for tidal stability of a fragment in the gravitational field of the central star. He estimated that planets formed this way would have masses of the order of 0.01 Solar masses ($M_\odot$) and that the disk would need a mass of $\approx 0.1\ M_\odot$. He retained von Weizsäcker's idea of turbulence in the disk, suggesting that *"Turbulence may be thought of as providing the initial density fluctuations and gravitational instability as amplifying them,"* an idea that has been revived in the modern theory of star formation in a turbulent interstellar cloud.

Safronov (1960) and Toomre (1964) rederived the Jeans condition in a flat disk, including differential rotation, gravity, and pressure effects. If the sound speed is $c_s$, the epicyclic frequency $\kappa$, and $\sigma$ the surface density of the disk (mass per unit area), then $Q = (c_s\kappa)/(\pi G\sigma) > 1$ for local stability to axisymmetric perturbations. Although Safronov was primarily interested in a flat disk of planetesimals, and Toomre was interested in a galactic disk of stars, and as stated, the derivation is valid only for axisymmetric perturbations, the "Toomre $Q$" is still a useful criterion even for stability to non-axisymmetric perturbations in gaseous disks. The critical value will depend on the equation of state and the details of the numerical code being used, but typically disks are stable if $Q > 1.5$.

Cameron (1969) later suggested that the protoplanetary disk, in the process of formation, could break up into axisymmetric rings which could then form planets by gravitational instability. There followed a series of evolutionary calculations for "giant gaseous protoplanets" which were assumed to have been formed by this mechanism (Bodenheimer, 1974; DeCampli and Cameron, 1979; Bodenheimer *et al.*, 1980). The general idea was that spherically symmetric condensations of approximately Jovian mass and Solar composition formed, in an unstable disk, with initial sizes of 1 to 2 AU, then contracted through an initial series of quasi-hydrostatic equilibria. The calculations involved the solution of the standard equations of stellar structure, including radiative and convective energy transport and grain opacities. The contraction phase lasts $2 \times 10^5$ yr for a protoplanet of 1.5 Jupiter masses ($M_J$) and $4 \times 10^6$ yr for a 0.3 $M_J$ protoplanet (Bodenheimer *et al.*, 1980). These times depend on the assumed grain opacities; interstellar grains were used in this particular calculation. Once the central temperature heats to 2000 K,

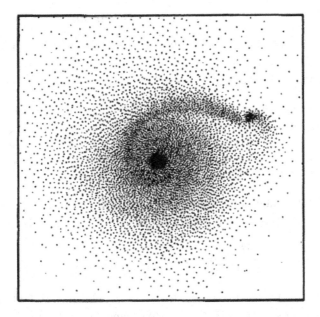

Fig. 1.1. Two-dimensional SPH simulation in the disk plane of gravitational insta-bility in an isothermal disk with mass equal to that in the central star. The particle positions are shown after slightly more than one disk rotation at its outer edge, which lies at about 100 AU from the star. Reprinted by permission from Adams and Benz (1992). ©Astronomical Society of the Pacific.

molecular dissociation sets in, leading to hydrodynamic collapse on a timescale of less than a year, with equilibrium regained at a radius only a few times larger than those of Jupiter and Saturn. DeCampli and Cameron (1979) were the first to make an estimate as to whether a solid core (deduced to be present in both Jupiter and Saturn now) could form during the early quasi-static equilibrium phase, finding that a 1 Earth mass ($M_\oplus$) core was possible only if the protoplanet mass was less than 1 $M_J$. This calculation of settling of solid material toward the center was followed up by Boss (1998b), who argued that a giant gaseous protoplanet of 1 $M_J$ could indeed form a core of a few $M_\oplus$, in line with present estimates of the core mass of Jupiter.

The first actual numerical simulation of gravitational instability in a gaseous disk in a situation relevant for planet formation was apparently done by Adams and Benz (1992), following a linear stability analysis by Shu *et al.* (1990). The calculation was done with a two-dimensional SPH code, the disk mass was equal to the mass of the central star, and the disk was assumed to be isothermal. The density was perturbed with an amplitude of 1% and an azimuthal wavenumber $m = 1$. The result was a one-armed spiral with a gravitationally bound knot (Fig. 1.1) of mass 1% that of the disk on an elliptical orbit; it was not determined whether the knot would survive for many orbits and evolve into a giant planet.

## 1.5 Core accretion: gas capture

Although the concept of a "planetesimal" had been discussed for a long time be-forehand (Chamberlin, 1903), Safronov (1969) was the first to give a fundamental and useful theory for the accretion of solid objects. He states:

"...despite the complexity of the accumulation process and the fact that fragmentation among colliding bodies was important, the process of growth of the largest bodies (the planetary 'embryos') can be described quantitatively in an entirely satisfactory manner if we assume that their growth resulted from the settling on them of significantly smaller bodies and that they were not fragmented during these collisions."

Thus his fundamental equation for the accretion rate of planetesimals onto a protoplanetary "embryo" was relatively simple. In its modern form,

$$\frac{\mathrm{d}M_{\mathrm{solid}}}{\mathrm{d}t} = \pi R_c^2 \sigma \Omega \left[ 1 + \left( \frac{v_e}{v} \right)^2 \right], \tag{1.2}$$

where $\pi R_c^2$ is the geometrical capture cross-section, $\Omega$ is the orbital frequency, $\sigma$ is the solid surface density in the disk, $v_e$ is the escape velocity from the embryo, and $v$ is the relative velocity of embryo and accreting planetesimal. The expression in brackets is known as $F_g$, the gravitational enhancement factor over the geometrical cross-section. An important requirement for a reasonable accretion timescale is that $F_g$ be large. However Safronov typically takes it in the range 7 to 11.

Safronov actually wrote the above equation as

$$\frac{\mathrm{d}M_{\mathrm{solid}}}{\mathrm{d}t} = \frac{4\pi (1 + 2\theta)}{P} \sigma_0 \left( 1 - \frac{m}{Q} \right) R_c^2, \tag{1.3}$$

where $\theta = (Gm)/(v^2 R_c)$ is known as the Safronov number, $m$ is the embryo mass, $Q$ is the present mass of the planet, $\sigma_0$ is the total initial solid surface density in the disk, and $P$ is the orbital period. In connection with the $m/Q$ factor he states:

"In the derivation of [the] formula ... for growth [times for terrestrial planets] it was assumed that the planetary zone was closed, or more precisely, that the total amount of solid material in the zone was conserved at all times and that its initial mass was equal to the present mass of the planet."

Thus effectively he has introduced the idea of the "minimum mass solar nebula" by requiring that the solid-surface density in the disk be just sufficient to correspond to the solid mass of the final planet.

Applying this assumption, he uses the equation to derive growth times. "Within 100 million years the Earth's mass must have grown to 98% of its present value," consistent with modern estimates of the growth time of the Earth. Detailed numerical calculations (Wetherill, 1980) of the formation of the terrestrial planets starting from roughly 100 lower-mass objects with low eccentricities spread out over the

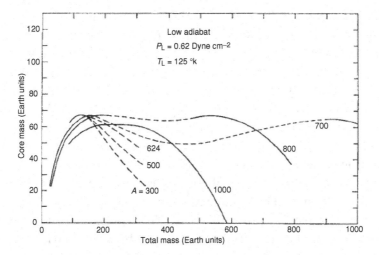

Fig. 1.2. The core mass as a function of total mass for a protoplanet consisting of a solid core plus a gaseous adiabatic envelope. Solid lines refer to hydrodynamically stable envelopes and dashed lines refer to unstable ones. $A$ is a parameter determined only by the distance of the planet from the central star, while $P_L$ and $T_L$ refer to the pressure and temperature, respectively, at the outer edge of the planet. Reprinted by permission from Perri and Cameron (1974). ©Academic Press.

terrestrial-planet zone gave this timescale, along with approximately the correct number of objects.

However Safronov also noted: *"It would appear ... that the distant planets (Uranus, Neptune, and Pluto) could not have managed to develop and use up all the matter within their zones within the lifetime of the solar system,"* a problem that is still not satisfactorily solved. In fact this statement actually led Cameron to pursue the gravitational instability hypothesis for the outer planets.

In connection with giant planet formation, Safronov mentions only briefly the process of gas capture: *"Effective accretion of gas by Jupiter and Saturn set in after they had attained a mass of about one to two Earth masses."* Cameron (1973) followed up on this remark with a statement to the effect that Jupiter could form by gas accretion once a solid core of about 10 $M_{\oplus}$ had been accumulated. He and Perri (1974) then made the first detailed calculation of a protoplanet consisting of a solid core and a gaseous envelope.

The Perri–Cameron model was assumed to be in strict hydrostatic equilibrium, to have a solid core of a given mass, and a gaseous envelope, assumed to be adiabatic, of Solar composition extending out to the Hill radius. The idea was to find structures that were dynamically unstable, implying that the gaseous envelope would rapidly collapse onto the core and that gas accretion would continue on a short timescale. They found that the structure was stable for values of the core mass up to a critical mass, above which it was unstable. Figure 1.2 shows the results for a particular choice of adiabat: the first maximum in the curve for a given value of the parameter $A$ corresponds to the core mass where the configuration becomes unstable. The parameter $A$ depends only on the distance from the Sun; relevant values of $A$ for the formation of Jupiter and Saturn are in the range 300 to 500. In this range the figure shows that the critical core mass is about 70 $M_\oplus$, too high as compared with current core mass determinations for Jupiter and Saturn. With a different reasonable choice for the adiabat, the critical core mass turns out to be even higher, about 115 $M_\oplus$. Nevertheless, the result does not depend sensitively on the distance, in approximate agreement with the properties of the giant planets.

This type of model was improved considerably by Mizuno (1980). He again constructed models in strict hydrostatic equilibrium, with a solid core and a gaseous envelope, extending outward to the Hill sphere. The boundary conditions for density and temperature at the outer edge were provided by a disk model. The energy equation was solved, given an energy source provided by planetesimals accreting through the envelope and landing on the core, at a fixed rate, for example $10^{-6}$ $M_\oplus$ per yr. The model was assumed to be in thermal equilibrium in the sense that the luminosity radiated was just equal to the rate at which the accreting planetesimals liberated gravitational energy. Also, energy transport by radiation and convection in the envelope was taken into account, with opacity assumed to be provided by the gas as well as grains. There were two main results from this calculation:

(1) He found a critical core mass, above which no model in strict hydrostatic equilibrium was possible. For interstellar grain opacity the value turned out to be 12 $M_\oplus$ (Fig. 1.3). It was assumed at the time that rapid gas accretion would follow, but later calculations (Pollack *et al.*, 1996) showed that in fact the gas accretion could be quite slow, particularly for core masses less than 10 $M_\oplus$. The models simply switched from strict hydrostatic equilibrium to quasi-hydrostatic equilibrium in which gravitational contraction of the envelope is of some importance.

(2) The value of the critical core mass turned out to be almost independent of the distance of the planet from the Sun. The fact that the value for the critical mass, as well as the independence on distance, agreed quite well with the estimates of the core masses of the giant planets at the time, put the core accretion–gas capture model on a fairly solid foundation.

## 1.6 Planet searches

The theories of planet formation that have just been discussed were formulated on the basis of the observed properties of the giant planets in the Solar System. Nevertheless, over the same time period that these theories were being developed, the search was on for extrasolar planets. The method used was primarily astrometry, in which an attempt is made to measure the periodic shift of the position of a star

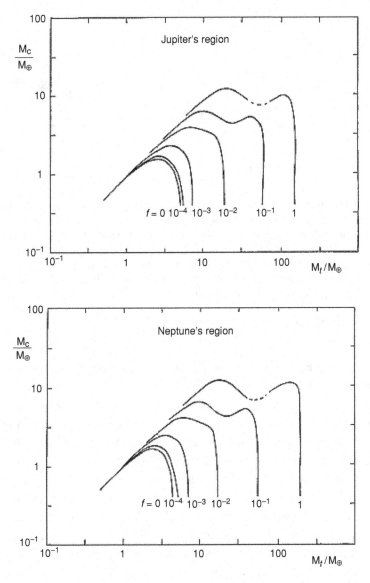

Fig. 1.3. The core mass as a function of total mass for a protoplanet consisting of a solid core plus a gaseous envelope calculated according to the equations of stellar structure. The masses are measured in units of $M_\oplus$, the Earth's mass. The parameter $f$ corresponds to the ratio of the assumed grain opacities to the interstellar values. Solid lines refer to hydrodynamically stable envelopes and dashed lines refer to unstable ones. *Top*: giant planet formation at 5 AU. *Bottom*: giant planet formation at 30 AU. The first maximum on each curve corresponds to the critical core mass. Reprinted by permission from Mizuno (1980). ©Progress of Theoretical Physics.

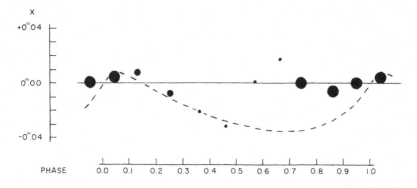

Fig. 1.4. The displacement of Barnard's star on the sky in the $x$-coordinate as a function of orbital phase, corresponding to a period of 24 years. The dashed line is a fit to the observations of van de Kamp (1963), while the filled circles are the measurements of Gatewood and Eichhorn (1973). The largest points have the highest degree of confidence. The solid line corresponds to zero displacement. Note that the deviation in $x$ that is looked for is only a few hundredths of a second of arc, while the image of the star on a photographic plate is more like 1–2 seconds across. Reprinted by permission from Gatewood and Eichhorn (1973). ©American Astronomical Society.

on the sky, caused by the gravitational effects of an orbiting planet. However as reported by Black (1991): *"The history of searches for other planetary systems is littered with published detections that vanish under further scrutiny."*

The most famous example is Barnard's star, named after the astronomer who discovered its large proper motion over the period 1894–1916. Peter van de Kamp measured, as accurately as he could, the position of the star on the sky, trying to deduce the presence of a low-mass companion by observing the periodic motion of the star around the center of mass of the system. His initial observational program ran from 1916 to 1962 and produced more than 2400 images. However, the apparent size of the star's orbit turned out to be only about 1/100 the size of the stellar images on a photographic plate. Nevertheless, he claimed (van de Kamp, 1963) to have found a planet, and the result was announced in the News and Views section of *Nature*. The companion was supposed to have a mass of 1.6 $M_J$, orbiting a star of 0.15 $M_\odot$ with a period of 24 yr, an eccentricity of 0.6, and a semimajor axis of 4.4 AU. However the system was measured independently by Gatewood and Eichhorn (1973) with a series of 241 plates, and they did not confirm the presence of the planet (Fig. 1.4). Adjustments to the telescope configuration at the Sproul Observatory were thought to have introduced spurious displacements of the star.

Nevertheless van de Kamp (1975) reanalyzed his data, using only that which was taken after the major adjustments had taken place, and now claimed the presence of two planets, with masses of 1 $M_J$ at a distance of 2.7 AU (11.5 yr period), and

0.4 $M_J$ at 4.2 AU (22 yr period), both on nearly circular orbits. However, because of the intrinsic errors in the astrometric observations, the astronomical community did not accept these results, and the presence of planets around this star is considered unconfirmed. On the other hand, it is not completely ruled out. Existing radial velocity measurements (Kürster *et al.*, 2003) do not reach the period range of 10–20 yr, but they do put an upper limit of 0.86 $M_J$ for any planet interior to 0.98 AU.

However, other techniques for planet detection have been developed. For example, a pulsar is a highly accurate clock. Thus a slight periodic variation in the clock period could represent the Doppler shift of the pulsar signal caused by its motion around the center of mass of the system, composed of itself plus an unseen low-mass companion. The quantity that is measured is actually the pulse-arrival time. This method also produced some initial false alarms. A famous example is the reported discovery of a planet in a circular orbit with a six-month period orbiting the neutron star PSR 1829 − 10 (Bailes *et al.*, 1991). Its mass was deduced to be about 10 $M_\oplus$.

Wolszczan (1993) comments:

*A potential of the pulse timing method to detect planets or planetary systems has been indicated in the past by reports on possible planet sized companions to the Crab pulsar ... and more recently to PSR 1829 – 10. In these and other similar cases, further studies have indicated that rather than originating from orbital motion, the observed periodic behavior of the pulse arrival times was caused by timing noise or by inaccuracies in the data analysis.*

In the case of PSR 1829 − 10, the error arose because the authors had failed to account for the small eccentricity of the Earth's orbit in reducing their observations of pulse arrival times.

Nevertheless, several theories, based on this announcement, were generated, whose purpose was to show how planets could still form in such an unexpected environment:

- Fallback of material from the supernova explosion that formed the neutron star results in a disk, rich in heavy elements out of which planets can form (Lin *et al.*, 1991).
- Two white dwarfs originally in a close binary system coalesce; much of the angular momentum remains in a low-mass disk, and the event results in the production of a neutron star (Podsiadlowski *et al.*, 1991).
- A neutron star collides with the central star of a pre-existing planetary system. The central star is disrupted and forms a short-lived extended envelope around the neutron star. The envelope then is dissipated and any inner planets are captured into orbits, which are circularized by interaction with the envelope, around the neutron star (Podsiadlowski *et al.*, 1991).
- The neutron star is in a binary system with a massive main-sequence companion star. The companion star expands to become a red giant and transfers some mass to a disk

around the neutron star. The companion later explodes in a supernova, the binary becomes unbound, but the disk remains around the neutron star. The outer parts of the disk cool and can form planets (Fabian and Podsiadlowski, 1991).

Although these turned out to be theories without a subject, a short time later another announcement was made (Wolszczan and Frail, 1992): they had discovered, from pulse timing with the Arecibo radio telescope, a system of two planets around the millisecond pulsar PSR1257 + 12. The planet masses were 3.4/sin $i$ and 2.8/sin $i$ $M_\oplus$, the distances from the pulsar were 0.36 and 0.47 AU, the orbital periods were 66.54 and 98.20 days, and the orbits were nearly circular with eccentricities in the range 0.020 and 0.024. Here $i$ is the unknown inclination of the plane of the orbit to the plane of the sky. In the discovery paper the authors suggested that the planets resulted from neutron star evolution in a low-mass binary system. The close-in low-mass main sequence companion, which was in orbit well inside the present orbits of the planets, could have been evaporated by the wind and radiation from the pulsar, and the ablated material would form a disk. The disk would evolve outwards and eventually dissipate, but meanwhile there would be the opportunity for the planets to form from disk material at 0.3–0.5 AU. This idea had been previously mentioned with regard to PSR1829 – 10 (Bailes *et al.*, 1991).

Freeman Dyson (1999) makes the following remarks concerning this first extra-solar planetary system:

*The news of his discovery was greeted by the community of astronomers in Princeton with considerable skepticism. I was lucky to be present when Wolszczan came to Princeton to confront the skeptics. This was a historic occasion. It provides an excellent example of the way the scientific establishment deals with young people who claim to have made important discoveries.*

*Alexander Wolszczan sat down to lunch with about fifty astronomers... The proceedings were informal and superficially friendly, but there was high tension in the air. Wolszczan had claimed to find planets orbiting the wrong kind of star ... The credibility of his discovery depended entirely on the credibility of his software programs ... he had to convince [his colleagues] that his software programs were free from bugs. ... Each astronomer who doubted the reality of Wolszczan's planets took a turn as prosecuting attorney, asking sharp questions about the details of the observations and trying to find weak points in [his] analysis. Wolszczan answered each question calmly and completely, showing that he had asked himself the same question and answered it long before. At the end of the lunch there were no more questions.*

*Wolszczan came through the ordeal victorious, and the skeptics gave him a friendly ovation.*

A short time later, the presence of the two planets was confirmed by observations at the National Radio Astronomy Observatory at Green Bank (Backer, 1993).

At least a dozen different theories for the origin of the pulsar planets were proposed almost immediately after the discovery, some of which were along the same lines as those which had been invented for PSR1829 − 10. Most of these are summarized by Podsiadlowski (1993) together with an estimate of their probability. The proposal mentioned above (Wolszczan and Frail, 1992) seems to be the most reasonable; however none of them have been followed through with detailed calculations covering both disk evolution and planet formation. The reason is probably that only a tiny fraction of stars could evolve in such a way to produce pulsar planets, and the conditions in disks for forming planets around pulsars undoubtedly differ physically, at least in some respects, from those in disks around solar-type stars. The ultimate goal is to understand the formation of planetary systems in situations where a potentially habitable zone exists, which means in the vicinity of long-lived main-sequence stars. Thus the discovery of the pulsar planets was an important first step, and it pointed the way to future discoveries. As Geoff Marcy remarked at the time, *"If bizarre stars like pulsars have planets, it seems other stars must."*

# 2

# The Formation and Evolution of Planetary Systems: placing our Solar System in context

Jeroen Bouwman

*Max-Planck-Institut für Astronomie, Heidelberg, Germany*

Michael R. Meyer, Jinyoung Serena Kim, and Murray D. Silverstone

*Steward Observatory, The University of Arizona, Tucson, AZ*

John M. Carpenter

*Astronomy, California Institute of Technology, Pasadena, CA*

Dean C. Hines

*Space Science Institute, Boulder, CO*

## 2.1 Introduction

Understanding the formation history of the Solar System is one of the major scientific goals within astronomy. Though many clues can be found within our Solar System, reconstructing its early history after 4.5 Gyr of often violent evolution is a difficult task. An alternative, and perhaps more accurate, complementary approach is to study nearby Solar-mass stars of different stellar ages and evolutionary phases. By tracing the evolution of star and planet formation by observing stars from just-formed young stellar objects (YSOs) with circumstellar disks, believed to be in the process of forming a planetary system, through older main-sequence stars with debris disks at various states of activity, valuable insights into the origin of the Solar System can be obtained.

Though the process of star formation is far from being completely understood, the generally accepted view on how stars form was put forward by Shu *et al.* (1987). Star formation can be divided into four distinct phases, after the initial formation of dense molecular cores (e.g. Evans, 1999) inside a giant molecular cloud, as, for instance, found in the Chameleon or Taurus regions. Following the classification first suggested by Lada and Wilking (1984), the different evolutionary stages of the protostar can be identified by the spectral energy distribution (SED) of the objects (Adams *et al.*, 1987), expanded with an additional class (Class 0), after the

*Planet Formation: Theory, Observation, and Experiments*, ed. Hubert Klahr and Wolfgang Brandner.
Published by Cambridge University Press. © Cambridge University Press 2006.

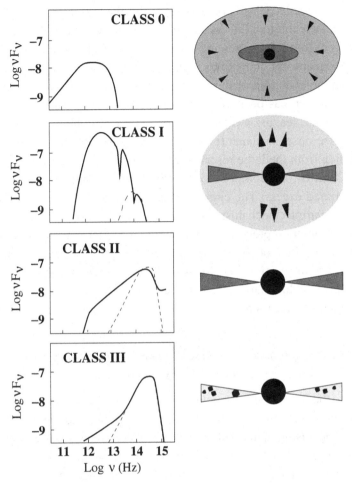

Fig. 2.1. Schematic overview of the various stages in the formation of a star evolving from a Class 0 to a Class III object (Natta, 1999). The left-hand panels show the typical SEDs (solid lines), where the dashed lines represent the emission from the central protostar. In the right-hand panels the system geometry is displayed showing the distribution of circumstellar gas and dust. The light and dark shaded areas depict low and high densities, respectively. The arrows indicate infall or outflow of material. In Class III objects, cometary-sized objects have formed in the flattened, gas depleted circumstellar disk.

discovery of a new group of objects by André *et al.* (1993) representing an earlier evolutionary phase. This schematic view of star formation can be seen in Fig. 2.1. The panels on the left show the typical SED of objects within a group, the panels on the right show the corresponding geometry or spatial distribution of material in the system. As the slowly rotating dense, molecular core collapses, a central protostar and surrounding disk will form (Class 0) still deeply embedded within an infalling envelope of gas and dust. At this stage the protostar is only visible at far-infrared and millimeter wavelengths. Class I objects represent the second phase as the central star accretes matter through the disk. A strong stellar wind can develop along the rotational axis of the system, creating a bipolar outflow. The star is still embedded, resulting in deep absorption features in its spectrum. Class II objects represent the third phase in the evolution of the protostar and its circumstellar disk, where the surrounding material has been largely dissipated by the stellar wind, terminating the infall of matter, leaving a protostar with an optically thick circumstellar disk. At this stage, the onset of planet formation is thought to occur, with massive giant planets forming from the large reservoir of gas present in the disk. The last group of objects, Class III, are comprised of (near) zero age main-sequence stars with almost no infrared excess. The circumstellar disk in these systems is largely dispersed through various mechanisms such as accretion on to the central star, photo-evaporation or radiation pressure, stellar winds or stellar encounters (Hollenbach *et al.*, 2000) or by incorporation into (proto)planetary objects. The dust seen in these systems originates most likely from collisions between planetoids, hence the name debris disk. In this chapter, we will focus on the evolution of the circumstellar disks starting at the transition period between Class II and Class III systems.

### 2.1.1 The formation of planets: from protoplanetary towards debris disk systems

The study into the formation of planetary systems took flight after the first detection of an extrasolar planet around a Solar-mass star (Mayor and Queloz, 1995). The continuously increasing number of planets discovered around main-sequence stars (e.g. Mayor *et al.*, 2004; Marcy *et al.*, 2000, see also Chapter 11 by Marcy) strongly supports the notion that planet formation is a common occurrence. These studies also show that there are substantial differences between the newly discovered planetary systems and the Solar System. To understand these differences, the conditions under which a system is formed needs to be known, i.e. one has to observe the birth of planets. Detecting planets, however, around pre-main-sequence stars, which are often still accreting and having outflows in the form of strong stellar winds or jets, is not possible with the radial-velocity techniques used to detect planets around main-sequence systems. We can, however, study the onset of planet formation by observing the dust and gas, the building-blocks of planets, in the circumstellar disks where they are born.

The circumstellar disks surrounding the pre-main-sequence T Tauri and Herbig Ae systems (Class II) are believed to be the site of ongoing planet formation (from hereon we will refer to these disks as protoplanetary disks). As most of the youngest (<1 Myr) Solar-mass stars have circumstellar disks (Strom *et al.*, 1989), with typical masses (Beckwith *et al.*, 1990) and sizes (McCaughrean and O'Dell, 1996; Dutrey *et al.*, 1996) comparable to the expected values for the primitive solar nebula, these disks are the natural candidates for the birth-sites of planets. The sub-micron sized dust grains present at the formation time in these disks can coagulate to form larger objects and eventually Earth-like planets (e.g. Weidenschilling, 1997a). The formation of giant gas planets can either follow the formation of a massive rocky core and a subsequent fast gas accretion (see Pollack *et al.*, 1996) or alternatively through gravitational instabilities and the subsequent fragmentation of the disk (Boss, 2001a). By deriving the composition of the circumstellar material, and identifying the processes governing the chemistry and coagulation, valuable insights can be gained in the workings of protoplanetary disks (e.g. Bouwman *et al.*, 2001, 2003; van Boekel *et al.*, 2003, 2004; Przygodda *et al.*, 2003) and thus the planetary-formation process. The results of these analyses can be compared directly with observations of Solar System objects like comets, meteorites and interplanetary dust particles (IDPs), preserving a record of the early phases in the evolution of the Solar System. Further, by determining the evolutionary timescales of circumstellar disks, like those for the dispersion of the circumstellar matter, constraints can be placed on the planet formation process and its timescales (e.g. Haisch *et al.*, 2001; Mamajek *et al.*, 2004).

Once the disk has been dispersed, only the circumstellar material which has been incorporated into larger asteroidal or planetary bodies remains, forming a so-called debris disk. The pre-main-sequence T Tauri and Herbig Ae stars are believed to be the precursors of such debris disks (Class III), or Vega-type (after the prototype) systems. In the gas-depleted disk, small dust grains can be removed efficiently by the stellar radiation field and wind. Any observed dust in these disks, therefore, must have been produced recently. The natural way to produce dust in a debris disk is by collisions between planetesimals (Weissman, 1984), not unlike those occurring in the asteroid and Kuiper-Belt objects. The discovery of systems having both planets and a debris disk (Pantin *et al.*, 2000; Beichman *et al.*, 2005) shows that many systems have a similar structure to that of the Solar System. The presence of a planet can influence and shape the structure of the debris disk (Kenyon and Bromley, 2004a, b). By determining the location and rate of dust production, constraints can be placed upon the dynamics of the planetesimal disk and the presence and influence of planets could be inferred. This will also provide insights into the final accretion phases of Earth-like planets, and the formation of asteroid- and Kuiper-like planetesimal belts.

### 2.1.2 The Spitzer Space Telescope and the formation and evolution of planetary systems legacy program

With the Infrared Astronomical Satellite (IRAS) and Infrared Space Observatory (ISO) missions, a tremendous advance in our knowledge of protoplanetary and debris disks could be achieved. However, these missions where limited to relatively

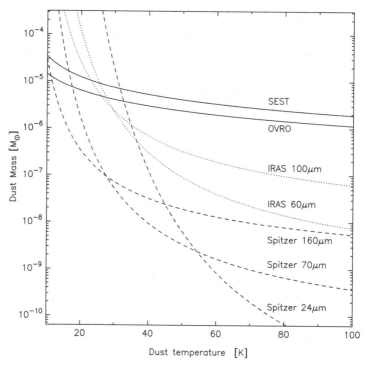

Fig. 2.2. Typical 5 $\sigma$ sensitivity limits to dust mass as a function of the dust temperature for the Owens Valley Radio Observatory and Swedish-ESO Sub-millimeter Telescope data (solid curves) compared with IRAS (dotted curves) and Spitzer (dashed curves). Dust masses were computed for a source at a distance of 50 pc, assuming that the emission is isothermal and optically thin, with $\beta = 1$ and $\kappa(1.3\ \text{mm}) = 0.02\ \text{cm}^2\ \text{g}^{-1}$. The ISO far-infrared photometric surveys (Habing *et al.*, 2001; Spangler *et al.*, 2001), not shown here for clarity, have sensitivities intermediate between IRAS and Spitzer. (Figure from Carpenter *et al.*, 2005.)

nearby and luminous A- and B-type sources, and provided only limited knowledge of the evolution of Solar-mass systems. With the launch of the Spitzer Space Telescope (Werner *et al.*, 2004), having a much higher sensitivity, also these less luminous systems are accessible to observations. The Spitzer telescope, an 85 cm cryogenic space observatory in Earth trailing orbit, was launched in August of 2003, and has an estimated minimum mission lifetime of 5 yr. The Spitzer Space Telescope carries three science instruments on board: the Infrared Array Camera (IRAC; Fazio *et al.*, 2004), the Infrared Spectrograph (IRS; Houck *et al.*, 2004), and the Multi-band Imaging Photometer for Spitzer (MIPS; Rieke *et al.*, 2004), which together provide imaging and spectroscopy in standard modes from 3.6 to 160 μm.

As an example of the capabilities of Spitzer, Fig. 2.2 shows the sensitivity to dust mass of the MIPS instrument relative to that of IRAS and the Owens Valley Radio

Observatory (OVRO) telescope and the Swedish-ESO Submillimeter Telescope (SEST). As shown in this figure, the mid- and far-infrared Spitzer observations are more sensitive to dust mass for dust warmer than ∼25 K. This makes Spitzer the ideal telescope to study the evolution of debris disk systems, typically showing spectral energy distributions dominated by the thermal re-emission from dust grains with temperatures of ∼50 K, peaking at ∼100 μm. The typical dust masses which can be detected in these systems are in the order of $5 \times 10^{-9}$ $M_{\odot}$ at a distance of 50 pc, assuming an opacity of $\kappa(1.3 \text{ mm}) = 0.02 \text{ cm}^2 \text{ g}^{-1}$ and $\beta = 1$.

To enable large-scale programs of broad scientific and public interest, and to provide access to uniform and coherent datasets as rapidly as possible in support of General Observer proposals, the so-called Legacy Science Program was established. The Formation and Evolution of Planetary Systems (FEPS) Spitzer Legacy Science Program is one of five such programs and builds upon the rich heritage of Spitzer's ancestors, the IRAS and ISO space observatories, complementing Guaranteed Time Observer programs also being pursued with Spitzer. The FEPS program will probe circumstellar dust properties around a representative sample of protoplanetary disks and debris disks, spanning the full range of circumstellar disk geometries and covering the major phases of planet-system formation and evolution. Our Legacy Program is designed to complement those of Guaranteed Time Observers studying both clusters and individual objects such that a direct link between disks commonly found surrounding pre-main-sequence stars (∼3.0 Myr), and the 4.56 Gyr-old Solar System can be made. Specifically, we will trace the evolution of planetary systems at ages ranging from:

(1) 3–10 Myr when stellar accretion from the disk terminates.
(2) 10–100 Myr when planets achieve their final masses via coalescence of solids and accretion of remnant molecular gas.
(3) 100–3000 Myr when the final architecture of solar systems takes form and frequent collisions between remnant planetesimals produce copious quantities of dust.

Our strategy is to obtain carefully calibrated spectral energy distributions assembled using all three Spitzer instruments for all stars in our sample to infer the radial distribution of dust. Our sample is comprised of 328 Sun-like stars distributed uniformly in log-age from 3 Myr to 3 Gyr. A more limited high-resolution spectroscopic survey of the younger targets amongst those in the dust disk survey will establish the gas content. In addition to insight into problems of fundamental scientific and philosophical interest, the FEPS Legacy Science Program will provide a rich database for follow-up observations with Spitzer, with existing and future ground- or space-based facilities.

In the following two sections, we will give a review of the results obtained within the FEPS Legacy Program, after the first year of observations.

## 2.2 From protoplanetary to debris disks: processing and dispersion of the inner dust disk

One of the major problems in understanding disk evolution and planet formation has been determining how primordial gas-rich accretions disks containing interstellar dust grains evolve into gas-depleted debris disks containing the products of collisions of large (proto)planetary and asteroidal bodies. Current near- and mid-infrared observations, tracing the dust content in the inner parts ($<10$ AU) of the disk, suggest that the dust in this region is dissipated, or that grains agglomerate into larger bodies, on timescales of 1 to 10 Myr, with 50% of low mass stars losing their inner dust disks within 3 Myr (Haisch *et al.*, 2001; Mamajek *et al.*, 2004). Though this loss of the smallest dust grains seems to be correlated with the cessation of material accreting from the disk onto the central star (Gullbring *et al.*, 1998), and would argue for a complete dispersion of the inner disk, it is still unclear if the loss of the dust disk also implies a dispersion of the gas (e.g. Najita *et al.*, 2003).

The disk dispersion timescales are most likely connected to the timescales on which planets form, and the question arises how these processes influence each other. Obviously, protoplanets have to be formed before the inner disk has been dispersed. Or, reversing the argument, it can be the formation of giant planets which causes removal of the inner disk. Once a giant planet has formed, it can effectively disconnect the inner disk from the large mass reservoir of the outer disk, resulting in the depletion of the inner disk through accretion onto the central star (see also Chapter 14 by Masset and Kley). Isotopic evidence from cometary material suggests that their parent bodies were formed within a few million years (see Chapter 5 by Trieloff and Palme), consistent with the dispersion timescales. Earth-like planets can reach their final masses at the later debris disk phases as indicated by measurements of terrestrial and lunar samples suggesting that the Earth–Moon system was 80 to 90% complete at an age of 30 Myr (Kleine *et al.*, 2002, 2003). If the disappearance of the dust disk also implies the removal of the gas, the dispersion timescales also impose constraints on the formation of giant planets. Current models of gas-giant planet formation suggest that Jupiter-mass planets orbiting at 3 to 10 AU can form within a few million years (see also Chapter 8 by Thommes and Duncan, Chapter 10 by Hubickyi, and Chapter 12 by Boss), consistent with the timescales for the disappearance of the inner dust disk.

To study the transition from primordial to debris disk, within the FEPS legacy program, a sample of 74 young (3–30 Myr) Sun-like stars (stellar mass ($M_{\star}$) = 0.8–1.2 $M_{\odot}$), selected without bias with respect to previously-known infrared excess, were observed with IRAC Channels 1 (3.6 μm), 2 (4.5 μm) and 4 (8.0 μm) (see further Silverstone *et al.*, 2006, for a detailed description of the sample and data reduction). Combining the IRAC measurements with Ks-band magnitude from

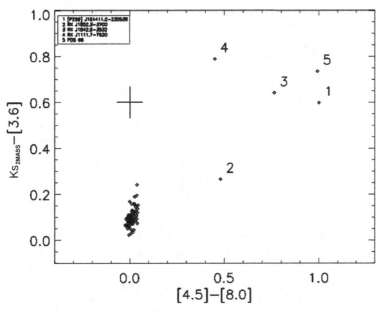

Fig. 2.3. The 2MASS Ks–IRAC 3.6 μm versus IRAC 4.5 μm–8 μm color-color diagram for 74 targets from the youngest age bins in the FEPS sample. Five apparent excess targets appears in the upper-right hand of this diagram. Plotted below the legend is a maximum typical error derived from the cluster of 69 sources with no apparent excess. (Figure from Silverstone *et al.*, 2006, see also for details on the IRAC observations.)

the Two-Micron All Sky Survey (2MASS) for the sample stars, we constructed a color-color diagram, to detect any excess emission with respect to the stellar photosphere. Fig. 2.3 shows the Ks [3.6 μm] versus [4.5 μm]–[8.0 μm] color-color diagram for the 74 stars in the sample. In the diagram the locus of points define objects having "bare" photospheres with no sign of any infrared excess. Clearly visible are five excess targets appearing at the upper right of this figure, having colors consistent with those expected from actively accreting classical T Tauri stars. Indeed, the spectral energy distributions of these stars are consistent with the emission from optically thick, massive dusty gas disks (Silverstone *et al.*, 2006; Carpenter *et al.*, 2005). This small fraction of excess sources shows that optically thick inner-circumstellar disks are rare surrounding Sun-like stars at ages greater than 3 Myr old. For the sources with no apparent excess, strong limits can be placed on the maximum amount of dust present in these systems. The measured IRAC fluxes imply dust-mass upper limits in the range of ∼0.1 to 3 $M_{Ceres}$ or $1.5 \times 10^{-5}$ to $4.3 \times 10^{-4}$ $M_{\oplus}$.

To determine how the near-infrared excess evolves with time, Silverstone *et al.* (2006) determined the fraction of systems observed with excess detected in the

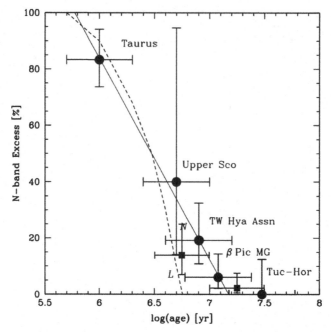

Fig. 2.4. The fraction of disk sources as a function of stellar age. The filled squares represent the fraction of IR excess sources based on the IRAC observations discussed in this section, for the 3–10 and 10–30 Myr age bins, the two youngest in the FEPS sample. These points represent the same statistic at all three IRAC bands. The black solid circles (and the solid-line fit was derived from) summaries of other 10 μm ground-based studies (summarized in Mamajek *et al.*, 2004), and the dashed line is a comparison to the L-band near-IR study by Haisch *et al.* (2001).

IRAC bands in two logarithmic age bins. The two age bins, between 3–10 Myr and 10–30 Myr, were chosen such that their widths are larger than the estimated 30% uncertainties in systemic ages, to allow for a proper statistical analysis. A total 29 sources are located in the youngest age bin and 45 in the oldest age bin. The fraction of systems with excess in each age bin is plotted in Fig. 2.4. The $1\sigma$ probability also plotted in this figure is estimated following the method of Gehrels (1986). The fraction of systems having a substantial inner disk decreases rapidly with time, with only 4 systems out of 29 with ages between 3 and 10 Myr, and 1 out of 45 targets with ages between 10 and 30 Myr exhibiting an infrared excess in one of the IRAC bands. These results agree well with previous studies in the L- and N-band (Haisch *et al.*, 2001; Mamajek *et al.*, 2004), which showed a similar rapid decrease of near-infrared excess in time. Looking at Fig. 2.3, there are no systems observed with colors intermediate between those due to an optically thick accretion disk and those with "bare" photospheres. This implies that the disk

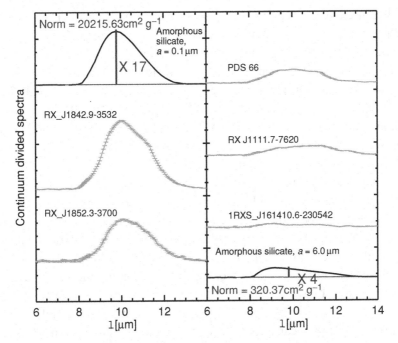

Fig. 2.5. Continuum divided IRS low-resolution spectra of the five excess sources identified in Fig. 2.3. The spectra are ordered (top to bottom, left to right) by decreasing relative band strength over continuum of the 10 μm silicate feature. Also plotted are the calculated continuum divided absorption opacities based on laboratory measurements of amorphous silicate glass with olivine stoichiometry (Dorschner *et al.*, 1995), for a grain size of 0.1 μm (top left) and 6 μm (bottom right), respectively. These calculations show that for increasing grain size, both the band strength over continuum as well as the absolute opacity of amorphous silicate decreases from a factor of 17 to 4, and from $2 \times 10^4$ cm$^2$ g$^{-1}$ to 320 cm$^2$ g$^{-1}$, respectively. Comparing the observed and calculated opacities, the observed trend of decreasing band strength can be explained as a change in grain size of the emitting dust grains.

dispersion process as shown in Fig. 2.5 is not a gradual and continuous process but must occur suddenly and at much shorter timescales than the observed systemic ages of the Silverstone *et al.* (2006) sample. One can estimate the timescale of this transition to be $< 7.5$ Myr$/29 = 0.26$ Myr using the youngest age bin and $< 20$ Myr$/45 = 0.45$ Myr using the oldest age bin.

Apart from the IRAC observations, IRS low-resolution (R~100) spectra were also obtained. While these spectra do not show any spectral features, indicative of small ($\leq 10$ μm) dust grains for those sources without a near-infrared excess, clear silicate dust emission features appear in the spectra of the five T Tauri excess stars. Fig. 2.5 shows the continuum divided IRS spectra around 10 μm for these sources. A clear trend can be observed, from the upper left to the lower right, in the relative strength over continuum of the 10 μm silicate feature. The silicate band strength varies from values similar to those observed for ISM grains to values almost identical to the underlying continuum. Comparing the strength and shape of the spectral features with calculated emission band profiles based on the measured optical properties of amorphous silicates, the observed trend can be identified with a change in the typical grain size from about 0.1 μm to ~5 to 10 μm radius. These latter values are about two orders of magnitude larger than those found in the ISM. This implies that a substantial grain growth must have taken place in the systems shown on the right-hand side of Fig. 2.5.

There appears to be no correlation between the near-infrared excess and the grain size of the amorphous silicate grain population as measured in the 10 μm region. One might expect such a correlation, as the grain opacity is directly related to grain size. The observed grain growth by two orders of magnitude will lower the opacity by the same amount, and in the case of an optically thin dust disk, also the observed infrared emission coming from such a disk. The absence of this correlation implies that the T Tauri stars with the largest grain sizes of the Silverstone *et al.* (2006) sample must still be optically thick. Also, within the sample of the five T Tauri stars, both the near-infrared excess and the measured grain size seem not to be correlated with systemic age. Though the sample of excess stars is too small to make any firm statistical conclusion, this latter point is consistent with other studies on larger samples (e.g. Bouwman *et al.*, 2001; Przygodda *et al.*, 2003). Interestingly, both the oldest T Tauri star (PDS 66, number 5) and the one with the largest grains ([PZ99] J161411.0-230536, number 1) show the largest near-infrared excesses.

Summarizing, the FEPS observations point towards a rapid dispersion of the inner disk, but the time at which the onset of this process occurs varies from system to system. The dispersion process, however, will occur within 10 Myr for 86% of the systems, and within 30 Myr for 98% of the young stellar systems. A possible explanation for the sudden disappearance of the inner disk and its infrared excess could be a runaway growth of the small-grain population forming cm- to m-sized objects, which would have too low an opacity to be detected. Practically all small grains, however, would have to be incorporated into these larger bodies. An alternative to this would be the formation of a massive gas planet, opening a gap in the accretion disk. This would effectively cut off the inner from the outer disk, resulting in a rapid emptying out of the inner disk due to viscous evolution. An observational test to distinguish between both scenarios would be the detection of gas in an inner disk depleted of small dust grains. In the first scenario a star would still be surrounded by its gas disk, while in the latter both dust and gas would disappear. The detection of gas signatures is part of the FEPS Legacy Program, results of which are to be expected in the second year of operations of Spitzer. As an alternative scenario to the above-mentioned mechanisms, a cessation of infall of material onto the outer disk from the (remnant) molecular parent cloud could also cause an emptying out of the inner accretion disk due to viscous evolution (Hollenbach *et al.*, 2000). By studying the surrounding and outer-disk evolution, one might hope to distinguish between this and the giant-planet model.

As to why certain systems disperse their disk sooner than others is still unclear. Questions like how grain growth is linked to disk properties such as size and density, and how the environment (binarity, stellar encounters, etc.) of the young stellar objects influences their evolution, need to be answered first before we can solve this problem. The combined observations of the FEPS Legacy and Guaranteed Time and General Observers Programs studying circumstellar disk evolution will be able to shed light on those questions.

## 2.3 Debris disks: Asteroid or Kuiper Belt?

After the dispersion of most of the gas and small dust grains, as discussed in the previous section, a remnant debris disk consisting of larger cometary- or asteroidal-type objects may remain. Dynamic stirring of such a disk would lead to collisions in which small dust grains can be produced, whose emission can be observed with Spitzer. Our present-day Solar System still has two major zones of debris: the Asteroid Belt at radii of 2 to 4 AU and the Kuiper Belt at radii between 30 and 50 AU. Collisions in these belts produce small dust grains observed as the zodiacal dust cloud. Understanding the dynamics of debris disks and the formation of Asteroid and Kuiper Belt-like structures, will be crucial in understanding the formation and evolution of planetary systems like our own.

While a large and increasing number of debris-disk systems has been detected, (e.g. Habing *et al.*, 2001; Decin *et al.*, 2003), the majority of the debris disks found so far are associated with more massive and luminous intermediate-mass (2–8 $M_\odot$) stars, as observatories such as ISO and IRAS did not have the sensitivity to detect debris disks around Solar-mass stars at distances beyond a few parsecs. The enhanced sensitivity of Spitzer will provide the means to greatly enhance our knowledge of the evolution of debris-disk systems around Solar-mass stars, and through which we can shed light on the evolution of the Solar System. The FEPS Legacy Program, from which the first results are presented in this section, will contribute significantly to this.

Similar to the study of the inner-disk evolution, discussed in the previous section, the FEPS Legacy Program will also provide a comprehensive study of the outer disk and its long-term evolution. As a first step, Carpenter *et al.* (2005) observed a large fraction of the FEPS target stars at millimeter or sub-millimeter wavelengths. Plotted in Fig. 2.6 is the long-term evolution of the cold outer disk as measured at these

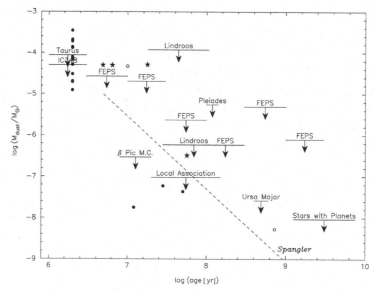

Fig. 2.6. Mean circumstellar dust mass derived from millimeter or sub-millimeter continuum observations (see Eq. 1 from Carpenter *et al.*, 2005) as a function of age for stars with stellar masses between 0.5 and 2 $M_\odot$. The stellar samples include stars observed as part of the FEPS legacy program, Taurus, IC 348, Lindroos binary stars, the $\beta$ Pic moving group and the Local Association, and stars with known planets from radial velocity surveys (see Carpenter *et al.*, 2005 for a discussion and references of the observations). Individual points indicate sources that have been detected at $S/N \sim 3$ in the FEPS sample (stars) and the other stellar samples represented in this figure (filled circles). The open circles show the location of TW Hya at 10 Myr and $\epsilon$ Eri at 730 Myr. The dashed line shows the mass–age relation derived by Spangler *et al.* (2001) from ISO observations. (Figure from Carpenter *et al.*, 2005.)

wavelengths. These observations show that while for the youngest systems disks are readily detected, no such massive disks are present in the older systems. The derived upper limits and few detections show that the typical disk masses for systems older than ~30 Myr are several orders of magnitude lower than those of young systems such as found in Taurus. This result is consistent with previous studies based on ISO observations that found a similar decline in disk mass (Spangler *et al.*, 2001, dashed line in Fig. 2.6). The observed decline in average disk mass most likely does not reflect a gradual dissipation of the outer disk, identical for all systems, but rather represents the equilibrium in a stochastic process, in which disks can have different epochs of dynamic heating of the planetesimal disk (e.g. Dominik and Decin, 2003).

The MIPS instrument on board Spitzer, providing photometric observations at 24 μm, 70 μm and 160 μm, will greatly improve our knowledge of the outer-disk

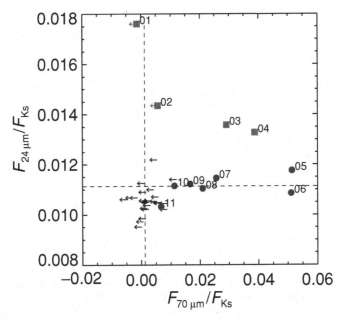

Fig. 2.7. Flux-density ratio plot of 70 μm/Ks versus 24 μm/Ks. Left arrows indicate $1\sigma$ upper limits of non-detections. The filled squares mark the positions of the warm debris disks (numbers 1 to 4, [PZ99] J161459.2-275023, HD 12039, HD 219498 and HD 141943, respectively; see also Fig. 2.8). The large filled circles mark the data points of the cold debris disks (numbers 5 to 11, HD 105, HD 8907, HD 122652, HD 145229, HD 150706, HD 6963 and HD 206374, respectively; see also Fig. 2.9). The dashed lines mark the positions of the expected flux ratios for a stellar atmosphere of a G0V main-sequence star with an effective temperature of 5850 K. Clearly visible is the separation between the warm and cold debris disks, the latter having 24 μm MIPS photometric fluxes consistent with stellar photospheres. Note also that for the brightest warm debris-disk sources no excess emission could be detected at 70 μm.

evolution, as it is ideally suited for the study of debris disks (see Fig. 2.2 and Section 1.1). Fig. 2.7 shows a compilation of the current results obtained with the MIPS instrument (Meyer *et al.*, 2004; Hines *et al.*, 2006; Kim *et al.*, 2005, this chapter). Plotted are the flux density ratios of 70 μm/Ks versus 24 μm/Ks. The numbered symbols mark the positions of sources showing an excess ($S/N > 3$) in one of the two MIPS bands. In this figure, the filled squares represent those systems with a relatively warm debris disk, meaning that they show an excess at 24 μm. The filled circles represent those systems which only show an excess at 70 μm, implying that the dust disk is colder and therefore does not emit at shorter wavelengths.

Though this sample is still too small for a proper statistical analysis a possible evolutionary trend is worth noticing. The systems with warm debris disks shown in Fig. 2.7, all located in the top half of the figure, are systematically younger than the

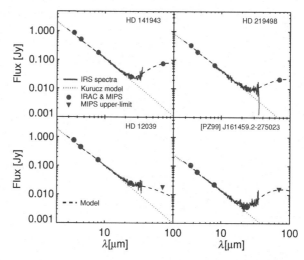

Fig. 2.8. Spectral energy distributions of four systems with a warm debris disk: HD 12039, HD 141943, [PZ99] J161459.2-275023 and HD 219498. Filled circles are IRAC and MIPS data points, and the solid lines are IRS low resolution spectra. The upside-down triangle marks MIPS 70 μm 3 σ upper limits. Dotted lines are Kurucz models. The dashed lines show current best-fit models to the dust excess.

stars surrounded by a cold debris disk. The systemic ages of the warm debris-disk systems are less than 150 Myr while the cold debris disk systems are closer to 1 Gyr. The two systems showing the largest 24 μm excess, [PZ99] J161459.2-275023 at an age of 5 Myr and HD 12039 at an age of 30 Myr, do not show evidence for any emission at 70 μm, implying that they can't have a substantial cold outer dust disk. The above observations are consistent with the idea that after the dissipation of most of the gas, a debris disk can be stirred by stellar encounters or a proto planet, leading to a dynamic heating of the planetesimal disk. This heating will lead to epochs of intense short-lived phases of dust producing collisions within the disk, moving from the inside out (Kenyon and Bromley, 2004a,b).

In this picture the disks with a 24 μm excess but without 70 μm excess would correspond to systems where the dynamic heating of the disk has just started, and the dust production is confined to a narrow region at the relative inner parts of the disk, similar to an asteroid belt. The cold debris disks would correspond to the more evolved systems where the main region of dust production is located at larger radii in structures similar to the Kuiper Belt. As these systems are substantially older, any dust produced in the inner regions in previous epochs would have been removed by radiation pressure or Pointing–Robertson drag. Within the cold debris-disk systems there is no apparent trend with age. This would be consistent with the notion of a stochastic process, where during the lifetime of the star–disk system, multiple epochs of enhanced dust production can occur. As the total reservoir of material

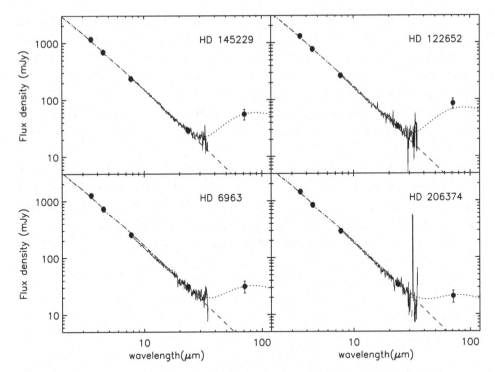

Fig. 2.9. Spectral energy distributions of four systems with a cold Kuiper-belt like debris disk. HD 145229, HD 122652, HD 206374 and HD 6963. Filled-circles are IRAC, and MIPS data points, solid lines are IRS low-resolution spectra. The dashed lines are Kurucz photospheric models, and the dotted lines show the current best-fit models. (Figures after Kim *et al.*, 2005.)

is expected to be larger in the outer disk and timescales longer, dust production producing larger observable quantities of dust could therefore be maintained for longer and multiple periods in time.

To determine the amount and distribution of the dust present in the systems shown in Fig. 2.7, and to quantify the above notion of dust production in Asteroid- or Kuiper-Belt structures, detailed radiative transfer modeling has been performed. Figs. 2.8 and 2.9 show examples of spectral energy distributions of warm and cold debris disks, respectively (Hines *et al.*, 2006; Kim *et al.*, 2005; Bouwman *et al.*, in preparation), together with a best model fit. Clearly visible in these figures is the excess above the stellar photosphere longward of $\sim$20 μm in the warm debris-disk systems and the 70 μm excesses in the cold debris-disk systems. For the HD 12039 and the [PZ99] J161459.2-275023 systems our modeling suggests that the dust emission originates from a narrow ring at about 5 AU and 15 AU, respectively. This would imply that the dust in these systems is located in an Asteroid Belt rather than a Kuiper Belt extending to larger radii. The total amount of dust in these rings is about $10^{-9}$ to $10^{-10}$ $M_\odot$. Our current best model fits constrain the inner radius of the dust disk in the HD 141943 and HD 219498 systems to be at least 10 to 15 AU, and to have a total dust mass of $\sim 10^{-8}$ $M_\odot$. This would imply that the dust we see in these systems most likely originates from a Kuiper Belt-like structure, but one that extents much closer in, by a factor of two, than is the case for the Solar System. The cold debris-disk systems shown in Fig. 2.9 typically have inner disk radii of 20 to 40 AU, very similar to the Kuiper Belt. The derived dust masses range between $10^{-8}$ and $10^{-9}$ $M_\odot$.

The exciting first results presented in this chapter are an excellent example of things to come, with the full sample of data of the FEPS program being analyzed in the near future. The Spitzer Space Telescope and the FEPS Legacy Program will considerably enhance our knowledge of planet-forming disks around Solar-mass systems, enabling us to place the Solar System in its proper context.

# 3

# Destruction of protoplanetary disks by photoevaporation

Sabine Richling

*Institut d'Astrophysique de Paris, Paris, France*

David Hollenbach

*NASA Ames Research Center, Moffett Field, USA*

Harold W. Yorke

*JPL, California Institute of Technology, Pasadena, USA*

## 3.1 Introduction

Planets form within circumstellar disks composed of a mixture of gas and dust grains. These disks result from the gravitational collapse of rotating molecular cloud cores. They are initially rather massive and consist of about 0.3 $M_\star$, where $M_\star$ is the mass of the central star (e.g. Yorke *et al.*, 1995). In contrast, the minimum mass required to build the planets of our Solar System is only about 0.01 Solar masses ($M_\odot$). Evidently, there are processes that redistribute the mass, transform the dust to larger particles, and disperse much of the gas and dust.

The processes which are responsible for the dispersal of the gas influence the formation of planets. For example, the timescale for gas dispersal as a function of the disk radius affects the composition of the resulting planetary system. As long as the dust particles are small enough to be tightly coupled to the gas, they follow the gas flow. If the gas is dispersed before the dust particles have had a chance to grow, all the dust will be lost and planetesimals and planets cannot form. Even if there is time for particles to coagulate and build sufficiently large rocky cores that can accrete gas (Pollack *et al.*, 1996; Hubickyj *et al.*, 2004), the formation of gas-giant planets like Jupiter and Saturn will be suppressed if the gas is dispersed before the accretion can occur. Furthermore, gas dispersal affects planet migration (e.g. Ward, 1997) and influences the orbital parameters of planetesimals and planets (Kominami and Ida, 2002).

Photoevaporation is a very efficient gas dispersal mechanism. Ultraviolet radiation heats the disk surface and the resulting pressure gradients can drive an

*Planet Formation: Theory, Observation, and Experiments*, ed. Hubert Klahr and Wolfgang Brandner.
Published by Cambridge University Press. © Cambridge University Press 2006.

expanding hydrodynamic flow which contains both gas and small dust particles. For the determination of the mass-loss rate of protoplanetary disks it is important to consider extreme-ultraviolet (EUV) photons with energy greater than 13.6 eV and far-ultraviolet (FUV) photons in the energy range 6–13.6 eV. EUV photons ionize hydrogen and heat the gas to a temperature of the order of $10^4$ K. FUV photons dissociate molecules, ionize carbon and heat the predominantly neutral gas generally via the photoelectric effect on dust grains. The most important cooling mechanisms in FUV-heated regions are generally the [OI] and [CII] fine-structure lines. The balance of heating and cooling results in temperatures in the range 100 K < $T$ < 5000 K. FUV-heated regions are also called photon-dominated regions (PDRs).

Two critical radii define the region where mass loss via an evaporating flow can be established. The inner critical radius is the gravitational radius

$$r_g = \frac{GM_\star}{kT} \sim 100 \text{ AU} \left(\frac{T}{1000 \text{ K}}\right)^{-1} \left(\frac{M_\star}{M_\odot}\right). \tag{3.1}$$

At $r_g$ the sound speed of the gas is equal to the escape speed from the gravitationally bound system. Early analytic models made the simple assumption that photoevaporation occurred for $r > r_g$, and that the warm surface was gravitationally bound for $r < r_g$. Recent results show that this was too crude an approximation. Adams *et al.* (2004) performed a streamline calculation that showed there is still significant mass loss even inside the gravitational radius for radii $r > 0.1 - 0.2 \, r_g$. This result is general and applies to both FUV and EUV heating and to both internal and external fields. Font *et al.* (2004) used a hydrodynamic code which also demonstrates this effect. At any rate, $\sim 0.15 \, r_g$ is a measure of the disk radius *outside of which* photoevaporation is important.

The other critical radius is the size $r_d$ of the disk itself. This outer radius may have been truncated by the pressure of an advancing ionization front, by stellar encounters, or by photoevaporation. In the case of external illumination, $r_d$ helps determine the mass-loss rate, since most of the mass loss is from $r_d$, where most of the disk surface area resides. In the case of illumination by the central star, most of the mass loss occurs near $r_g$, and $r_d$ plays less of a role, except as a mass reservoir.

Photoevaporating disks have been investigated theoretically under different geometries and different assumptions about the ultraviolet field. The main cases are external and internal illumination as well as FUV and EUV radiation. Table 3.1 gives a chronological overview of recent work indicating which cases have been analyzed. Johnstone *et al.* (1998) and Störzer and Hollenbach (1999) present a good summary of when EUV and when FUV photons dominate the photoevaporative mass loss for the case of external O stars illuminating the disks of nearby low mass

Table 3.1. *Summary of recent models on photoevaporating disks*

| Reference | external | | internal | |
|---|---|---|---|---|
| | EUV | FUV | EUV | FUV |
| Scally and Clarke (2001) | X | X | | |
| Clarke *et al.* (2001) | | | X | |
| Matsuyama *et al.* (2003) | X | X | X | |
| Ruden (2004) | | | X | |
| Font *et al.* (2004) | | | X | |
| Lugo *et al.* (2004) | | | X | |
| Adams *et al.* (2004) | | X | | |
| Throop and Bally (2005) | X | | | |
| Richling and Yorke (in preparation) | X | X | X | X |

stars. The dominating ultraviolet band is a function of both the distance to the O star as well as $r_d$. In the case of external illumination by B stars, FUV photons nearly always dominate.

## 3.2 Photoevaporation and other dispersal mechanisms

Other potential gas-dispersal mechanisms are accretion onto the central star, tidal stripping due to close stellar encounters and the entrainment of disk surface gas and dust into an outflowing shear layer driven by protostellar winds. Hollenbach *et al.*, (2000) discussed these mechanisms and found that viscous accretion is the dominant dispersal mechanism for the inner ($r \ll r_g$) part of the disk. With the possible exception of isolated low-mass stars with little external field (see discussion below), the destruction of the outer parts of the disk is mainly dominated by photoevaporation.

The remaining mechanisms are in most cases of minor importance. Tidal stripping due to stellar encounters is considered to be relatively rare and affects only the very outermost parts of the disk. The disk is typically stripped to a radius about one-third of the distance of closest approach. The paucity of significant events is confirmed by N-body simulations of very compact stellar aggregates like the Orion Nebular Cluster (Scally and Clarke, 2001). These authors find that the distribution of the minimum stellar encounter separation peaks at 1000 AU. Only about 4% of the stars have encounters closer than 100 AU.

The mass entrainment of disk surface material by a wind which rips across its surface is likely an inefficient mechanism to disperse material out of the system (Hollenbach *et al.*, 2000), but further calculations are warranted. Elmegreen (1979) suggests that the impacting wind could be absorbed by the disk, and the resultant angular momentum transfer would actually enhance the accretion of disk material onto the star.

## 3.3 Photoevaporation by external radiation

Johnstone *et al.* (1998) investigated the photoevaporation of protostellar disks by an external radiation field using an analytical model. They distinguish between two qualitatively different flow patterns which arise within the ionized envelope of a photoevaporating disk, i.e. EUV- and FUV-dominated flows. For the case of O stars, EUV-dominated flows occur close to the star, whereas FUV-dominated flows tend to apply for moderate-sized ($r_g < r_d < 100$ AU) photoevaporating disks at larger distances.

In the case of FUV-dominated flows, the neutral flow launches supersonically from the disk surface, passes a shock front behind which it slows down and becomes subsonic. The subsonic region extends up to the ionization front. From there the ionized flow is able to escape supersonically again (see also Fig. 3.1). The mass-loss rate of FUV-dominated flows in high FUV fields,

$$\dot{M}_{\mathrm{FUV}} \sim 6 \times 10^{-8} \, \mathrm{M_{\odot} \, yr^{-1}} \left( \frac{N_d}{5 \times 10^{21} \, \mathrm{cm^{-2}}} \right) \left( \frac{r_d}{30 \, \mathrm{AU}} \right), \qquad (3.2)$$

depends mainly on the column density $N_d$ of neutral gas that is heated by FUV and on the radius of the disk $r_d$. It is surprisingly insensitive to the (high) FUV flux. This insensitivity arises because the heated gas temperature and $N_d$ are only weakly dependent on the FUV flux. It should be noted that this equation only holds if $r_d > 0.2 \, r_g$. Since high FUV fields heat gas to roughly 1000 K, this translates to $r_d >$ roughly 20 AU.

An improved model which considers the structure of the PDR and the flow speed off the disk surface in a more consistent manner (Störzer and Hollenbach, 1999) shows that for high FUV fields there is a weak dependence of the mass-loss rate on the distance $d$ to the UV source (or on the FUV flux) $\dot{M}_{\mathrm{FUV}} \propto d^{\beta}$ with $\beta < 1$. The default value for the column density inside the FUV-heated region, $N_d = 5 \times 10^{21} \, \mathrm{cm^{-2}}$ given in Eq. 3.2, yields best fits to observations of photoevaporating disks in the Orion Nebula.

In the case of EUV-dominated flows, the PDR between the disk surface and the ionization front is small in spatial extent and the neutral flow is prevented from becoming supersonic. In this case, the mass-loss rate,

$$\dot{M}_{\mathrm{EUV}} = 7 \times 10^{-9} \, \mathrm{M_{\odot} \, yr^{-1}} \left( \frac{\Phi_{\mathrm{EUV}}}{10^{49} \, \mathrm{s^{-1}}} \right)^{0.5} \left( \frac{d}{10^{17} \, \mathrm{cm}} \right)^{-1} \left( \frac{r_d}{10 \, \mathrm{AU}} \right)^{1.5}, \qquad (3.3)$$

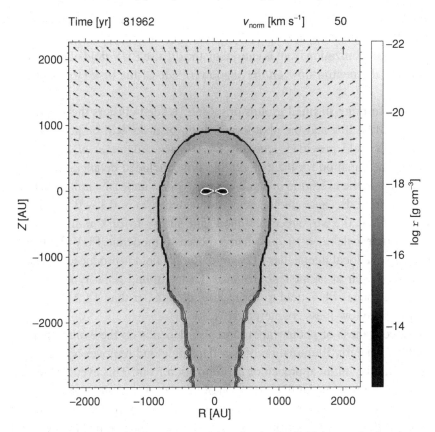

Time [yr]   81962          $v_{norm}$ [km s$^{-1}$]   50

Fig. 3.1. Density distribution (greyscale) and velocity field (arrows) of a photo-evaporating disk due to EUV and FUV radiation from an external source located outside the computational domain in the positive $z$-direction. The velocity normalization is given at the top right. The white line indicates the ionization front of carbon and approximately traces the inner border of the PDR. The outer border of the PDR is the hydrogen ionization front (black line) (Richling and Yorke, 2000).

is characterized by the EUV photon rate $\Phi_{EUV}$ and $d$. The dependence on the disk size $\dot{M}_{EUV} \propto r_d^\alpha$ is stronger for EUV-dominated flows ($\alpha_{EUV} = 1.5$) than for FUV-dominated flows ($\alpha_{FUV} = 1$).

Numerical calculations considering both EUV and FUV radiation (Richling and Yorke, 2000) confirm that the extent of the PDR and the possibility of the formation of a shock front in the neutral region depends on the relative importance of the EUV and FUV photon flux as well as on the dust properties which define the heating ability of the dust grains via the photoelectric effect. The dependence of the mass-loss rate on the disk radius is between the idealized cases with $\alpha_{FUV} < \alpha < \alpha_{EUV}$. Fig. 3.1 displays a result of the two-dimensional radiation hydrodynamic simulations. The cool disk is embedded in a warm PDR. The hydrogen ionization front shows a head–tail structure which is a typical feature of observed photoevaporating disks in dense clusters, where both the external EUV and FUV fluxes are strong. The tail-shaped ionization front is caused by the diffuse EUV field of the HII region illuminating the neutral flow on the shadowed side of the disk (Johnstone *et al.*, 1998).

For stars forming in small groups or more extended clusters the external EUV radiation will be low and the photoevaporation of disks will be governed by the external FUV field provided generally by B stars. Adams *et al.* (2004) solved the streamline equation for the evaporating flow, included detailed chemistry and concentrated on the evolution of small ($r_d < r_g$) disks under the influence of an external FUV field. Because photoevaporation extends to $\sim 0.1 - 0.2\,r_g$, low-mass stars with $M_\star < 0.5\,M_\odot$ are evaporated and shrink to disk radii smaller than 15 AU on timescales $t < 10^7$ yr when exposed to moderate FUV fields a factor of $10^3$ times the local interstellar FUV field. Such a field is produced, for example, by an early B star at about 1 pc distance. Disks around Solar-type stars require a 10-times stronger FUV field to shrink to radii smaller than 15 AU on the same timescale as the low-mass counterparts.

The effectiveness of photoevaporation also depends on the evolutionary status of the disk. An early illumination of a collapsing cloud removes a fraction of the material with the initial approach of the ionization front. This event leaves behind a smaller mass reservoir for the formation of the star–disk system. Stars with lower mass and disks with smaller radii are the result. The consequences are strongest if the onset of illumination is before one free-fall time of the parental molecular cloud core (Richling and Yorke, in preparation). This mechanism of an early photoerosion of prestellar cores can also explain the formation of free-floating brown dwarfs or planetary-mass objects (Whitworth and Zinnecker, 2004).

## 3.4 Photoevaporation by the central star

Photoevaporation of disks by the central star is most effective in the case of high-mass stars, where the central star itself produces a high rate of EUV photons up to $10^{50}$ photons s$^{-1}$ for the most massive O-type stars (Sternberg *et al.*, 2004). This case was investigated with semi-analytical models by Hollenbach *et al.* (1994). Their analysis yields a mass-loss rate due to EUV photons of

$$\dot{M}_{\mathrm{EUV}} = 4 \times 10^{-10}\,\mathrm{M_\odot\,yr^{-1}} \left( \frac{\Phi_{\mathrm{EUV}}}{10^{41}\,\mathrm{s^{-1}}} \right)^{0.5} \left( \frac{M_\star}{\mathrm{M_\odot}} \right)^{0.5}. \tag{3.4}$$

This result applies for both high- and low-mass central stars, and is valid for a weak stellar wind. For high-mass stars, $\dot{M}_{\mathrm{EUV}}$ can be as high as $10^{-5}\,\mathrm{M_\odot\,yr^{-1}}$, so that the outer ($r > 0.2\,r_g$ with $r_g \sim 50$ AU) portions of disks around high-mass stars would evaporate in $\sim 10^5$ years. The effect of a strong stellar wind is such that the ram pressure reduces the scale height of the atmosphere above the disk and the EUV photons are allowed to penetrate more easily to larger radii. This slightly increases the mass-loss rate from the outer parts of the disk. Nevertheless, the most important parameter is the EUV photon rate $\Phi_{EUV}$. It is noteworthy that the diffuse EUV field, caused by recombining electrons and protons in the disk's ionized atmosphere inside $r_g$, controls the EUV-induced mass-loss rates (Hollenbach *et al.*, 1994). Radiation hydrodynamic simulations (Yorke and Welz, 1996; Richling and Yorke, 1997) find a similar power-law index for the dependence of the mass-loss rate on the EUV photon rate of the central star.

For high-mass stars this mass-loss rate results in disk lifetimes which are long enough to explain the rather high abundance (and implied $10^5$ yr lifetimes) of ultra-compact HII regions (UCHII regions). Diagnostic radiative transfer calculations (Kessel *et al.*, 1998) of the hydrodynamic results show that the photoevaporating-disk model can explain many properties of UCHII regions, e.g. the spectral energy distribution (SED) and the spatial appearance in high-resolution radio maps in the case of the bipolar source MWC 349 A. A more recent work using a parametric description of the evaporating wind (Lugo *et al.*, 2004) also finds that the SED of MWC 349 A can be very well fitted by a photoevaporating disk model.

Is photoevaporation by the central star also important in the case of Solar-mass stars? To date, the only published models of photoevaporation by a central low-mass star rely on the very uncertain EUV luminosity of the central star. Shu *et al.* (1993) showed that with an EUV luminosity of $10^{41}$ photons s$^{-1}$, the early Sun could have photoevaporated the gas beyond Saturn before the cores of Neptune and Uranus formed, leaving them gas poor.

Clarke *et al.* (2001) assumed similar EUV luminosities and find that after $\sim 10^6$ to $10^7$ years of viscous evolution relatively unperturbed by photoevaporation, the viscous accretion inflow rates fall below the photoevaporation rates at $r_g$. At this point, a gap opens up at $r_g$ and the inner disk rapidly (on an inner disk viscous timescale of $\sim 10^5$ yr) drains onto the central star or spreads to $r_g$ where it evaporates. In this fashion, an inner hole is rapidly produced extending to $r_g$. The outer disk then evaporates on a longer timescale. These authors find the rapid transition from classical T Tauri stars to weak-line T Tauri stars in concordance with observations.

Matsuyama *et al.* (2003) pointed out that if the EUV luminosity is created by accretion onto the star, then, as the accretion rate diminishes, the EUV luminosity drops and the timescale to create a gap greatly increases. Alexander *et al.* (2004a) further pointed out that the EUV photons created by accretion are unlikely to escape the accretion column to irradiate the outer disk. However, Alexander *et al.* (2005) presented indirect observational evidence that an active chromosphere may persist

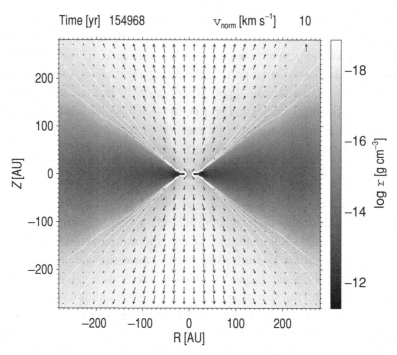

Fig. 3.2. Density distribution (greyscale) and velocity field (arrows) of a photo-evaporating disk due to FUV radiation from the central star. The velocity normalization is given at the top right. The white contour lines indicate the degree of ionization of carbon and are given for $x_C = 0.2, 0.4, 0.6$ and $0.8$ (Richling and Yorke, in preparation).

in T Tauri stars even without strong accretion, and that EUV luminosities of $> 10^{41}$ photons s$^{-1}$ may persist in low-mass stars for extended periods to illuminate their outer disks. Ruden (2004) provided a detailed analytic analysis which describes the evolution of disks in the presence of viscous accretion and photoevaporation and compares his results favorably with these two groups. Only if the EUV luminosity remains high due to chromospheric activity does EUV photoevaporation play an important role in the evolution of disks around isolated low-mass stars.

Models are needed of the photoevaporation of disks around low-mass stars caused by the FUV and X-ray luminosities of the central star. Radiation hydrodynamic simulations which include FUV as well as EUV radiation from the central star (Richling and Yorke, in preparation) indeed show that EUV photons are completely quenched by the material close to the star. In the absence of prolonged chromospheric activity which provides high levels of EUV, only FUV photons contribute to the photoevaporation of the disk and mass-loss rates of the order $\dot{M} \sim 10^{-7}\,\mathrm{M_\odot\,yr^{-1}}$ for $\Phi_{FUV} \sim 10^{44} - 10^{45}\,\mathrm{s^{-1}}$ are obtained. Fig. 3.2 shows an example from these calculations. In contrast to Fig. 3.1, the hydrogen ionization front (black line) does not appear at all, because the EUV photons are absorbed in the innermost cell. FUV radiation alone drives the evaporating flow.

Alexander *et al.* (2004b) studied the possibility that X-rays from the central star heat the disk surface. In comparison to UV radiation, the X-ray emission from T Tauri stars is well known from observations. They used a simple two-dimensional model in order to derive an upper limit for the mass-loss rate and find a lower mass-loss rate than in the case of UV radiation. In the presence of a significant UV field, X-ray driven disk winds seem to be unlikely to play a significant role in the evolution of disk winds around low-mass stars.

## 3.5 Photoevaporation and dust evolution

The previous sections solely addressed the evolution of the gas and its dispersal due to photoevaporation. However, when considering planet formation, it is also important to consider the evolution of the dust, which ultimately provides the solid material for rocky planets and the cores of gas giants. Once the disk has formed, the dust grains begin to coagulate and settle into the midplane (e.g. Weidenschilling, 1997b). The importance of photoevaporation during this stage of planetesimal formation depends on the ratio of the coagulation timescale to the evaporation timescale. Adams *et al.* (2004) found that in the case of external FUV illumination coagulation tends to take place more rapidly than mass loss for radii less than a cut-off radius of about 100 AU, even in relatively harsh environments. Thus, the formation of Kuiper-Belt objects should be possible around most stars.

Once coagulated dust has concentrated in the midplane, the roughly centimeter-sized particles can grow further by collisions or by local gravitational instability (Goldreich and Ward, 1973; Youdin and Shu, 2002). A numerical model by Throop and Bally (2005) investigated this phase of planetesimal formation. They followed the evolution of gas and dust independently and considered the effects of vertical sedimentation and external photoevaporation. Their results show that when dust particles grow and settle toward the midplane, the outer layer of the disk becomes dust depleted leading to dust-depleted evaporating flows. In contrast, the dust-to-gas surface-density ratio in the disk midplane grows until it meets the gravitational instability criteria of Youdin and Shu (2002), indicating that kilometer-sized planetesimals could spontaneously form. The criteria may be met at later times or not at all in the case of disks which are not exposed to significant UV radiation. Thus, these results imply that photoevaporation may induce a rapid formation of planetesimals.

Later on, planetesimals collide and interact gravitationally to build terrestrial planets and the cores of giant planets. Gas-giant planets will only form at radii where large enough cores form to gravitationally accrete the remaining gas before the gas is completely evaporated. If our Solar System formed in the presence of a strong external FUV radiation field, photoevaporation could explain why Neptune and Uranus in our Solar System are gas poor, whereas Jupiter and Saturn are relatively gas rich (Adams *et al.*, 2004). Alternatively, as discussed above, the EUV or FUV photons from the central star may be sufficient to cause the gas-poor nature of Neptune and Uranus.

## 3.6 Conclusions

Photoevaporation is an important gas dispersal mechanism and affects the formation of planets. Early models made the simple approximation of zero mass loss inside of the characteristic radius $r_g$. For FUV-induced photoevaporation, $r_g$ was of order 50–100 AU for low-mass stars, and therefore FUV photoevaporation would appear to be negligible in the planet-forming region. However, recent results show that substantial photoevaporation occurs to radii as small as $0.1\,r_g$, or 5–10 AU, and therefore FUV photoevaporation can play a significant role in planet formation.

External illumination is an important dispersal mechanism for disks around low-mass stars in clusters that harbor O or B stars. For example, photoevaporating-disk models can explain the mass-loss rates of $10^{-7}$–$10^{-6}\,M_\odot\,\mathrm{yr}^{-1}$ which have been obtained from observations for the photoevaporating disks in the Orion Nebula (Henney and O'Dell, 1999). These rates imply lifetimes as short as $10^5$ years, at least in the outer $r > 0.1\,r_g$ regions. This could not only stop gas-giant formation in these regions, but even the formation of Kuiper-Belt objects.

Internal illumination can also trigger significant photoevaporation of a disk around a central illuminating star. High-mass stars rapidly ($\sim 10^5$ yr) photoevaporate the outer ($r > 5$ AU) portions of their disks, leaving little time for planet formation there. Assuming viscous timescales in the inner disks are also of that magnitude, there is little time for planet formation at all around high-mass stars. It appears that photoevaporation by the EUV produced by a low-mass star may not significantly affect its disk evolution, which is largely controlled by viscous accretion. The only caveat to this conclusion is that significant effects may occur if the EUV luminosity remains elevated, at $\sim 10^{41}$ photons $\mathrm{s}^{-1}$ levels, for extended periods after accretion has diminished. The effects of the X-rays from the low-mass star appear to cause insignificant photoevaporation. The effects of the FUV from the low-mass star is under investigation. Preliminary hydrodynamic models suggest rather high mass-loss rates, which will have important effects on planet formation in the outer portions of disks.

## Acknowledgments

SR acknowledges the financial support provided through the European Community's Human Potential Programme under contract HPRN-CT-2002-00308, PLANETS. DH acknowledges support by the NASA Origins and Astrophysical Theory Programs and HWY acknowledges support by NASA under grant NRA-03-OSS-01-TPF. Portions of the research described here were conducted at the Jet Propulsion Laboratory (JPL), California Institute of Technology.

# 4

# Turbulence in protoplanetary accretion disks: driving mechanisms and role in planet formation

Hubert Klahr

*Max-Planck-Institut für Astronomie, Heidelberg, Germany*

Michał Różyczka

*N. Copernicus Astronomical Center, Warsaw, Poland*

Natalia Dziourkevitch

*Max-Planck-Institut für Astronomie, Heidelberg, Germany*

Richard Wünsch

*N. Copernicus Astronomical Center, Warsaw, Poland and Astronomical Institute, Academy of Sciences of the Czech Republic, Prague, Czech Republic*

Anders Johansen

*Max-Planck-Institut für Astronomie, Heidelberg, Germany*

## 4.1 Introduction

The observed characteristics of molecular clouds from which stars form can be reproduced by simulations of magnetohydrodynamic (MHD) turbulence, indicating the vital role played by magnetic fields in the processes of star formation. The fields support dense cloud cores against collapse, but they cannot do so indefinitely, because only charged particles couple to the field lines while neutral atoms and molecules can freely slip through. Through this process, called ambipolar diffusion, the cores slowly contract. The recombination rate in denser gas increases, causing the ionisation degree of the core to decrease. According to available observational data, once the core has contracted to $\sim 0.03$ pc it decouples from the magnetic field and enters the dynamic collapse phase. During the collapse the angular momentum is locked into the core and remains unchanged (Hogerheijde, 2004).

### 4.1.1 Protostellar collapse and formation of disks

The typical specific angular momentum of a core on the verge of dynamic collapse, $j_c$, amounts to $\sim 10^{21}$ cm$^2$ s$^{-1}$, and is many orders of magnitude larger than the

*Planet Formation: Theory, Observation, and Experiments*, ed. Hubert Klahr and Wolfgang Brandner.
Published by Cambridge University Press. © Cambridge University Press 2006.

typical specific angular momentum of a star (Hogerheijde, 2004). The inevitable conclusion is that the protostellar object resulting from the collapse must be surrounded by a large, rotationally supported disk (hereafter, protoplanetary disk) in which the original angular momentum of the core is stored. The outer radius of the disk, $r_d$, may be roughly estimated based on Kepler's law. Since

$$\Omega^2(r_d) = \frac{j_c^2}{r_d^4} = \frac{GM_c}{r_d^3}, \tag{4.1}$$

where $\Omega$ is the Keplerian angular velocity and $M_c$ is the mass of the core, we have

$$r_d = 500 \left( \frac{j_c}{10^{21} \, \text{cm}^2 \, \text{s}^{-1}} \right)^2 \left( \frac{M_c}{M_\odot} \right)^{-1} \text{AU}, \tag{4.2}$$

and indeed, disks with radii of a few hundred AU have been directly imaged around young stars (McCaughrean and O'Dell, 1996; Dutrey *et al.*, 1996).

Numerical simulations show that the disks resulting from the dynamic collapse can contain up to 30–40% of the original mass of the core; nevertheless they have nearly Keplerian rotational profiles (Yorke and Bodenheimer, 1999). Since the expected mass of the final planetary system is negligible compared to the final mass of the star, a very efficient mass loss from the disk must occur after the collapse. The mass can be lost via accretion onto the protostar and via outflows, which, at least in the early evolutionary phases of the disk, may take the form of well-collimated jets (Bacciotti *et al.*, 2003). The processes responsible for the outflows have been explored rather poorly. As they operate at the surface and/or at the very centre of the disk, whereas the planets are thought to form in the midplane of the disk, we usually neglect them when considering planet formation. We should remember however, that they may cause a significant redistribution of angular momentum within the disk, thus modifying the accretion rate and the conditions at the midplane.

### 4.1.2 Observations of accretion in protoplanetary systems

Interferometric observations of the CO line emission demonstrate that the circumstellar material around young stars has a flattened structure and is in Keplerian rotation (Simon *et al.*, 2000). Dust grains suspended in the gas scatter the stellar light, often revealing the disk-like geometry. Direct images of these disks have been obtained by the Hubble Space Telescope and adaptive optics systems on ground-based telescopes (e.g. Weinberger *et al.*, 2002). Continuum images in the millimeter range suggest that most of the mass is located at rather large distances from central objects ($r \geq 30$–50 AU). Using a gas-to-dust ratio of $10^2$, analyses of the dust emission indicate that the total (dust + gas) masses are in the range 0.001 to 0.1 $M_\odot$.

However estimating the disk mass is difficult and the present estimates are rather uncertain (Dutrey *et al.*, 2004).

The lifetimes of protoplanetary disks range from $10^6$ to $10^7$ yr (Haisch *et al.*, 2001). While similar estimates result from the meteoritic evidence obtained in the Solar System (Russell *et al.*, 1996), the large dispersion in the disk lifetimes remains puzzling. The accretion rate $\dot{M}$ decreases by several orders of magnitude within the lifetime of a disk (Hartmann *et al.*, 1998). The youngest, highly embedded systems accrete at a few $10^{-5}$ $M_\odot$ yr$^{-1}$, and in systems which undergo FU Orionis-type outbursts $\dot{M}$ can reach up to a few $10^{-4}$ $M_\odot$ yr$^{-1}$ (Hartmann and Kenyon, 1996). Presently, the most reliable estimates of $\dot{M}$ are based on measurements of the excess emissions superimposed onto the intrinsic photospheric spectrum of the central object. It is generally accepted that this excess arises from the accretion shock formed as disk material falls onto the photosphere (either directly or along magnetic-field lines). Characteristic of this type of flow are emission lines, whose equivalent widths decrease as the system ages. Best studied are low-mass objects ($0.1$ $M_\odot < M < 1$ $M_\odot$), which, depending on the equivalent width of the $H_\alpha$ line, are classified as either classical or weak-line T Tauri stars (CTTS or WTTS). Based on a large sample of CTTS with known $\dot{M}$ it was found that these objects typically accrete at a rate of $10^{-8}$–$10^{-7}$ $M_\odot$ yr$^{-1}$ (Calvet *et al.*, 2004).

Because of their weaker infrared excesses and emission lines, indicating weak or non-existent accretion activity, WTTS are usually considered as descendants of CTTS. However, a significant fraction of the WTTS have ages not much different from those of CTTS (Calvet *et al.*, 2004). The low percentage of stars with properties intermediate between CTTS and WTTS indicates that the disk dispersal time is of the order of only $\sim 10^5$ yr. Taken together, these observations suggest that the transition from the evolutionary phase with an active protoplanetary disk to the phase in which the disk is essentially undetectable is very rapid (Duvert *et al.*, 2000). Based on a positive correlation observed between $\dot{M}$ and the rate of mass loss from the system, one may infer that a vital role is played here by stellar and/or disk winds (the latter being powered or at least aided by irradiation).

### 4.1.3 Self-gravity and the early evolution of disks

A disk whose mass is significant compared to the mass of the protostar may develop gravitational instabilities which operate on a timescale of a few orbits of the unstable region (usually located in the outer part of the disk). The susceptibility of a thin disk to its own gravity can be estimated with the help of the Toomre parameter,

$$Q \equiv \frac{c_{s,m}\kappa}{\pi G \Sigma} = 60 \left(\frac{M_\star}{M_\odot} \frac{T_m}{100 \text{ K}}\right)^{0.5} \left(\frac{r}{\text{AU}}\right)^{-1.5} \left(\frac{\Sigma}{10^3 \text{ g cm}^{-2}}\right)^{-1}, \qquad (4.3)$$

where $c_{s,m}$ and $T_m$ are the sound speed and temperature at the midplane, $\kappa$ is the epicyclic frequency (in a nearly Keplerian disk $\kappa \approx \Omega$), $M_\star$ is the mass of the central object, $\Sigma$ is the column density, and it is assumed that the disk is composed of pure molecular hydrogen. If $Q \leq 1$, the disk is locally linearly unstable and may fragment.

Equation (4.3) holds for disks that are *locally isothermal* (the disk may be regarded as locally isothermal if in a thin annulus located at a distance $r$ from the central object the temperature $T$ varies with $z$ in such a way that the bulk of mass has a temperature nearly equal to $T_m$). For such disks the equation of hydrostatic equilibrium in $z$ takes the form

$$c_{s,m}^2 \frac{d\rho}{dz} = -\rho\Omega^2 z, \tag{4.4}$$

from which an approximate relation

$$H = c_{s,m}/\Omega \tag{4.5}$$

follows, relating the scale height of the disk, $H$, to local sound speed and angular velocity. Inserting (4.5) into (4.3) yields

$$Q \simeq \frac{H}{r_d} \frac{M_\star}{M_d}, \tag{4.6}$$

where $M_d \equiv \pi r_d^2 \Sigma$ is the approximate mass of the disk. The ratio $H/r_d$ observed in protoplanetary systems typically ranges from 0.1 to 0.2, indicating that disks with $M_d/M_\star \gtrsim 0.2$ can be unstable.

There is little doubt that disks with $Q < 1$ do indeed fragment on an orbital timescale, but the final outcome of the non-linear evolution of disks with $Q \approx 1$ remains controversial (Pickett *et al.*, 2003). The issue of whether a disk formed from a collapsing molecular core can be prone to fragmentation due to self-gravity is also confused. At the end of the dynamic collapse the disk is rather hot, and it is unlikely that $Q$ could approach 1 anywhere within it. Radiative cooling reduces $Q$; however instead of fragmenting, the disk could regain stability by reducing $\Sigma$ in the unstable region and/or by heating itself up. Heating can occur directly due to trailing spiral shocks that are excited as $Q$ decreases (Laughlin and Różyczka, 1996), and/or through the dissipation of "gravito-turbulence" (Gammie, 2001). Both spiral shocks and gravito-turbulence increase the effective viscosity in the disk, thus increasing the accretion rate. It is conceivable that an equilibrium between radiative cooling and accretional heating can be reached for $Q > 1$.

Another, frequently forgotten, effect of self-gravity is that it modifies the vertical structure of the disk. The modifications can be significant even when $M_d$ is small compared to $M_\star$ (Hure, 2000).

### 4.1.4 Viscous evolution

The evolution of a gravitationally stable disk is usually described in terms of viscous diffusion. A viscous disk decays on a *viscous timescale*,

$$\tau_v = R_d^2 \nu^{-1}, \tag{4.7}$$

where $\nu$ is the effective viscosity coefficient. Given the sizes of protoplanetary disks, a viscosity

$$\nu = 2 \times 10^{17} \left(\frac{r}{500 \text{ AU}}\right)^2 \left(\frac{\tau_d}{10^7 \text{ yr}}\right)^{-1} \text{ cm}^2 \text{ s}^{-1} \tag{4.8}$$

is needed for $\tau_v$ to be comparable to the observed lifetimes. The source of such a large viscosity (many orders of magnitude exceeding the microscopic viscosity of the gas) remains unknown. Various candidates, among which the magnetorotational instability (MRI) is widely regarded as the most promising one, are discussed in the following sections of this chapter. Here we focus on a simple approach, originally proposed by Shakura and Sunayev (1973), on which most models of the disks have been based. In this approach it is assumed that the viscosity originates from turbulent motions, and the viscosity coefficient is defined by

$$\nu = \alpha c_{s,m} H, \tag{4.9}$$

where $\alpha$ is a dimensionless parameter and $H$ (the disk scale height) is a natural upper limit for the size of turbulent eddies. Adopting (4.9), one effectively assumes that the characteristic velocity associated with the largest eddies is $\alpha c_{s,m}$. As the turbulent motions are most probably subsonic (otherwise they would quickly dissipate due to shocks), $\alpha$ must be smaller than 1.

According to the Definition (4.9), the viscosity coefficient depends on $r$ only. Using (4.5), one obtains an alternative expression,

$$\nu = \alpha c_{s,m}^2 \Omega^{-1}. \tag{4.10}$$

Equation (4.10) is often cast into a more general form (Bell *et al.*, 1997), e.g.

$$\nu = \alpha c_{s,m}^2 \Omega^{-1} = \frac{\alpha P}{\Omega \rho}, \tag{4.11}$$

which allows $\nu$ to vary both in $r$ and $z$, making it suitable for two-dimensional modelling (Hure, 2000; Klahr *et al.*, 1999). Note that Eq. (4.11) implies a shear stress proportional to the pressure.

In the absence of external torques the viscous disk conserves its angular momentum. While most of its mass *loses* angular momentum and is accreted onto the central body, a small amount of mass *gains* angular momentum and moves away from the central body. The orbital energy of the accreted matter is transformed into heat. A thin stationary disk is heated at a rate

$$Q_v = \frac{9}{4} \left( \frac{GM}{r^3} \right)^{0.5} \nu \Sigma \tag{4.12}$$

per unit area (for the derivation see e.g. Spruit, 2001), and cooled at the same rate by the radiative flux emerging from its surface. The vertically integrated enthalpy divided by $Q_v$ defines the *thermal timescale*, $\tau_{th}$, on which substantial changes of temperature distribution occur. Neglecting factors of order unity we have

$$\tau_{th} \approx c_s^2 \Sigma / Q_v = (\alpha \Omega)^{-1} . \tag{4.13}$$

The orbital or *dynamical* timescale $\tau_d \equiv \Omega^{-1}$ is shorter than $\tau_{th}$, which in turn is shorter than the viscous time scale $\tau_v$ (note that $\tau_v = r_d^2/(\alpha c_s H) = \tau_{th}(r_d/H)^2$).

Modeling and observations suggest that in protoplanetary disks $\alpha \approx 10^{-3}$–$10^{-2}$. It must be remembered, however, that the "$\alpha$-approach" is a rather rough research tool which provides but a superficial insight into the workings of the disks. It fails, for example, to reproduce the aforementioned rapid dispersal of protoplanetary disks. An "$\alpha$-disk" evolves on a single, viscous timescale, which in order to re-produce the observed lifetimes has to be of the order of $10^6$–$10^7$ yr. Based on such a slow dispersal one would expect to find a substantial class of objects with properties intermediate between CTTS and WTTS (Armitage *et al.*, 2003), in clear disagreement with the observations. Moreover, observational evidence, numerical simulations and theoretical arguments indicate that $\alpha$ is a free function of local parameters of the disk rather than a constant (Hure, 2000; Fleming *et al.*, 2000).

## 4.2 Magnetohydrodynamic turbulence

Magnetic fields might be assumed to be weak and not important for astrophysical disks. Such an assumption would make the physics much easier, but the angular momentum problem can be hardly solved without the additional viscosity driven by magnetic instabilities. The situation where magnetic fields might not be important is characterized by the inequality

$$\frac{\langle B \rangle^2}{8\pi} \ll \frac{\rho \langle u \rangle^2}{2}, \tag{4.14}$$

where $\langle B \rangle$ and $\langle u \rangle$ are the mean magnetic and velocity fields.

Keplerian disks are hydrodynamically stable configurations according to the Rayleigh criterion,

$$\frac{d j^2}{dr} > 0, \tag{4.15}$$

where $r$ is the radius and $j = r^2\Omega$ is the angular momentum per unit mass.

Nevertheless, differentially rotating gaseous disks with sufficiently high conductivity are unstable under the influence of a weak magnetic field. The magneto-rotational instability (MRI) has been known for more than three decades (Velikhov, 1959; Chandrasekhar, 1960, 1961), but the importance for accretion disks was first pointed out by Balbus and Hawley (1991).

The two important conditions for MRI are firstly that there is a Keplerian rotational profile in the disk and secondly, that fluid elements are connected via magnetic field lines so they can exchange momentum. Now consider the following similar situation: there are two satellites orbiting the earth and chasing each other on the same orbit. This situation is stable with respect to little perturbations in the satellite orbit. But now connect them with a rubber band and the situation changes; if one ever so slightly kicks one of the satellites towards a lower orbit it will accelerate its angular velocity due to the conservation of angular momentum (Kepler profile). Thus, the distance between both satellites increases and the rubber band becomes stretched (MHD field lines become bent). The rubber band is now breaking the first satellite and putting a torque on the second satellite. More precisely, the first satellite loses angular momentum which is put onto the second satellite. Because now the inner satellite has too little angular momentum to maintain its orbit it further declines and the second satellite with excess angular momentum moves radially outward (Kepler profile). Thus the tension of the rubber band between both satellites does not bring them closer together, but separates them even more, and the instability loop is closed. Finally one satellite will drop into the Earth's atmosphere and the other can be expelled or put in a larger orbit, at least until the rubber band is torn apart.

Now replace satellites with gas parcels and the rubber band with magnetic fields and you have an idea how the MRI works.[1]

### 4.2.1 Non-ideal magnetohydrodynamics

The presence of a weak initial magnetic field in the background can always be expected, but how will the idealised MRI action be changed in the case of the low conductivity expected in protoplanetary disks (see e.g. Semenov *et al.*, 2004)?

---

[1] Warning: it would be wrong to identify the magnetic-field lines with the rubber bands themselves, as a magnetic field line does not pull on electrons moving along with the field lines but those moving perpendicular to the field (Lorentz force). The real MRI situation is thus slightly different in so far as a vertical field line connects two fluid parcels at the same radius but differents heights above the midplane. This field line can now exert a force on the fluid elements when they move apart, because the field line bends.

The basic dynamic equations of ideal MHD are:

$$\frac{D \ln \rho}{Dt} + \nabla \cdot u = 0, \tag{4.16}$$

$$\frac{Du}{Dt} + \frac{1}{\rho} \nabla (P + \Psi) - \frac{1}{4\pi\rho} B \times [\nabla \times B] = 0, \tag{4.17}$$

$$\frac{\partial B}{\partial t} - \nabla \times (u \times B) = 0. \tag{4.18}$$

The notation $D/Dt$ indicates the Lagrangian derivative, $\Psi$ is an external (point-source) gravitational potential, and $P$ is pressure. Other symbols have their usual meanings ($u$ for velocity fields, $B$ for magnetic fields, and $\rho$ for density).

Non-ideal effects appear if the gas has a non-vanishing resistivity and is only partly ionized. In this case the induction equation changes to

$$\frac{\partial B}{\partial t} = \nabla \times [u_e \times B - \eta \nabla \times B], \tag{4.19}$$

where $u_e$ is the velocity of electrons and $\eta$ is the magnetic diffusivity. In ideal MHD the velocity of electrons, ions and the neutral gas is identical, but not so now. Under the assumption that all atoms are ionized only once (i.e. we ignore the continuous ionization and recombination processes), we can expand the electron velocity in terms of the neutral gas velocity and the relative velocity with respect to the ions $u_i$:

$$u_e = u + (u_e - u_i) + (u_i - u) = u - \frac{J}{n_e e} + \frac{J \times B}{c \gamma \rho_i \rho}. \tag{4.20}$$

Substituting the last relation into Eq. (4.19) leads to the appearance of new terms on the right-hand side of the induction equation,

$$\frac{\partial B}{\partial t} = \nabla \times \left[ u \times B - \frac{4\pi \eta J}{c} - \frac{J \times B}{n_e e} + \frac{(J \times B) \times B}{c \gamma \rho_i \rho} \right], \tag{4.21}$$

where $n_e$ is the electron number density, $\rho_i$ is the ion density and $\gamma$ is a drag coefficient between ions and neutrals. The terms on the right-hand side are called, correspondingly, inductive (I), Ohmic (O), Hall (H) and ambipolar diffusion (A) terms. The relative importance of non-ideal terms in the induction equation can be represented in the following way (Balbus and Terquem, 2001):

$$\frac{I}{O} \sim \frac{1}{Re_m}, \quad \frac{H}{O} \sim \frac{\omega_{ce}}{\nu_{en}}, \quad \frac{A}{H} \sim \frac{\omega_{ci}}{\gamma \rho}, \tag{4.22}$$

where $Re_m$ is the magnetic Reynolds number, and $\omega_{ce}$ and $\omega_{ci}$ are the electron and ion cyclotron frequencies. The magnetic Reynolds number is a dimensionless measure of the relative importance of resistivity,

$$\mathrm{Re_m} = \frac{V_A h}{\eta},$$  (4.23)

with the Alfenic speed $V_A = B/\sqrt{4\pi\rho}$ and $h$ as a characteristic length (e.g. the scale height $H$ of a disk of radius $r$ with $H/r \ll 1$). The magnetic Reynolds number is the ratio of dynamical scales to diffusive timescales. In systems with $\mathrm{Re_m} < 1$ the dissipation will kill all MHD processes.

### 4.2.2 Ohmic dissipation

In general, Ohmic dissipation effectively puts a restriction on the wave numbers of the unstable modes. The turbulence with the smallest wavelengths is dissipated, while the modes with larger wavelengths will have lower growth rates. When the smallest unstable wavelength is larger or equal to the disk thickness then the MRI cannot develop any more. In the work of Balbus and Terquem (2001) this was demonstrated to be the case for $\mathrm{Re_m} < 100$.

### 4.2.3 Ambipolar diffusion

The ambipolar diffusion term (A) can be important in outer parts of the protostellar disks. Here the ions interact with the magnetic field, which then couple the magnetic field to the neutrals via ion-neutral collisions. Ambipolar diffusion acts as a field-dependent resistivity. It leads to the heating of the gas. The growth rate of MRI with ambipolar diffusion is smaller compared to the ideal MHD case (Padoan *et al.*, 2000).

### 4.2.4 Hall term

The Hall term, on the other hand, is in fact an additional electromotive force and can basically act as an additional boost for the MHD turbulence pushing the critical Reynolds numbers down by one or two orders of magnitude. The dependence of the saturation level of Maxwell and Reynolds stresses on the presence and strength of the Hall effect (together with the Ohmic term) was analysed with help of two- and three-dimensional simulations (Sano and Stone, 2002a, b).

The value of $\mathrm{Re_{m,crit}}$, in general, depends on the field geometry (Fleming *et al.*, 2000); if there is a mean vertical field present, it is about 100, otherwise, it is much larger, of the order $10^4$. However, there are indications that effects of Hall electromotive forces result in smaller $\mathrm{Re_{m,crit}}$ (Wardle, 1999; Balbus and Terquem, 2001).

For a protoplanetary disk the Ohmic dissipation will dominate in every dust-rich region. Only when dust is settled down to the midplane, the Hall term dominates in the upper layers over the Ohmic dissipation for $r < 10$–$100$ AU, which still depends on the strength of magnetic field. Inside of $r < 1$–$5$ AU, Ohmic dissipation suppresses all MRI modes, except for $r < 0.1$ AU where thermal ionization becomes important and the dust is evaporated.

We conclude that the stability of the disk is determined by the local ionization structure and that some parts of the disk can be perfectly stable while some other regions produce strong turbulence. We will elaborate on that in the following section.

## 4.3 Layered accretion

Both numerical simulations (Hawley *et al.*, 1995) and analytical work (Balbus and Papaloizou, 1999) show that the MHD turbulence leads to a viscosity consistent with the description in Eq. (4.10) however, $\alpha$ may vary both in space and time. A good coupling between the gas and the magnetic field is necessary for the MRI to operate. Since protoplanetary disks are weakly ionized, in some regions the resistivity of the disk may be so high that the magnetic field decouples from the gas and the MRI decays. The MRI-free region is usually referred to as a *dead zone*, whereas the remaining part of the disk is referred to as *active*.

Assuming that electrons are the main current carriers, which is true if the density of small grains is low (Gammie, 1996), the resistivity is connected to ionization degree $x_e = n_e/n_n$ via the relation (Hayashi, 1981)

$$\eta = 6.5 \times 10^3 x_e^{-1} \text{ cm}^2 \text{ s}^{-1}. \tag{4.24}$$

Inserting Eqs. (4.24) and (4.9) into (4.23) and using $H = c_s/\Omega$, where $\Omega = \sqrt{GM/r^3}$ is the Keplerian rotation frequency, we obtain

$$\text{Re}_m = 7.4 \times 10^{13} x_e \alpha^{1/2} \left(\frac{r}{AU}\right)^{3/2} \left(\frac{T}{500 \text{ K}}\right) \left(\frac{M}{M_\odot}\right)^{-1/2}, \tag{4.25}$$

where $G$ is the constant of gravity, $M$ is the mass of the central star and $r$ is the radius. Equation (4.25) yields the threshold value of ionisation degree $x_e$; the MRI can operate if $\text{Re}_m > 100$ which corresponds approximately to $x_e \geq 10^{-12}$.

The most important sources of free electrons are collisional (thermal) ionization (important in the inner region with a high temperature), and the ionization by cosmic rays and X-rays emitted by the central star (important in the surface layers of the disk). The ionization due to the radioactive decay of $^{40}$K and $^{26}$Al is also sometimes considered, but the resulting ionization degree is several orders of magnitude lower than necessary for MRI (Stepinski, 1992).

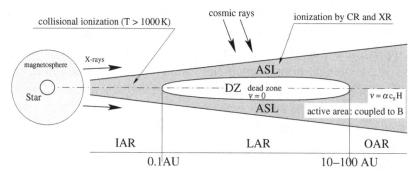

Fig. 4.1. The schematic structure of the layered disk: the inner active region (IAR), the layered accretion region (LAR) with two active surface layers (ASL) and the dead zone (DZ), and the outer active region (OAR).

This led Gammie (1996) to propose a model of a layered disk with the following structure (see Fig. 4.1): near the star there is an inner active region where the high temperature is responsible for sufficient ionization. At medium radii there is a layered accretion region, with active surface layers ionized by cosmic rays and X-rays, and the dead zone sandwiched between them near the midplane of the disk. The dead zone disappears at large radii where the disk surface density drops, and the cosmic rays and X-rays are able to penetrate the whole disk.

### 4.3.1 Ionization structure

In the case of thermal ionization, the main source of free electrons is the alkali metals potassium and sodium. Umebayashi (1983) determined the ionization degree $x_e$ as a function of density and temperature assuming a complete chemical equilibrium. He found that $x_e$ rises steeply from $10^{-16}$ to $10^{-11}$ near $T \simeq 1000$ K, mainly due to the ionization of potassium. In relevant density and temperature regions the Saha equation can be approximated as

$$x_e \equiv \frac{n_e}{n_n} = 6.47 \times 10^{-13} \left(\frac{a}{10^{-7}}\right)^{1/2} \left(\frac{T}{10^3}\right)^{3/4}$$
$$\times \left(\frac{2.4 \times 10^{15} \text{ cm}^{-3}}{n_n}\right)^{1/2} \frac{\exp(-25\,188/T)}{1.15 \times 10^{-11}}, \qquad (4.26)$$

where $n_e$ and $n_n$ are the electron and neutral number densities, respectively, and $a$ is the potassium abundance relative to hydrogen (Fromang *et al.*, 2002; Balbus and Hawley, 2000). Taking into account the nearly exponential dependence of $x_e$ on the temperature and relatively weak temperature dependence of other factors in

Eq. (4.25), it is usually assumed (Gammie, 1996; Fromang *et al.*, 2002) that material with a temperature above 1000 K is active.

The cosmic rays ionization rate is given by

$$\zeta_{CR} \simeq 10^{-17}\,\mathrm{s}^{-1} \tag{4.27}$$

(Spitzer and Tomasko, 1968). Cosmic rays are able to penetrate the surface layers of the disk down to a column density of $\sim 100\,\mathrm{g\,cm}^{-2}$ (Umebayashi and Nakano, 1981). However, the low-energy particles mainly responsible for the ionization may be excluded by the magnetized stellar winds (Fromang *et al.*, 2002) making ionization by cosmic rays ineffective.

T Tauri stars are strong sources of X-rays (e.g. Glassgold *et al.*, 2000) which are probably emitted by hot electrons in a magnetosphere. Their X-ray luminosities are $10^{29}$–$10^{32}$ erg s$^{-1}$ for photon energies between 1 and 5 keV (Feigelson *et al.*, 1993; Casanova *et al.*, 1995). The X-ray-ionization structure in the protoplanetary disk was determined by Glassgold *et al.* (1997). They used a simple, vertically isothermal model of the accretion disk with the mass of the minimum Solar Nebula. The X-ray source was modeled as an isothermal corona with temperature $T_X$, radius $r_X$ and total X-ray luminosity $L_X$. This model was extended by Igea and Glassgold (1999) taking into account the photon scattering using Monte Carlo simulations and by Fromang *et al.* (2002) taking the vertical temperature structure of the disk into account.

The ionization rate due to X-rays (Glassgold *et al.*, 1997; Fromang *et al.*, 2002) at a radius $r$ is

$$\zeta_{XR} = \frac{L_X/2}{4\pi r^2 k T_X}\sigma(kT_X)\frac{kT_X}{\Delta\varepsilon}J(\tau), \tag{4.28}$$

where $k$ is the Boltzmann constant. An interaction of X-ray photons with an atom or molecule produces a fast photo-electron which collisionally generates secondary electrons. Here $\Delta\epsilon = 37$ eV is the average energy necessary for the secondary ionization, while $\sigma(kT_X)$ is the ionization cross-section which may be approximated by a power-law,

$$\sigma(kT_X) = 8.5 \times 10^{-23}\left(\frac{kT_X}{\mathrm{keV}}\right)^{-n}\mathrm{cm}^2, \tag{4.29}$$

where the power-law index $n = 2.81$ is valid for the case where heavy elements are depleted due to grain formation. The dimensionless integral $J(\tau) = \int_{x_0}^{\infty} x^{-n}\exp(-x - \tau x^{-n})\,\mathrm{d}x$ comes from integrating the product of the cross-section and the attenuated X-ray flux over the X-ray spectrum above a threshold energy $x_0$. $\tau$ is the optical depth at an energy $kT_X$. For large $\tau$, $J(\tau)$ can be approximated as

$$J(\tau) \simeq A\tau^{-a} \exp(-B\tau^{-b}), \qquad (4.30)$$

where $A = 0.686$, $B = 1.778$, $a = 0.606$, and $b = 0.262$.

A description of recombination processes is necessary to translate the ionization rates $\zeta_{CR}$ and $\zeta_{XR}$ into electron fractions. It is usually assumed that the dust settles rapidly towards the midplane and the dominant recombination process is molecular dissociative recombination (Gammie, 1996). Then the ionization rate is given as

$$x_e = \sqrt{\frac{\zeta}{\beta n_n}}, \qquad (4.31)$$

where $\beta \sim 8.7 \times 10^{-6}\, T^{-1/2}\, \mathrm{cm^3\, s^{-1}}$ is the dissociative recombination coefficient (Glassgold *et al.*, 1986). For a more detailed treatment of recombination processes, including charge exchange between molecular ions and metal atoms, and radiative recombination due to capturing of electrons by heavy-metal ions we refer to Fromang *et al.* (2002).

Combining Eqs. (4.27) and (4.31) yields the ionization degree marginally sufficient to couple the gas to the magnetic field. So, provided that they are not removed by the stellar wind, cosmic rays are able to ionize the surface layers of the disk, and to couple the gas to the magnetic field there. The thickness of the active surface layer is given by the stopping column density of cosmic rays, for which a value of $\sim 100\, \mathrm{g\, cm^{-2}}$ is usually adopted.

Models of disk ionization by X-rays (Glassgold *et al.*, 1997; Fromang *et al.*, 2002) show that the column density of the active surface layer strongly depends on model parameters: actual properties of the X-ray source, the value of $Re_{m,crit}$, and the abundance of heavy metals in the disk (because of their relatively slow recombination). The outer radius of the dead zone varies from several AU to $\sim 100$ AU. For some models with low $Re_{m,crit}$ (due to the Hall term) or high heavy-metal abundances the dead zone may even disappear completely.

It is assumed in Eq. (4.31) that the medium is in chemical equilibrium. The plausibility of this assumption was tested by Semenov *et al.*, (2004), who computed the disk ionization degree using a network of chemical reactions that take place in the gas–dust medium. They found that the equilibrium approach is appropriate in most parts of the disk. However, in some places – especially in weakly ionized ones, the ionization degree can differ by more than an order of magnitude from the equilibrium value. Another problematic region is the intermediate layer between the surface layer and the dead zone. In the dead zone, most species are frozen out on grain surfaces, while in the surface layer and the disk atmosphere, they are destroyed by the high-energy stellar radiation. In the intermediate layer, the chemistry is very complex, and more than a hundred species and reactions have to be included to determine the ionization degree with reasonable accuracy.

### 4.3.2 Layered disk evolution

An analytical model of the surface active layer similar to the standard model for an accretion disk was obtained by Gammie (1996). He assumed that the main source of the ionization in the surface layer is the cosmic rays providing an active layer with a constant column density $\Sigma_a \simeq 100 \text{ g cm}^{-2}$. An interesting consequence of such an assumption is that the accretion rate in the layer, $\dot{M}$, is a function of $r$, implying that the mass accumulates in the dead zone at a rate of

$$\dot{\Sigma}_{DZ} = \frac{1}{2\pi r} \frac{d\dot{M}}{dr}. \tag{4.32}$$

Such accumulation cannot last forever – when the surface density of the disk reaches a value sufficient to maintain the temperature above $\sim 1000$ K by the accretion process, any perturbation can make the dead zone active. When this happens, the accretion rate and the disk luminosity dramatically increase, which led Gammie (1996) to suggest this mechanism as an explanation for the FU Orionis outbursts. This scenario was further studied by Armitage *et al.* (2001), who assumed that the dead zone becomes active due to the gravitational instability. Their model shows repeating periods of strong accretion separated by quiescent intervals lasting $\sim 10^5$ yr, which is in agreement with the observed properties of FU Orionis objects (see e.g. Hartmann and Kenyon, 1996).

On the other hand, the surface layer ionized by stellar X-rays exhibits a lower mass-accumulation rate, since its surface density decreases with the radius. The time for the accumulation of mass necessary for the gravitational instability is comparable to the lifetime of the protoplanetary disks (Fromang *et al.*, 2002).

Two-dimensional axially symmetric radiation hydrodynamic models of layered disks were obtained by Wünsch *et al.* (2005). The authors show that variations of the thickness of the dead zone influence the structure of the surface active layer, leading to growing perturbations of the mass accumulation rate. As a result, the dead zone splits into rings. The model also shows that radiation from the inner disk heats the dead zone and makes it active shortly after mass accumulation has started. These periodic "ignitions" of the inner part of the dead zone, that occur on a timescale of $\sim 100$ yr, continuously remove material from the dead zone. Therefore, the accumulation of the large amounts of mass necessary for the FU Orionis type event does not seem to be likely.

## 4.4 Alternative instabilities in the dead zone

One might ask the following questions: How "dead" is the "dead zone"? What happens if the entire disk is too low in ionization for MHD effects to be important? Is there a pure hydro instability in accretion disks?

Barotropic Keplerian disks are centrifugally stable as predicted by the Rayleigh criterion. Balbus *et al.* (1996) did extensive numerical and analytical investigations on this matter. On the other hand, some authors still claim the existence of a non-linear shear instability operating in accretion disks (e.g. Richard and Zahn, 1999). This is mainly motivated by the experiments by Wendt (1933), who found non-linear instability in a Rayleigh stable Taylor–Couette flow for the case of Reynolds numbers of the order $10^5$. However Schultz-Grunow (1959) demonstrated that the results of Wendt were due to flaws in the experimental setup and found the Taylor–Couette flow to remain stable at comparable Reynolds numbers. Nevertheless, the debate on non-linear instability is still open (Richard, 2003; Hersant *et al.*, 2005).

Anyway, if there is no self-gravity working, basically no ionization present and no shear instability possible, the disk would develop no turbulence, transport no angular momentum, release no accretion energy, and would cool down to the ambient temperature, while orbiting the star on a time-constant orbit. In such a disk, timescales for the growth of planetesimals would be much longer than in the standard scenarios, as the Brownian motion of the micron-sized dust is smaller and there is no turbulence acting on the dust in the centimeter to meter size range.

Sedimentation of the grains to the midplane could create a shear instability (Cuzzi *et al.*, 1993) which then generates turbulence. Cuzzi *et al.* (1996) argue that the Solar Nebula at one time must have been turbulent in order to explain the size segregation effects during the formation of chondrites. In any case we want to stress that it is of major relevance for planet formation to know whether protoplanetary disks are turbulent or not.

Klahr and Bodenheimer (2003) presented numerical simulations of a purely hydrodynamic instability that works in accretion disks, namely a baroclinic instability, similar to the one responsible for turbulent patterns on planets, for example, Jupiter's red spot, and the weather patterns of cyclones and anti-cyclones on Earth. Cabot (1984) investigated the possibility and efficiency of a baroclinic instability for cataclysmic variables (CVs) in a local fashion, concentrating on local vertical fluctuations. He found that this instability produces insufficient viscosity to explain CVs, but nevertheless enough viscosity ($\alpha \approx 10^{-2}$) for protoplanetary disks.

Ryu and Goodman (1992) considered linear growth of non-axisymmetric disturbances in convectively unstable disks. They used the shearing-sheet approximation in a uniform disk and found that the flux of angular momentum was inwards. Lin *et al.* (1993) performed a linear stability analysis of non-axisymmetric convective instabilities in disks but allowed for some disk structure in the radial direction. Their disk model includes a small radial interval in which the entropy has a small local maximum of 7% above the background but with a steep drop with an average slope of $K \sim R^{-3}$ ($K$ is the polytropic constant; see below). Such a situation could be baroclinically unstable, and in fact that region is found to be associated with outward transport of angular momentum. Lovelace *et al.* (1999) and Li *et al.* (2000) investigated the stability of a strong local entropy maximum in a thin Keplerian disk and found the situation to be unstable to the formation of Rossby waves, which transported angular momentum outward and ultimately formed vortices (Li *et al.*, 2001). Sheehan *et al.* (1999) studied the propagation and generation of Rossby waves in the protoplanetary nebula in great detail, but they had to assume some turbulence as a prerequisite.

Recently Klahr (2004) and Johnson and Gammie (2005) performed a local linear stability analysis for accretion disks with radial stratification, i.e. under the influence of a global radial entropy gradient. As a result the linear theory predicts a transient linear instability that will amplify perturbations only for a limited time or up to a certain finite amplification. This can be understood as a result of the growth time of the instability being longer than the shear time which destroys the modes that are able to grow. So only non-linear effects can lead to a relevant amplification. This could help to explain the observed instability in numerical simulations (Klahr and Bodenheimer, 2003) as an ultimate result of a transient linear instability.

We can conclude from all these investigations that there are so far no non-MHD instabilities proven to reliably operate in accretion disks. One might argue that this is also the case for the MRI and related MHD instabilities with respect to the uncertainties in the magnetic Reynolds numbers present in real protoplanetary disks. Nevertheless, MHD turbulence is simple to reproduce in a three-dimensional numerical simulation and thus the ideal working horse to be exploited for investigations about the influence of turbulence on the planet-formation process.

## 4.5 Transport by turbulence

Knowledge of the transport properties of particles and molecules embedded in a turbulent gas medium is of importance in many aspects of protoplanetary-disk modeling. If the spatial number density distribution of e.g. dust grains in a disk is required for the model, one must know the effect of turbulent diffusion on the dust grains.

*Vertical diffusion* – the distribution of tiny dust grains, with radii smaller than around 100 μm, determines the observability of protoplanetary disks around young stellar objects through their contribution to the infrared parts of the spectrum. An interesting observational effect of turbulent diffusion is its influence on the vertical settling of dust grains. The settling affects the spectral energy distribution of protoplanetary disks, since flaring disks, i.e. where the scale height of the gas density increases with radial distance, have a much stronger mid- to far-infrared excess than self-shadowing disks, where the scale height after a certain distance from the protostar begins to fall with radial distance (e.g. Dullemond, 2002). Recent model calculations by Dullemond and Dominik (2004) show that the vertical settling of dust grains towards the midplane of the disk can change an initially flaring disk to a partially self-shadowing disk, thus affecting the observability of the disk (see also Chapter 7 by Henning *et al.*). These calculations depend among other things on the strength of the turbulence in the disk (the turbulent viscosity) and on the turbulent diffusion coefficient of dust grains in the direction perpendicular to the disk midplane. Also recently Ilgner *et al.* (2004) considered the effect of vertical mixing in protoplanetary disks on the distribution of various chemical species and found the distribution to be greatly influenced by mass-transport processes, again underlining the importance of vertical turbulent diffusion in the modeling of protoplanetary disks.

*Radial diffusion* – crystalline silicate dust grain features observed in comet spectra are often attributed to radial mixing in the Solar Nebula (e.g. Hanner, 1999). Silicate dust grains are formed primarily in amorphous form, but they can become crystalline if exposed to temperatures above $\sim 1000$ K. Such a heating can obviously have occurred in the hot inner parts of the Solar Nebula, whereas comets are expected to have formed in the cold outer regions of the Nebula, so in this picture some radial mixing must take place between the inner and outer Nebula. Bockelée-Morvan *et al.* (2002) considered disk models where crystallization of silicates happens in the inner, hot parts of the disk. They calculate that in a few times $10^4$ yr the crystalline silicate fraction reaches a uniform value outside the crystallization region due to radial turbulent diffusion and that the value can approach unity for realistic disk parameters. From high-resolution observations of three protoplanetary disks van Boekel *et al.* (2004) found that the inner 1–2 AU of these disks contain a higher crystalline silicate fraction than the outer 2–20 AU, supporting the theory that crystalline dust grains form in the hot inner disk and are subsequently transported to the outer disk by turbulent gas motion.

### 4.5.1 Dust dynamics

Dust grains moving through the gas in a protoplanetary disk are affected by a friction force proportional to the velocity difference between dust and gas and in the opposite direction. The strength of the friction is determined by the friction time $\tau_f$. For sufficiently small dust grains at subsonic velocity the friction time is proportional to the radius of the dust grains and independent of the relative velocity between dust and gas (Weidenschilling, 1977a).

It is possible (with a few reasonable assumptions) to solve algebraically for the equilibrium dust velocity as a function of gas velocity and density. For not too large objects (e.g. <1 m) it can be assumed that the dynamic timescale of the gas is much longer than the friction time, an assumption valid, for example, for stationary disk solutions and also in turbulence when the friction time is much shorter than the turbulent eddy lifetime. Under those conditions one yields from the combined equations of motion for dust (velocity $w$) and gas (velocity $u$):

$$w = u + \tau_f \frac{1}{\rho} \nabla P . \tag{4.33}$$

One can interpret the short friction-time approximation, Eq. (4.33), as expressing a tendency for dust grains to climb up the local gas pressure gradient $\nabla P$ (see Johansen and Klahr, 2005, for details).

As an example of the use of the short friction-time approximation one can consider a laminar Keplerian disk where the vertical gravity is balanced by a pressure gradient due to stratification. Hydrostatic equilibrium in the vertical direction means that $0 = -\rho^{-1}\partial P/\partial z - \Omega_0^2 z$, so Eq. (4.33) gives the equilibrium dust velocity as $w_z = -\tau_f \Omega_0^2 z$. This can be either interpreted as dust grains falling due to vertical gravity, or in the short friction-time approach as dust grains climbing up the global pressure gradient (since there are no local pressure gradients in this laminar model).

Another example involves a radial pressure gradient $\partial \ln P/\partial \ln r = \alpha$. For an isothermal equation of state the pressure gradient force is given by $-\rho^{-1} \nabla P = -c_s^2 \nabla \ln \rho$, where $c_s$ is the constant isothermal sound speed. Using the short friction-time expression one gets $w_x = \tau_f(c_s^2/r)\alpha$. Thus for a pressure that falls radially outwards (negative $\alpha$), the dust grains will migrate inwards, a well-known problem in planet formation (Weidenschilling, 1977a).

### 4.5.2 Dust-trapping mechanisms

In the two previous examples only a global pressure gradient appears. But in a turbulent disk, local fluctuating regions of high and low pressure are expected to occur. Then dust grains continuously move up the local pressure gradient, potentially trapping dust grains in regions of high pressure and expelling them from regions of low pressure. An example of this is shown in Fig. 4.2. The plot is from Johansen *et al.* (2006) and shows on the left the gas column density in a magneto-rotational turbulence simulation. The right plot shows the corresponding dust

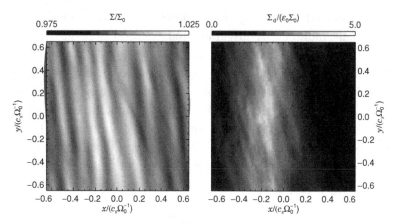

Fig. 4.2. Example of pressure-gradient trapping. Fluctuating regions of high gas density (left plot) lead to an enormous concentration of meter-sized protoplanetesimals (right plot). Notice that the gas column density is only increased by a few %, while the dust column density at the same place is around 400% above the average.

column density. The meter-sized dust particles have been concentrated up to 400% by the very modest gas-density fluctuations (part of the dust concentration is due to a global gas-pressure gradient forcing dust particles to drift inwards). Similar concentrations of meter-sized particles were found by Rice *et al.* (2004) in the overdense spiral arms of self-gravitating disks. The concentrations of dust particles in radial over-density regions has also been explored by Klahr and Lin (2001) and by Haghighipour and Boss (2003).

For larger dust grains, where the friction time becomes comparable to the orbital time of the disk, the short friction-time approximation is no longer valid. Here acceleration effects can become important. Any rotationally supported structure in the gas is kept in place due to an equilibrium between the global Coriolis force from the rotating disk and the centrifugal force of the rotating structure. Larger dust grains experience a slow acceleration (due to friction with the gas) as they enter such a structure. Rotating initially with the gas, but at a slightly slower rate, the Coriolis force dominates over the centrifugal force, and the dust grains are sucked into the eddy. This can be seen as the result of the mismatch between the epicyclic frequency of the particle (which does not feel any pressure) and the rotational frequency of the vortex, which is in pressure equilibrium.

This vortex-force trapping was proposed by Barge and Sommeria (1995), and has since then been the subject of much theoretical investigation (e.g. Tanga *et al.*, 1996; Chavanis, 2000; Johansen *et al.*, 2004). The conclusions are that vortices are extremely efficient at trapping dust grains, and this efficiency may even explain how gas planets are formed before the dispersion of the gas disk (Klahr and Bodenheimer, 2005). Vortex trapping seems to be much more efficient at trapping dust grains in turbulent features than pressure-gradient trapping (Johansen *et al.*, 2004), but it requires the presence of long-lived vortex structures (see Section 4.4).

In the absence of the Coriolis force, vortex trapping cannot occur. Here the centrifugal force of a rotating structure must always be balanced by a radially inwards-pointing pressure-gradient force. Rotating regions are thus low-pressure regions that can expel dust grains because of "inverse" pressure-gradient trapping. Concentrations between vortices have been proposed by Cuzzi *et al.* (2001) to give a preferential size-sorting in the Solar Nebula that can explain the size distribution of chondrules (millimeter-sized solid inclusions found in primitive meteorites; see e.g. Norton, 2002). Contrary to these concentrations between vortices one finds vertical convection cells in protoplanetary disks that concentrate dust particles, which again can be interpreted as a special case of pressure-gradient trapping (Klahr and Henning, 1997).

None of these local trapping mechanisms occur in a laminar disk, which shows how important gas turbulence is for the conditions under which planetesimals form.

### 4.5.3 Turbulent diffusion

Turbulent transport is often described mathematically as a diffusive process. Here the turbulent flux of dust number density $n$ is assumed to be proportional to the gradient of the dust-to-gas ratio (Dubrulle *et al.*, 1995). The effect of turbulent diffusion on the dust number density can be parametrized as

$$\frac{\partial n}{\partial t} = -\nabla \cdot \left[ w n - D_t \rho \nabla \left( \frac{n}{\rho} \right) \right].$$

(4.34)

This diffusion equation for dust can be solved for grains falling towards the midplane with velocity $w_z = -\tau_f \Omega_0^2 z$, as given by the short friction-time approximation. Solving now for the stationary state $\partial n / \partial t = 0$, one obtains an ordinary differential equation in the dust-to-gas ratio $\epsilon = \rho_d / \rho$:

$$\frac{\partial \epsilon}{\partial z} + \frac{\tau_f \Omega_0^2}{D_t} z \epsilon = 0$$

(4.35)

with the general solution $\epsilon = \epsilon_1 \exp[-z^2/(2H_\epsilon^2)]$. Here the scale height is given by the expression $H_\epsilon^2 = D_t/(\tau_f \Omega_0^2)$, so a vertically settled dust layer's ability to undergo gravitational instability, and thus the disk's ability to form planetesimals via self-gravity (Safronov, 1955; Goldreich and Ward, 1973), depends strongly on the value of the turbulent diffusion coefficient.

The diffusion description of particle transport does not provide any value for the turbulent diffusion coefficient $D_t$. This now leaves two important issues to consider: firstly what the value of the turbulent diffusion coefficient is compared to the turbulent viscosity, and secondly how it depends on particle size.

It is often assumed simply that $D_t \equiv \nu_t$. This argument is based on the fact that Reynolds stresses, which give rise to the turbulent viscosity, and diffusion stresses, which give rise to diffusion, are linked together by their common dependence on the velocity field. Thus fluctuating velocity fields should give around the same value of $\nu_t$ and $D_t$ (Tennekes and Lumley, 1972). Recently Johansen and Klahr (2005) did numerical simulations of dust grains embedded in magnetorotational turbulence. They found that radial diffusion is indeed as strong as turbulent viscosity, but since the major contribution to the turbulent viscosity is magnetic stresses, this actually means that dust diffusion is very much more effective than diffusion of angular momentum by Reynolds stresses. In magnetorotational turbulence the vertical velocity fluctuations are much smaller than the radial fluctuations (Brandenburg *et al.*, 1995). In accordance with this, Johansen and Klahr (2005) found that the turbulent diffusion coefficient is anisotropic with a diffusion weaker by a factor of two in the vertical direction.

A lot of analytical work has been devoted to determining the dependence of the diffusion coefficient on grain size (Safronov, 1955; Cuzzi *et al.*, 1993; Dubrulle *et al.*, 1995; Schräpler and Henning, 2004). According to Schräpler and Henning (2004), ignoring here the effect of the mean motion of the dust grains, the diffusion coefficient can be written as

$$D_t = \frac{D_0}{1 + St} . \qquad (4.36)$$

Here St is the Stokes number, and the factor $1/(1 + St)$ determines the variation of diffusion coefficient with particle radius. The Stokes number is defined as the ratio of the friction time to the turn-over time $\tau_c$ of the largest eddies. Defining the rotation speed of the largest eddies as $\nu_e = \alpha_t^q c_s$ and choosing $q = 0.5$, one can (following Schräpler and Henning, 2004) arrive at the expression

$$D_t = \frac{D_0}{1 + 4^{-1}\pi \Omega_0 \tau_f} . \qquad (4.37)$$

Thus for large grains with $\Omega_0 \tau_f > 1$, the diffusion coefficient falls rapidly with grain radius.

## 4.6 Conclusions

Turbulence in protoplanetary accretion disks is most likely similar to the weather on Earth. Sometimes it is stormy and sometimes there is not even a breeze. And while there is a hurricane in Florida, one can have sunshine in Heidelberg. The only difference might be that the storms in the disk can have a magnetic nature, which nevertheless will not change the local appearance of turbulence. The importance of turbulence for the planet-formation process is obvious. Turbulence mixes and transports the grains and molecules in the disk. Larger grains can even be segregated and concentrated in turbulent flow features like vortices and pressure maxima. Again there is another analogoly from meteorology: the largest hail stones are produced only in the strongest thunder storms, because they can stay airborne for a longer time and continue to grow. This clearly stresses the importance of turbulence for any kind of particle growth, thus also for the growth of planetesimals.

# 5

# The origin of solids in the early Solar System

Mario Trieloff

*Mineralogisches Institut, Heidelberg, Germany*

Herbert Palme

*Institut für Mineralogie und Geochemie, Köln, Germany*

## 5.1 Introduction: geoscience meets astronomy

Much of our knowledge about the formation of planets in the Solar System and in particular concepts and ideas about the origin of the Earth are derived from studies of extraterrestrial matter. Meteorites (Sears, 2004; Lauretta *et al.*, 2006; Krot *et al.*, 2006) were available for laboratory investigations long before space probes were sent out for *in situ* investigations of planetary surfaces, or Moon rocks were brought back to Earth. Meteorite studies provided such important parameters as the age of the Earth and the time of formation of the first solids in the Solar System (Chen and Wasserburg, 1981; Allegre *et al.*, 1995; Amelin *et al.*, 2002), as well as the average abundances of the elements in the Solar System (Anders and Grevesse, 1989; Palme and Jones, 2004). Traditionally, the study of rocky material requires techniques that fundamentally differ from astronomical techniques. While electromagnetic radiation from stars is analyzed by spectroscopy, the solid samples of aggregated cosmic dust and rocky matter from planetary surfaces require the use of laboratory instruments that allow the determination of their chemical and isotopic composition. Planetary surface materials are present in polymineralic assemblages. Formation conditions and thermal stability of individual minerals provide important boundary conditions for the genesis and history of the analyzed materials. Such studies require a thorough mineralogical background. The abundances and properties of the rock-forming elements, the focus of geo- and cosmochemical research, are, however, not necessarily of major concern to astrophysicists. The most abundant elements in the rocky planets and in meteorites are – in view of the whole Solar System – only minor elements and are thus insignificant in many objects of astrophysical studies. The fundamental circumstance that we live on a terrestrial planet implies

*Planet Formation: Theory, Observation, and Experiments*, ed. Hubert Klahr and Wolfgang Brandner.
Published by Cambridge University Press. © Cambridge University Press 2006.

that the cosmochemically most abundant elements (H, He, C, N, O) are heavily depleted, but these elements are dominant in the cosmos. While the light elements carry most of the mass, solid phases in rocky planets are the result of condensation and/or accretion of the remaining material, in astronomical terms often referred to as the "dust" component. Nevertheless, dust forms a major element in almost all models of protoplanetary disks since the dust provides virtually all of the opacity, and the dust component in protoplanetary disks is even easier to observe for astronomers than the gas (see also the review by Henning *et al.* in Chapter 7).

After the first protoplanetary disks were observed and the first extrasolar planet was discovered in the mid-nineties (Mayor and Queloz, 1995; McCaughrean and O'Dell, 1996), the astrophysical community rediscovered their interest in planetary sciences. Within the upcoming decade observations of extrasolar planets – most probably also Earth-like planets – and studies of protoplanetary disks will significantly increase our knowledge of the origin of planetary systems in general. As much as astrophysically oriented observational and theoretical advances in the last decade were achieved by new instrumental and computational capabilities, progress in the techniques of solid-matter research in the geosciences now allow precise elemental, isotopic and structural analyses of sub-micrometer grains of meteorites. With the most advanced instruments it is now possible to determine the isotopic composition of some elements in an area of less than 0.1 $\mu m^2$ (Messenger *et al.*, 2003). In addition, the total mass of available extraterrestrial material has increased significantly with the discovery of a huge number of meteorites from cold and hot deserts, and the collection of extraterrestrial dust has delivered additional material of possibly cometary origin. Thus cosmochemists will meet the renewed "planetary" interest of the astronomical community by providing a wealth of new discoveries and ideas concerning the formation and evolution of the accretion disk of our Solar System, the Solar Nebula 4.6 Gyr ago. Accordingly, we can expect in the forthcoming years an increasingly fruitful exchange of ideas and closer collaborations between the two scientific communities that will allow us to place our own Solar System into a completely new context (e.g. how unique is it?). The results that are available from our early Solar System will in turn have important consequences for theoretically and observationally constrained astronomical models of star and planet formation in general.

In this review we will provide an update of some of the fundamental concepts envisioned by cosmochemical studies on the timing of accretion of planetesimals and planets in the early Solar System, and we will consider fundamental processes that lead to compositional (isotopic and/or elemental) variability of early Solar-System matter. Detailed reviews on meteorite properties and their significance for the early Solar-System evolution are available in past and forthcoming literature (Papike, 1998; Sears, 2004; Lauretta *et al.*, 2005; Krot *et al.*, 2005).

## 5.2  Meteorites: remnants of planetesimal formation 4.6 billion years ago in the asteroid belt

Orbital reconstructions of a number of precisely observed meteorite falls (Lauretta *et al.*, 2005) demonstrate that most of the >28 000 meteorites in terrestrial collections are fragments of small bodies from the asteroid belt between Mars and Jupiter (except for some 60 Martian and Lunar meteorites). Asteroids are important in the origin of planets because modern theories of planet formation imply a hierarchical growth from dust to planets. Asteroids are the remnants of a swarm of planetesimals <1000 km in diameter. These planetesimals did not accrete to form a regular planet, most probably hindered by the early presence of Jupiter, which today still shapes the dynamic structure of the asteroid belt. Meteoriticists infer that ≈100 different planetesimals are needed to account for the present population of meteorites available for research (Sears, 2004; Lauretta *et al.*, 2005; Krot *et al.*, 2005). The major arguments supporting this scenario are:

(1)  Elemental and isotopic differences among meteorite classes preclude their origin from a single large planet and require distinct parent bodies.
(2)  Most meteoritic matter is from undifferentiated planetesimals and thus unaffected by major parent-body heating that occurred during or shortly after formation. This favors a multitude of small bodies incapable of retaining heat effectively.

Indeed, the major process affecting meteorite parent bodies after their formation was shock metamorphism (Stöffler *et al.*, 1991), caused by hypervelocity collisions in the asteroid belt (Michel *et al.*, 2001) resulting in shock waves with pressures of up to several 100 kbar. Impacts induced local heating effects, disturbed primary textures, and in extreme cases led to melting and occasionally to evaporation. Nevertheless a comparatively large number of undisturbed rocks are among the meteorites that escaped impact metamorphism and have retained signatures of their formation in the very early Solar System.

Other small bodies from the outer Solar System are comets, which formed in colder parts of the protoplanetary disk than asteroids. Their primitive character is ascertained by their high proportion of *volatile* elements. Their composition carries – as do asteroids for the region between Mars and Jupiter – information on the conditions in the early outer Solar System. However, the investigation of cometary matter via laboratory studies is limited to a very small amount of material and extremely small sample sizes: some cometary particles may have been identified among a certain class of interplanetary dust particles, which typically are 10 μm in size, but are at this level still heterogeneous, composed of many sub-particles, some only 100 nm in size. Analyses of cometary matter are technically challenging, but they will allow completely new insights into processing of matter in outer protoplanetary disks (Brownlee *et al.*, 1996; Bradley *et al.*, 1999; Jessberger *et al.*, 2001; Messenger *et al.*, 2003).

## 5.3 Calcium-aluminum-rich inclusions and chondrules: remnants from the earliest Solar System

Unlike the larger differentiated terrestrial planets that have metallic Fe–Ni cores, and silicate mantles and crusts, most meteoritic matter is from small, undifferentiated parent bodies. These meteorites have approximately Solar abundances of Fe and Ni, where Fe is partly present as metal, and partly oxidized residing in silicates. The term "undifferentiated" also refers to the side-by-side occurrence of phases formed at low and at high temperatures. Low-temperature phases are concentrated in the fine-grained matrix that is composed of µm-sized minerals and typically rich in volatile elements and compounds. High-temperature phases are rare, cm-sized calcium-aluminum-rich inclusions (CAIs) and abundant sub-mm to mm-sized droplets, crystallized from liquid silicates. From these so-called chondrules undifferentiated meteorites received the name chondrites (Fig. 5.1). These high-temperature components predate accumulation of planetesimals and record high temperature events in the early Solar Nebula. The two major classes of undifferentiated meteorites are ordinary chondrites – the most abundant class of meteorites delivered to Earth – and carbonaceous chondrites. Carbonaceous chondrites can have significant amounts of water and carbon and contain more matrix (30–100%; Table 5.1) and fewer chondrules than ordinary chondrites. Matrix often contains *hydrous minerals* resulting from ancient interaction of liquid water and primary minerals.

Calcium-aluminum-rich inclusions (Fig. 5.1) are composed of highly *refractory* elements (Al, Ca, Ti, and trace elements Zr, Hf, U, Th, rare earth elements, and also refractory metals W, Ir, Os, etc.). The composition is reflected in the mineralogy[1] with Ti, Al-rich pyroxene, melilite, spinel and anorthite (calcium-rich feldspar). This is compatible with a high-temperature origin close to 2000 K. Calcium-aluminium-rich inclusions are often interpreted as the first Solar Nebula

---

[1] Minerals in meteorites are frequently considered in terms of a condensation sequence: refractory minerals containing Ca and Al comprise oxides (corundum $Al_2O_3$, hibonite $Ca(Al,Ti,Mg)_{12}O_{19}$, spinel $MgAl_2O_4$) and silicates (melilite – solid solution of akermanite $Ca_2MgSi_2O_7$ and gehlenite $Ca_2Al_2SiO_7$). They condense at higher temperatures than metal or the Mg-silicates forsterite ($Mg_2SiO_4$) and enstatite ($Mg_2Si_2O_6$) – the Mg-rich endmembers of solid solutions olivine and pyroxene. At lower temperatures, Ca and Al minerals can be transformed into diopside ($CaMgSi_2O_6$, a Ca-rich pyroxene) and anorthite ($CaAl_2Si_2O_8$, the Ca-rich endmember of feldspar solid solution), and Fe can enter Mg-silicates. Volatile elements condense in the most recent compounds/minerals. Aqueous alteration processes occur mainly on carbonaceous chondrite parent bodies and produce hydrous minerals – e.g. smectite, serpentine $(Mg,Fe)_3Si_2O_5(OH)_4$ – or carbonates, e.g. calcite $CaCO_3$. Mineral formation processes can be dated if they fractionate parent and daughter elements of radiometric systems. Important for radiometric dating are minerals in which parent nuclides are concentrated, e.g. phosphates for U–Pb–Pb and $^{244}$Pu fission-track dating, and feldspar for K–Ar and $^{26}$Al–$^{26}$Mg dating.

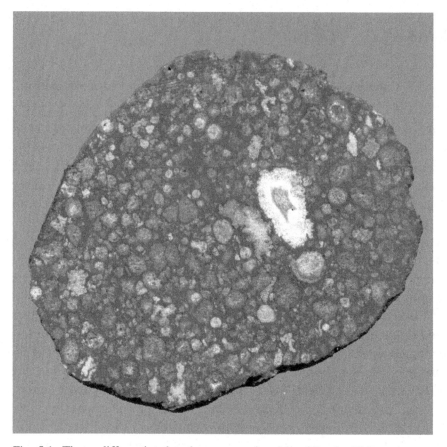

Fig. 5.1. The undifferentiated carbonaceous chondrite Allende. This rock was probably never heated to temperatures above 500 °C. It preserves a variety of primitive, pre-accretionary components, such as chondrules, and Ca-Al-rich inclusions, the oldest known material in our Solar System. Meteorite collection, Dieter Heinlein, Augsburg, Germany.

condensates, others think they are residues of early heating processes. In any case they are the oldest dated objects in our Solar System with an age of $4567.2 \pm 0.6$ million yr (Amelin *et al.*, 2002), as obtained by *radiometric dating* using the U–Pb–Pb method based on the decay of long-lived parent nuclides $^{238}$U and $^{235}$U into $^{206}$Pb and $^{207}$Pb (Chen and Wasserburg, 1981; Allègre *et al.*, 1995). The formation of chondrules (Fig. 5.1) requires temperatures of 1600 K or more depending on the extent of melting. Most chondrules appear to be 1 to 4 million yr younger than CAIs (Russell *et al.*, 1996; Kita *et al.*, 2000; Amelin *et al.*, 2002), although some chondrules approach the ages of CAIs (Bizzarro *et al.*, 2004).

It is not clear how CAIs or chondrules fit with observations or models of coagulation of dust particles (see also the reviews by Wurm and Blum in Chapter 6 and Henning *et al.* in Chapter 7). Some researchers believe that CAIs formed close to the protoSun and were then transported by X-winds to larger radial distances, where they were incorporated into locally formed rocks (Shu *et al.*, 1996).

A large number of models for chondrule formation is discussed in the literature (Scott *et al.*, 1996), ranging from lightning, collisions between first generation planetesimals (e.g. Lugmair and Shukolyukov, 2001; Sears, 2004), or shock waves in the Solar Nebula (Desch and Connolly, 2002). Chondrules cooled faster than CAIs (100–2000 $Kh^{-1}$ versus 10 $Kh^{-1}$), reflecting different thermal regimes. The ubiquitous presence of chondrules in all chondritic meteorites, except CI-chondrites (see Section 5.4), suggests that the chondrule-forming process was an important and widespread process in the early Solar System. However, it is uncertain if all matter in planets and asteroids has passed through the chondrule-forming process. Recent studies suggest a chemical complementarity in some chondrites between chondrules and matrix implying that chondrules and matrix formed from the same reservoir, essentially Solar composition for non-volatile elements. This excludes models where chondrules formed in a different part of the Nebula and combined accidentally with some fine-grained material (Klerner and Palme, 2000; Klerner, 2001).

## 5.4 Compositional variety of chondrites: planetesimal formation occurred at a variety of conditions in the protoplanetary disk

In general, chondrites represent undifferentiated, chemically "primitive" material with approximately Solar element abundances of non-volatile elements. In one group, the CI chondrites, element abundances match the spectroscopically determined Solar compositions to better than 10% (Anders and Grevesse, 1989; Palme and Jones, 2004) for most elements heavier than oxygen (except for the extremely volatile noble gases). Indeed, the Solar abundances of many trace elements that are difficult to determine in the Sun are now taken from meteorites. However, the various chondrite groups have characteristic compositional differences (Table 5.1; Figs. 5.2, 5.3, 5.4 and 5.5) and they deviate from CI element abundances, e.g. by having lower abundances of volatile elements. These differences do not only imply the origin of chondritic asteroids from several independent small precursor planetesimals, they also indicate that formation conditions of meteoritic material in the protoplanetary disk varied – most probably with radial distance from the early Sun and/or with time – and that material was incompletely mixed during or shortly before planetesimal formation (though not necessarily during earlier evolutionary stages, see below).

### 5.4.1 Metal abundance and oxidation state

Fig. 5.2 shows the different chondrite groups in a Urey–Craig diagram, where reduced Fe (in metal or sulfide) is plotted versus oxidized Fe (in silicate). Only CI

Table 5.1. Properties of chondrite groups

| Chondrite group | | EH | EL | H | L | LL | R[1] | CI | CM | CR[2] | CO | CK | CV |
|---|---|---|---|---|---|---|---|---|---|---|---|---|---|
| Chondrule abundance[3] | [vol%] | 20–40 | 20–40 | 60–80 | 60–80 | 60–80 | >40 | <<1 | 20 | 50–60 | 35–40 | 15 | 35–45 |
| Chondrule mean diameter[3] | [mm] | 0.2–0.6 | 0.8 | 0.3 | 0.7 | 0.9 | 0.4 | – | 0.3 | 0.7 | 0.2–0.3 | 0.7 | 1.0 |
| Tot. molten chondrules[4] | [%] | 17 | 17 | 16 | 16 | 16 | ≈8 | – | ≈5 | <1 | 4 | <1 | 6 |
| Refractory inclusions[3] | [vol%] | 0.1–1 | 0.1–1 | 0.1–1 | 0.1–1 | 0.1–1 | <1 | <<1 | 5 | 0.5 | 10–15 | 4 | 6–12 |
| Matrix abundance[3] | [vol%] | <5 | <5 | 10–15 | 10–15 | 10–15 | 36 | >99 | 70 | 30–50 | 30–40 | 75 | 40–50 |
| FeNi metal[3,5] | [vol%] | 8? | 15? | 10 | 5 | 2 | <0.1 | 0 | 0.1 | 5–8 | 1–5 | <0.01 | 0–5 |
| $Fe_{metal}/Fe_{total}$[6] | | 0.76 | 0.83 | 0.58 | 0.29 | 0.11 | 0.00 | 0 | 0 | 0.22 | 0–0.2 | NO | 0–0.3 |
| Fa-content of Olv[6] | [mol%] | | | 16–20 | 23–26 | 27–32 | 37–40 | | | | | | |
| Mg/Si[6] | | 0.77 | 0.83 | 0.96 | 0.93 | 0.94 | 0.90 | 1.05 | 1.05 | 1.06 | 1.05 | 1.13 | 1.07 |
| Ca/Si[6] | | 0.035 | 0.038 | 0.050 | 0.046 | 0.049 | 0.051 | 0.064 | 0.068 | 0.060 | 0.067 | 0.068 | 0.084 |
| Fe/Si[6] | | 0.95 | 0.62 | 0.81 | 0.57 | 0.52 | 0.74 | 0.86 | 0.80 | 0.81 | 0.77 | 0.83 | 0.76 |
| (Refract. El./Mg)$_{CI}$[7] | | 0.87 | 0.83 | 0.93 | 0.94 | 0.90 | 0.95 | 1.00 | 1.15 | 1.03 | 1.13 | 1.21 | 1.35 |
| Petrologic types[6] | | 3–5 | 3–6 | 3–6 | 3–6 | 3–6 | 3–6 | 1 | 2 | 2 | 3 | 3–6 | 3 |
| Max. metamorph. temp.[6] | [K] | 1020 | 1220 | 1220 | 1220 | 1220 | 1220 | 430 | 670 | 670 | 870 | 1220 | 870 |

[1] see Berlin (2003) and references therein; [2] CR group without CH chondrites; [3] Sears (2004) and Papike (1998); [4] Rubin and Wasson (1995); [5] in matrix; [6] Sears and Dodd (1988) and Sears (2004); [7] Scott et al. (1996)

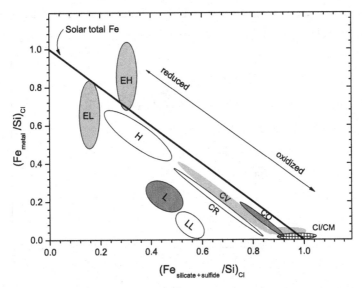

Fig. 5.2. Compositional variety of chondrites in a Urey–Craig diagram (after Sears, 2004): total metal content as well as oxidation state varies significantly, indicating different formation conditions. For definition of chondrite groups see Section 5.4.1 and Table 5.1.

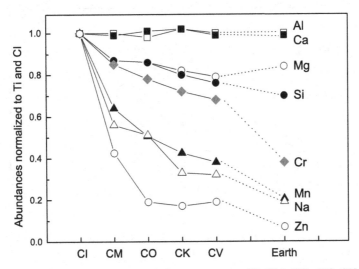

Fig. 5.3. Carbonaceous chondrites define a sequence CI–CM–CO–CK–CV of decreasing volatile abundances (and/or increasing refractory abundance). The decrease in condensation temperatures from Al to Zn is accompanied by decreasing abundances. CV chondrites are most depleted with a composition similar to the Earth.

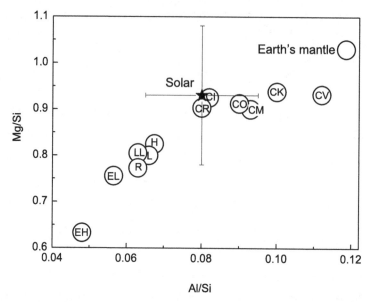

Fig. 5.4. In carbonaceous chondrites (CI,CR,CM,CO,CK,CV) Mg/Si ratios are constant, but Al/Si ratios are variable; ordinary (H,L,LL,R) and enstatite chondrites (EH,EL) are fractionated in both ratios relative to CI, which is identical to the Solar photospheric ratio (modified from O'Neill and Palme, 1998).

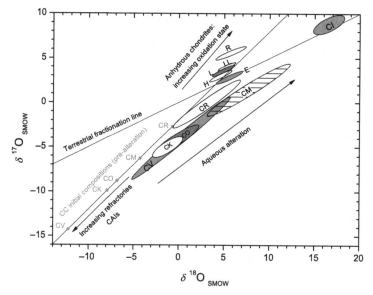

Fig. 5.5. Compositional variety of chondrites in an oxygen three-isotope diagram: refractory phases represented by CAIs have the highest enrichment in $^{16}$O. Trends of individual meteorite classes parallel to the terrestrial fractionation line with a slope of 0.5 are due to mass-dependent fractionation processes, e.g. igneous processes. Effects of aqueous alteration and oxidation state are also indicated.

chondrites have – completely oxidized – Fe with a Solar Fe/Mg ratio (indicated by the solid line). The most reduced groups are EH and EL enstatite chondrites, named after the Fe-free Mg-endmember enstatite of the mineral group pyroxene. Here, Fe is not present in an oxidized form in silicates, it occurs mainly in metals and sulfides which are more abundant in EH (high Fe) than in EL (low Fe) chondrites. Ordinary chondrites are classified according to the total abundance of Fe, and the ratio of reduced (metallic) Fe to oxidized Fe (Fig. 5.2; Table 5.1): H (high total iron), L (low total iron) and LL (low total iron, low metal). Carbonaceous chondrites are classified according to their contents of volatile elements and some other parameters. They are named after a prominent member of the respective group: CI (Ivuna), CM (Mighei), CO (Ornans), CK (Karoonda), CV (Vigarano), CR (Renazzo). Fig. 5.2 indicates very different degrees of oxidation for the various groups, and different $Fe_{total}/Si$ ratios. However, note that the plotted ranges represent average values of individual members of the respective groups. Primitive, unequilibrated meteorites do not record a single oxygen fugacity. They are mixtures of very oxidized and very reduced components.

### 5.4.2 Ratio of refractory to volatile elements

The dominant factor controlling the chemical composition of chondrites is the volatility of an element. Quantification of volatility is done by calculating condensation temperatures. This is the temperature at which 50% of an element is in the solid phase during cooling of a nebular gas (Grossman and Larimer, 1974; Larimer, 1979; Palme, 2001; Allègre *et al.*, 2001). The first phases to condense are Ca and Al oxides and trace elements such as rare earth elements, Hf, Sc, V, Ti, Sr, with condensation temperatures between 1400 and 1850 K at a total pressure of $10^{-3}$ bar. Refractory metals like W, Os, Ir, Re condense as metal alloys under the reducing conditions of the Solar Nebula. The refractory component is physically associated with CAIs that constitute up to 13% of CO chondrites (Table 5.1) and – more rarely – Mg-rich olivines (forsterites). Fig. 5.3 shows elemental abundances in various types of carbonaceous chondrites normalized to the refractory element Ti. The figure demonstrates that abundances of the refractory elements Ca, Al and Ti (i.e. the composition of the refractory component) are virtually uniform. However, moderately volatile elements with lower condensation temperatures (640–1230 K) such as Mn, Na, and Zn (but also Rb, K, F) are gradually depleted in the sequence CI–CM–CO–CK–CV. The more volatile an element, the lower its concentration. Ordinary chondrites do not fit with this classification. Although they are also depleted in volatile elements their pattern is very different from that of carbonaceous chondrites. In Fig. 5.3 we have plotted, for comparison, compositional data of the upper mantle of the Earth. Similarities with carbonaceous chondrites are apparent.

### 5.4.3  Major element fractionations: Mg, Si, Fe

The major rock-forming elements Mg and Si and metallic Fe are moderately volatile with intermediate condensation temperatures between 1250 and 1350 K (at $10^{-3}$ bar). While carbonaceous chondrites have constant Mg/Si ratios (Fig. 5.4, reflecting a similar degree of depletion in Mg and Si in Fig. 5.3), ordinary and enstatite chondrites show comparatively large variations in Mg/Si ratios (Fig. 5.4). This is indicative of *fractionation* of forsterite (Mg-rich olivine). Forsterite is the first Mg-silicate to condense and it has a Mg/Si ratio of two whereas the Solar Nebula Mg/Si ratio is about one. Removal of early formed refractory forsterite will make the residue Si richer (Larimer, 1979; Nagahara and Ozawa, 1996; Palme, 2001). Similarly, the variable Al/Si ratios in Fig. 5.4 can be explained in terms of fractionation of a refractory component such as that represented by CAIs, i.e. removal of such material from enstatite or ordinary chondrite precursors, or addition to carbonaceous chondrites.

The variable Fe/Si ratios in chondrite groups in Fig. 5.2, where meteorites plot on lines parallel to the line designated "Solar total Fe", may also be volatility related, however, incomplete sampling of condensed metal may also be achieved by other fractionation mechanisms in the early Solar Nebula, e.g. involving magnetic fields.

### 5.4.4  Oxygen isotopes

One of the most fundamental parameters in classifying chondritic meteorites is the oxygen isotopic composition (Clayton, 1993) as shown in Fig. 5.5 (prepared using METBASE 2004 data). The $^{18}O/^{16}O$ and $^{17}O/^{18}O$ ratios are given as permil deviations (using the *delta notation* $\delta^{18}O$ and $\delta^{17}O$) from the standard composition of standard mean ocean water (SMOW). The terrestrial fractionation line (TFL) serves as a reference line: terrestrial samples (with very few exceptions, e.g. stratospheric ozone) have oxygen isotope compositions that plot along this line with a slope of about 0.5, a result of mass-dependent fractionations which occur during physical processes such as evaporation or condensation. Variations in $^{18}O/^{16}O$ ratios are twice those of $^{17}O/^{18}O$ ratios, corresponding to the atomic-mass differences.

Several meteorite groups plot on fractionation lines parallel to the TFL (Fig. 5.5), indicating similar fractionation processes but different initial oxygen isotopic compositions. Compared to ordinary and enstatite chondrites with a limited range of oxygen isotopes, carbonaceous chondrites show a significantly larger range and variations are not restricted to lines with slope 0.5. The high variability is partly due to the presence of water, which reacted with anhydrous (water-free) phases of chondrites (Franchi *et al.*, 2001; Young, 2001), a peculiar $^{16}O$ enrichment of refractory phases, and the highly unequilibrated (i.e. chemically heterogeneous) character of most carbonaceous chondrites. The likely initial composition before aqueous alteration plots on the line connecting CAI-like composition and ordinary chondrites (gray line in Fig. 5.5). This line has a slope of one and indicates the addition of a component with pure $^{16}O$ hosted by refractory phases. The origin of this $^{16}O$-component is presently not understood (see Franchi *et al.*, 2001).

In summary, it appears that on a larger scale oxygen isotopes are much more uniform than on a local scale: large objects, such as the Earth, the Moon and Mars show comparatively small variations in oxygen isotopes very close to the terrestrial fractionation line. Smaller planetesimals such as parent bodies of ordinary chondrites and carbonaceous chondrites have larger variations, and individual components of unequilibrated meteorites have the largest variations in oxygen isotopes.

The total variations in the oxygen isotopic composition of components in the CM chondrite Murchison varies from $\delta^{17}O = -41.07$ for spinel to $\delta^{17}O = +16.67$ for calcite (Clayton and Mayeda, 1984) and thus exceeds by far the variations in bulk meteorites (compare with Fig. 5.5). Gas–solid exchange reactions of individual meteoritic components are considered to be essential in establishing the variations in oxygen isotopic compositions in meteorites and components of meteorites (Clayton, 1993). The exact origin of the various oxygen isotopic components is still unclear, particularly the origin of the $^{16}O$-rich component in refractory components. However, the association of this component with refractory material that formed within the Solar System (e.g. CAIs), or the correlation with oxidation state in ordinary and enstatite chondrites – i.e. a property established in the Solar System – leads most researchers to the conclusion that the $^{16}O$-rich component also originated in our Solar System, and that it is not an inherited presolar component.

## 5.5 Isotopic homogeneity of Solar-System materials

Although the oxygen isotopic composition of Solar-System objects is heterogeneous, with percent variations on small spatial scales, and permil variations on a planetary scale, this variation is the exception rather than the rule, probably related to the fact that the major fraction of oxygen in the inner Solar System is gaseous – even the most volatile rich meteorite Orgueil (CI-type) contains only about half of the Solar oxygen endowment. Indeed, most elements have within 0.01% the same isotopic composition in samples from Earth, Moon, Mars and more than 100 different meteorite parent bodies. Examples are isotopic compositions of K (Humayun and Clayton, 1995) and Si, but also elements that are used for chronology, such as Sr and Nd (Palme, 2001). This indicates that Solar-System materials were in general well mixed or chemically homogenized before or at formation of the Solar System, and that heterogeneities (e.g. variations of refractory and volatile element proportions, or oxygen isotopes) were established in the early Solar System itself.

Isotopic homogeneity as a dominant feature of Solar-System materials appears somewhat surprising when considering the fact that the condensable matter consisting of heavy nuclei were contributed by very different stellar sources feeding the interstellar medium. In the following subsection we discuss that such heterogeneities do indeed exist – in very rare cases and on a very small spatial scale – but that these are unlikely to influence bulk compositions, even on a μm or ng scale.

### 5.5.1 Heterogeneity inherited from the interstellar medium: restricted to rare individual grains

One possibility for variable chondrite compositions is heterogeneity inherited from the interstellar medium: tiny μm-sized grains of silicon carbide, graphite, corundum, spinel, and nm-sized diamond grains with strong isotope anomalies occur in all groups of chondritic meteorites and were recovered as residues after acid treatment (Anders and Zinner, 1993; Huss and Lewis, 1995; Hoppe and Zinner, 2000; Ott, 2001; Nittler, 2003). These presolar grains condensed in stellar outflows, e.g. in asymptotic giant branch (AGB) stars or supernovae, and their extremely variable isotopic signatures are used to constrain nucleosynthetic processes of their parental stars. Recently, silicate grains of interstellar origin have also been found in meteorites (Nguyen and Zinner, 2004) and also in interplanetary dust particles of presumed cometary origin (Messenger *et al.*, 2003). These grains survived destructive processes in the interstellar medium (e.g. supernova shock waves) or in the early Solar Nebula; thermal processing like evaporation or annealing before planetesimal formation, or thermal and aqueous alteration on parent bodies. However, isotope anomalies from these grains are not expected to show up in chondrite bulk compositions, as large interstellar grains are rare and most of the recovered grains are tiny: large grains, e.g. 10 μm-sized silicon carbide grains, have abundances as low as $\approx$14 ppm (Ott, 2001). Smaller, several 100 nm-sized, silicate subgrains within interplanetary dust particles of cometary origin – with large anomalies in the oxygen isotopic composition – have abundances between 450 and 5500 ppm (Floss and Stadermann, 2004; Messenger *et al.*, 2003), silicates in very primitive meteorite matrices (Nguyen *et al.*, 2004) have an abundance of $\approx$40 ppm. A further significant fraction of such small silicate subgrains may also have preserved smaller (not yet detectable) isotope anomalies inherited from their stellar sources. However, it is possible that isotopic homogenization of most interstellar medium

(ISM) grains was achieved by grain destruction and recondensation or other processes in the interstellar medium (Jones, 2001). In any case, isotopic homogeneity
on a μm scale was achieved for most Solar-System materials: typical sample sizes
for precise isotopic analyses require material of – at least – an aggregate of 30 μm
in size, corresponding to $3 \times 10^{-5}$ mg, and consisting of 27 000 grains of 1 μm size
or 27 000 000 grains of 100 nm size. Consequently, anomalies in elemental or isotopic composition inherited from the interstellar medium are quite unlikely to cause
measurable variations of bulk compositions when presently available techniques
are applied.

### 5.5.2 Heterogeneous or homogeneous distribution of short-lived nuclides: mixed evidence

A remarkable discovery was the finding of the decay products of short-lived radioactive nuclides in meteorites. After the first demonstration of excess $^{129}$Xe from
$^{129}$I (half-life, $T_{1/2} = 16$ Myr) in stone meteorites (Jeffery and Reynolds, 1961) and
excess $^{26}$Mg from $^{26}$Al ($T_{1/2} = 0.73$ Myr) in CAIs (Lee *et al.*, 1976), several other
short-lived radioactive nuclei were identified by the analysis of isotopes of their
daughter nuclides: $^{53}$Mn $\rightarrow$ $^{53}$Cr, $T_{1/2} = 3.7$ Myr (Birck and Allègre, 1985), $^{60}$Fe
$\rightarrow$ $^{60}$Ni, $T_{1/2} = 1.5$ Myr (Shukolyukov and Lugmair, 1993; Tachibana and Huss,
2003), $^{10}$Be $\rightarrow$ $^{10}$B, $T_{1/2} = 1.5$ Myr (McKeegan *et al.*, 2000). Due to their short
half-lives, their nucleosynthesis must have taken place shortly before incorporation
into the Solar System. Either they were injected into the early Solar System by
ejection from nearby stars, e.g. from the local cluster or the region where the Sun
formed (Wasserburg *et al.*, 1998; Boss and Vanhala, 2001; Hartmann, 2001) or
they were produced by strong Solar proton irradiation of solid matter close to the
early Sun (Lee *et al.*, 1998), or both. In the first case they were probably uniformly
distributed as they participated in the elemental and isotopic homogenization of
the bulk Solar-System materials. In the second case their abundance should have
varied with distance from the Sun. Such a non-uniform distribution would preclude
nuclei such as $^{26}$Al being used for chronology (see below). Some isotopes such as
$^{10}$Be are probably produced by nuclear reactions induced by Solar-particle radiation, implying heterogenous distribution, others like, for example, $^{60}$Fe can only
be produced in sufficient abundance by nucleosynthetic processes in a supernova
(Wasserburg *et al.*, 1998; Lee *et al.*, 1998; Russell *et al.*, 2001). It is now generally
assumed that $^{26}$Al was homogeneously distributed within the early Solar System:
firstly, the good agreement of the $^{26}$Al–$^{26}$Mg chronometer with absolute U–Pb–Pb
ages in different types of meteorites requires homogeneous distribution (Amelin
*et al.*, 2002; Zinner and Göpel, 2002); second, $^{26}$Al is abundant in the interstellar
medium fed by evolved stars' outflows (Knödlseder *et al.*, 2002).

## 5.6 Dating accretion, heating, and cooling of planetesimals

Although the majority of the meteoritic mass collected on Earth is from undiffer-entiated parent asteroids, a comparatively large fraction of meteorites come from small bodies that suffered melting and differentiation in a similar fashion as the larger terrestrial planets (Papike, 1998; Lugmair and Galer, 1992): various classes of iron meteorites and basaltic meteorites, partial melts from the interior of small planetesimals (hence the term basaltic achondrites). Apparently, heating that trig-gered differentiation did not only occur on large terrestrial planets, but also on some small asteroids. However, whereas heating and subsequent melting of the larger terrestrial planets was caused by large collisions in the final stages of plane-tary accretion, small planetesimals were heated by the decay of $^{26}$Al and/or $^{60}$Fe, as discussed by Lee *et al.* (1976), Miyamoto *et al.* (1981), Wood and Pellas (1991), and Trieloff *et al.* (2003). The effects of heating were not only seen in rare differentiated meteorites, but also in chondritic meteorites where heating has led to the equilibra-tion of texture and minerals (Dodd, 1969). These effects (e.g. uniformity of mineral compositions, size of secondary minerals, visibility of primary chondrules, etc.) are quantified by the petrologic type ranging from 1 to 6 (see Table 5.1; van Schmus and Wood, 1967): the higher the number the higher the maximum metamorphic temperature. Most carbonaceous chondrites are of petrologic type between 1 and 3 – type 1 experienced the mildest heating, but it suffered the highest degree of aqueous alteration. Ordinary chondrites were metamorphosed more strongly, up to type 6. Further heating would lead to melting. Members of all three groups (H, L, LL, representing three different parent bodies) were heated to variable degrees, between petrologic types 3 and 6: for example, an H4 chondrite that still displays recognizable primary textures like matrix and chondrules was heated up to max-imum temperatures of ≈970 K. An H6 chondrite was heated up to ≈1220 K, so that primary textures were erased and replaced by ≈100 μm-sized recrystallized secondary minerals (e.g. feldspar and phosphates that can be used for radiometric dating, see below).

Fig. 5.6 shows model calculations for an ordinary chondrite parent body as-suming internal heating by $^{26}$Al decay (Trieloff *et al.*, 2003). A layered structure with increasing temperature versus the center of the planetesimal is obtained. The right panel in Fig. 5.6 shows cooling curves calculated after a model by Miyamoto *et al.* (1981) for an asteroid of 100 km radius: the initial $^{26}$Al/$^{27}$Al ratio cho-sen is sufficiently low to avoid melting and achieve the maximum metamorphic temperatures in the central regions of the parent planet that are required for re-producing heating effects in type 6 chondrites. The corresponding $^{26}$Al/$^{27}$Al ratio is about one order of magnitude lower than that found in CAIs from carbona-ceous chondrites ($^{26}$Al/$^{27}$Al $= 5 \times 10^{-5}$; Lee *et al.*, 1976). Thus the H-chondrite

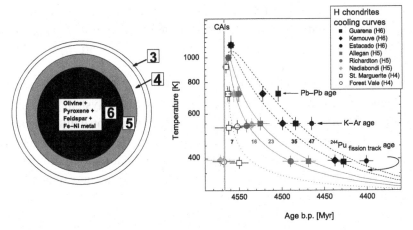

Fig. 5.6. Ordinary-chondrite parent bodies were heated more strongly (up to 1220 K) than carbonaceous chondrites, and only a small fraction escaped severe heating due to rapid cooling at shallow layers (types 3 and 4). Cooling curves determined for a number of unshocked chondrites with undisturbed radiometric record of the early history are key evidence to demonstrate cooling in such an onion-skin-layered parent asteroid, whereby strongly heated central regions cooled about 150 Myr longer than more shallow layers. Curves were calculated assuming an asteroid with 100 km radius. Numbers correspond to depths in km (after Trieloff *et al.*, 2003).

parent body would have accreted ≈2 Myr after CAIs. Earlier accretion would have led to melting with subsequent differentiation, in later accretion heating effects would have been insufficient to cause the temperature increase required to produce the thermal metamorphism typical of ordinary chondrites of petrologic type 6.

The calculated heating and cooling curves in Fig. 5.6 can be tested by *radionuclide chronometry*: The data points in Fig. 5.6 are radiometric ages obtained by various chronometers: U–Pb–Pb (Göpel *et al.*, 1994), $^{40}$K–$^{40}$Ar and $^{244}$Pu fission tracks (Trieloff *et al.*, 2003). Here it is important to mention a basic principle of radionuclide chronometry, the concept of *cooling ages*. A radiometric age does not necessarily correspond to the point of time when the rock crystallized from the melt – it rather reflects the time when the temperature reached the *closure temperature* where diffusive equilibration of daughter isotopes from radioactive nuclei becomes ineffective and the radiometric clock is started. For example, the U–Pb–Pb age of the phosphate mineral apatite defines the time when the temperature reached about 720 K (Cherniak *et al.*, 1991), and diffusion of Pb in apatite ceased and the exchange of Pb isotopes with neighboring minerals stopped (i.e. did not *equilibrate*). The corresponding temperature for the K–Ar system in chondritic feldspar is 550 K (Trieloff *et al.*, 2003; Pellas *et al.*, 1997; Turner *et al.*, 1978), when Ar could no more diffuse out of individual feldspar grains and was thus quantitatively retained by this mineral. It should be noted that closure temperatures additionally depend on mineral grain size and cooling rates, and the above values are typical values for equilibrated ordinary chondrites.

Two major conclusions can be derived from the data in Fig. 5.6: first, the data clearly suggest that parent-body heating in the early Solar System was caused by $^{26}$Al heating (Trieloff *et al.*, 2003). Nevertheless, minor contributions from other sources appear possible, e.g. $^{60}$Fe or accretional heating involving porous rocks (Kunihiro *et al.*, 2004); second, it can be recognized that many radiometric ages do not date the time of solid formation or accretion, but define extended cooling histories in asteroid-sized parent bodies.

## 5.7  A timescale of early Solar-System events

Fig. 5.7 shows the timescale of the first few Myr in the early Solar System, based on precise U–Pb–Pb ages, combined with ages based on short-lived nuclide chronometries[2] ($^{26}$Al–$^{26}$Mg, $^{129}$I–$^{129}$Xe, $^{53}$Mn–$^{53}$Cr) – short-lived nuclides such as $^{26}$Al can be used for radiometric dating (if homogeneous distribution in the early Solar System is assumed, see above), and due to their short half-life the time-resolution of these systems is better than for long-lived chronometers (Gilmour and Saxton, 2001). Age data are shown for Allende CAIs, chondrules (Amelin *et al.*, 2002), and metamorphic minerals (phosphates and feldspars) of the H4 chondrites Forest Vale and St. Marguerite (Göpel *et al.*, 1994; Zinner and Göpel, 2002). These objects and minerals cooled fast (within <1 Myr), allowing comparison of U–Pb–Pb ages and $^{26}$Al–$^{26}$Mg, $^{129}$I–$^{129}$Xe, and $^{53}$Mn–$^{53}$Cr ages neglecting closure temperatures. The $^{26}$Al–$^{26}$Mg, $^{129}$I–$^{129}$Xe (Brazzle *et al.*, 1999), and $^{53}$Mn–$^{53}$Cr ages (Lugmair and Shukolyukov, 1998; Polnau and Lugmair, 2001) are normalized using a variety of tie points (also plotted in part in Fig. 5.7), similar to a recent approach by Gilmour *et al.* (2004); for details see footnote. The timescales give coherent results, indicating that $^{26}$Al, $^{129}$I and $^{53}$Mn were indeed distributed homogeneously in the early Solar System (or in the particular region where these samples come from) and that they can be used as chronometers. The data in Fig. 5.7 show that CAIs

[2]  Short-lived nuclide chronometries were calibrated by least square fitting time scales with different tie points (not single tie points for each scale, as often done previously): Forest Vale (FV), St. Marguerite (SM), Richardton (RI), Kernouve (KER), Acapulco (ACA) and LEW86010 (LEW). First, the optimum calibrations of Mn–Cr versus Pb–Pb (using SM, FV, RI, ACA, LEW), I–Xe versus Pb–Pb (using SM, KER, RI, ACA) and Mn–Cr versus I–Xe (SM, RI, ACA) were evaluated. The final calibration was performed by choosing one consistent parameter set for all three systems for which the deviation from these optimum calibrations was minimized. This led to absolute ages of 4562.5 Myr for I–Xe of Shallowater enstatite and 4556.6 Myr for Mn–Cr of LEW86010 ($^{53}$Mn/$^{55}$Mn = 1.25 × 10$^{-6}$). Al–Mg versus Pb–Pb (using CAIs, FV, SM) yielded 4567.6 Myr for Al–Mg of CAIs ($^{26}$Al/$^{27}$Al = 5 × 10$^{-5}$). Contrary to Gilmour *et al.* (2004) we did not use *oldest* chondrules as tie points, taking into account the variation of chondrule ages.

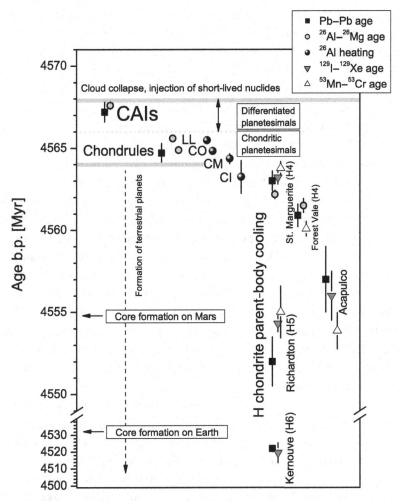

Fig. 5.7. Timescale of events in the early Solar System derived from the meteoritic record. Calcium-aluminium-rich inclusion (CAIs) formation was followed by sequential planetesimal accretion in distinct zones within <4 Myr. Planetesimals that accreted within <2 Myr after CAIs incorporated $^{26}$Al sufficient for intense heating, melting and differentiation. Undifferentiated chondritic parent bodies accreted subsequently, as indicated by experimentally determined initial average $^{26}$Al abundances in chondrules (of LL ordinary and CO carbonaceous chondrites) that correspond to the observed degree of heating of their parent bodies. In each zone local chondrule formation and subsequent planetesimal formation occurred rapidly (<1 Myr), although a minor proportion of older chondrules may have been incorporated as well. Pb–Pb and other short-lived nuclide-based ages are shown for some key samples, including metamorphic minerals in H chondrites that reflect extended cooling in the parent asteroid.

are the oldest dated solids, followed by chondrules about 2 Myr later, with chondrules from ordinary chondrites (type LL chondrites, $^{26}Al/^{27}Al = 7.4 \times 10^{-6}$; Kita *et al.*, 2000) slightly older than those of carbonaceous (CO) chondrites ($^{26}Al/^{27}Al = (3.8 \pm 0.7) \times 10^{-6}$; Kunihiro *et al.*, 2004). However, also note that chondrules from CV Allende (not plotted here) have a larger spread in ages (Bizzarro *et al.*, 2004). While chondrules must have formed as individual melt droplets in space, metamorphic minerals (phosphates and feldspars) of St. Marguerite and Forest Vale require parent-body accretion, heating and cooling within 2–3 Myr after chondrule formation. Such a short time can only be achieved by $^{26}Al$ heating, if parent-body accretion followed chondrule formation almost instantaneously.

Such a scenario is independently required by the chemical complementarity between chondrules and matrix mentioned above (Klerner and Palme, 2000; Klerner, 2001). Rapid parent-body accretion after chondrule formation is furthermore demanded by (or consistent with) the fact that the time of chondrule formation corresponds to exactly the time when $^{26}Al$ had the appropriate abundance for providing the heat required to produce the observed maximum metamorphic temperatures for both ordinary and carbonaceous chondrites. The $^{26}Al$ model ages derived from the metamorphic temperatures are shown as a separate time scale in Fig. 5.7 (calculations after Miyamoto *et al.*, 1981, for a 100 km radius asteroid, using parameters after LaTourrette and Wasserburg, 1998; Young, 2001, obtained consistent ages for carbonaceous chondrites).

The consistent timescales suggest that parent bodies of ordinary chondrites accreted slightly earlier and experienced correspondingly higher temperatures than those of carbonaceous chondrites. It furthermore implies that chondrule formation was strongly related in time and space to accretion in the respective radial zones or – in other words – conditions in certain nebular regions were appropriate for chondrule formation and accretion. Furthermore, this scenario quite naturally explains the apparent strange 2 Myr age "gap" between CAI and chondrule formation (Russell *et al.*, 1996; Kita *et al.*, 2000; Gilmour and Saxton, 2001; Amelin *et al.*, 2002): planetesimals that accreted earlier than chondritic parent bodies incorporated correspondingly large amounts of $^{26}Al$ and inevitably differentiated. This conclusion is supported by recent analyses of differentiated meteorites. Radiogenic $^{182}W$ (see below) in Fe meteorites led Kleine *et al.* (2005) to the conclusion that iron meteorites are older than chondritic parent bodies. Crust formation on basaltic achondrite parent bodies <3 Myr after CAIs is implied by $^{26}Al$–$^{26}Mg$ dating (Bizzarro *et al.*, 2005). These early planetesimals may have contained chondrules, but because they suffered complete melting, evidence for chondrules is erased. It should be noted that this scenario does not exclude that some chondrules are as old as Allende CAIs, as found by Bizzarro *et al.* (2004) – it rather confirms that chondrule-forming processes were active since CAI formation, but that most older chondrules had been consumed in earlier-formed planetesimals that experienced melting because of higher $^{26}$Al contents.

However, it should also be cautioned that this scenario needs further, independent evidence. For example, some Fe meteorites have younger ages than chondrules (Kleine *et al.*, 2005). In addition, recent studies of chondrules from CO chondrites yielded $^{26}$Al/$^{27}$Al ages as high as chondrule ages from LL chondrites, depending on chondrule type (Kita *et al.*, 2004) and a significant fraction of chondrules from the CV chondrite Allende display ages similar to CAIs (Bizzarro *et al.*, 2004). Radiometric ages may not date chondrule formation in all cases: for example, ages could have been reset later during parent, body metamorphism, as, for example, observed for the I–Xe system (Gilmour and Saxton, 2001; Brazzle *et al.*, 1999), or they could date the age of an older precursor protolith rather than the chondrule-forming process, e.g. if Mg isotopes are determined in whole chondrules rather than in chondrule minerals (Bizzarro *et al.*, 2004).

## 5.8 Formation of terrestrial planets

In model calculations for the accretion of the terrestrial planets (Wetherill and Stewart, 1993; Kokubo and Ida, 2000), planetary embryos form by gravitational "cleaning" of the planetesimals feeding zones ("runaway growth"). This is achieved within one million yr. In a second step that takes 10 to 100 million yr gravitational scattering and random collisions lead to the formation of the five terrestrial planets (including the Moon). From a cosmochemical viewpoint it is not a trivial task to constrain the formation time interval of the large terrestrial planets. Although we have rocks from Earth and Mars (and the Moon), large planets are – contrary to small asteroids – "geologically active" for a much longer period of time than required for accretion, implying continuous cycles of rock destruction and formation. For example, large-scale mantle convection on Earth produces new oceanic crust at diverging plate boundaries, while old oceanic crust (always younger than 200 Myr) is continuously recycled back into the mantle by subduction at converging plate boundaries. The oldest rock units of the continental crust are 3.8 Gyr old; only few relict mineral grains (found in younger rock units) record ages of up to 4.4 billion years (Wilde *et al.*, 2001), still 150–200 Myr after the formation of the Solar System.

As there are no terrestrial rocks that survived early processing during planetary formation and early planetary differentiation, radiometric ages are used to date formation of the Fe–Ni core and the proto-(silicate) mantle. In this way, Pb isotopes have been used to determine an age of the Earth since the mid-1950s yielding typically values of 4450±50 Myr (Allègre *et al.*, 1995). Recently, the decay of $^{182}$Hf (half-life: 9 Myr) to $^{182}$W (Halliday *et al.*, 2001) was used to infer, from measurements of meteorites, terrestrial rocks and Martian meteorites, times to complete core formation within 33 Myr for Earth and 13 Myr for Mars (Kleine *et al.*, 2002). Assuming an exponentially decreasing rate of accretion, 63% of the planetary masses could have been present at 10–12 Myr and 2–4 Myr, respectively (Yin *et al.*, 2002). These results agree with dynamic timescales of terrestrial planet formation (Wetherill and Stewart, 1993; Kokubo and Ida, 2000).

## 5.9 Disk dissipation, Jupiter formation and gas–solid fractionation

Up to this point, we have reviewed basic isotopic and chemical properties of chondritic parent bodies, and the timescales of hierarchical growth from small aggregates like CAIs and chondrules to km-sized planetesimals, and finally the growth of the terrestrial planets by collisions of planetary embryos. We have concluded that both differentiated and undifferentiated (chondritic) asteroid-sized planetesimals formed within the first 4 Myr of the Solar System. This agrees well with average lifetimes of protoplanetary disks observed by astronomers on the basis of excess infrared radiation from fine dust (Haisch *et al.*, 2001): at 3 Myr after formation of the protostar, 50% of young stellar objects display a dust disk. Indeed, the striking agreement of the two values may simply indicate that the disappearance of fine dust is caused by coagulation and planetesimal formation (Throop *et al.*, 2001; see also Chapter 7 by Henning *et al.* and Chapter 6 by Wurm and Blum). Such a fast accretion of planetesimals also implies the early presence of potential building blocks for terrestrial planets. Hence, although final assembling of large terrestrial planets lasts significantly longer than the commonly inferred disk lifetimes of a few Myr, terrestrial planets can well be expected in extrasolar planetary systems, because their building blocks – be these chondritic or differentiated planetesimals – are present within a few Myr.

Furthermore, constraints on the timescale of Jupiter formation may be obtained here: if Jupiter inhibited planet formation in the early asteroid belt, it should have been present before asteroids accreted to a larger planet with, for example, the size of Mars, that accreted a significant mass fraction within a few Myr (Kleine *et al.*, 2002; Yin *et al.*, 2002). A further important constraint on the time of Jupiter formation is dissipation of the gaseous disk, as Jupiter must have grown in the presence of nebular hydrogen-dominated gas, whether this occurred "bottom up" (see Chapter 10 by Hubickyj) or "top down" (Boss, 1997, see also Chapter 12; Mayer *et al.*, 2002). However, in this case we may not directly use disk lifetimes inferred from infrared excess observations, as the disappearance of fine dust may just be caused by coagulation. The gas itself may stay significantly longer in protoplanetary disks. Direct astronomical observations of the gas ($CO$, $H_2$) are intrinsically difficult and are rarely done. Presently lifetimes of nebular gas clouds are not well constrained and they may be somewhat longer than those of the dust component by a few tens of Myr (Briceño *et al.*, 2001; Thi *et al.*, 2001). However, Briceño *et al.* (2001) also looked for H-alpha emission which is a signature of gas accretion onto the protostar and found that it was correlated with the infrared excess from dust grains. This can be interpreted as disappearance of disk gas in the inner disks within a few Myr. Another essential point in such discussions are observations of extrasolar giant planets very close to their parent stars: while migrations via planet–disk interactions can be modelled (see Chapter 14 by Masset and Kley), it is not clear what stops these planets from migrating into the star – inner disk gas dispersal would be an obvious solution for this problem.

Cosmochemical constraints on disk gas dissipation in the early Solar System can potentially be derived from irradiation effects recorded by meteoritic matter, e.g. Solar-wind ions implanted into early planetesimal surfaces, because Solar-wind implantation requires irradiation settings that are not shielded by gas (or fine dust). Many meteorites (so-called regolith breccias) contain a record of exposure on asteroid surfaces by, for example, the presence of noble gases (He, Ne) implanted as Solar-wind ions, with a characteristic isotopic composition. However, it is difficult to decide if very early irradiation records exist, e.g. if Solar-type noble gases were implanted by a Solar wind in the early Solar System when small bodies accreted to asteroid-sized objects, rather than during late stages when asteroids were progressively fragmented into smaller objects. Trieloff *et al.* (2000, 2002) demonstrated that the Earth contains Solar-wind implanted noble gases, possibly implanted on the surface of small precursor planetesimals (<1 km). Evidence from meteorites indicates that m-sized planetesimals were irradiated very early by the Solar wind (Goswami and Lal, 1979; Goswami and McDougall, 1983; Nakamura *et al.*, 1999). At this time dust grains in the nebula must have aggregated to larger objects to allow penetration of Solar radiation. This result would imply irradiation of matter at least in some meteorite-forming regions and the inner Solar System where the terrestrial planets formed. If irradiation took place in the midplane of the Solar Nebula (and not off-disk, e.g. into single grains settling later to the midplane) this would mean that gas in the inner disk was dissipated over a timescale of <4 Myr, and also sets a tight constraint for Jupiter formation.

Another independent observation that could indicate progressive gas dispersal when meteorite parent bodies were assembled is the fact that chondrules and also CAIs formed under much more oxidizing conditions than would be expected from typical gas/dust ratios in protoplanetary disks (of about 100). Many observations, such as the existence of melts, and FeO-bearing oxidized silicates imply that gas/dust ratios were enhanced by a factor of 1000, although it is not clear if this was due to locally enhanced dust/gas ratios or the general depletion or loss of gas from the disk.

An important aspect of disk dissipation is the potential of this process to fractionate gas from solids, and to produce the observed fractionations of refractory and volatile components. Remarkably, the sequence CI–CM–CV which is a sequence of increasing volatile depletion (Fig. 5.3; Palme, 2001) is also a sequence of increasing proportions of Solar-wind-implanted Ne, indicating a possible causal link between disk clearing, gas-solid fractionation by loss of volatiles, and Solar-wind implantation (Trieloff *et al.*, 2002). Direct radiometric dating of disk dissipation may be possible by interpreting the $^{53}$Mn–$^{53}$Cr system of carbonaceous chondrites and the Earth as fractionation of volatile Mn and refractory Cr in the inner Solar System a few Myr after CAI formation (Palme, 2001; Halliday *et al.*, 2001). However, it remains to be clarified if reasons other than loss of disk gas can be excluded as fractionation mechanisms. For example, condensed refractory components may also be removed, if early formed solids decouple from the gas and spiral into the Sun.

## 5.10 Summary

Here we summarize the origin of first solids and the growth of planetary bodies in our Solar System from a geo-scientific point of view, and within the astrophysical context of star and planet formation:

(1) Star formation starts with the collapse of an interstellar cloud of gas and dust ($\approx 1\%$ per mass). This cloud contains mainly sub-$\mu$m-sized grains, partly from evolved stars with significant mass outflows (Gail, 2003), supernovae and other sources. Grains of silicon carbide and graphite, or oxides and silicates, carry distinct isotopic fingerprints inherited from the corresponding nucleosynthetic processes (Hoppe and Zinner, 2000; Ott, 2001; Nittler, 2003). An unknown – probably large – fraction of these grains may have been reprocessed (evaporated, condensed) during hundreds of Myrs residence time in the interstellar medium (Jones, 2001). In any case, even if a considerable fraction of grains is unprocessed and isotopically anomalous (which remains to be investigated with improved nano-SIMS techniques), any volume of dust (and gas) with a sufficiently large number of grains will contain very close-to-cosmic (or Solar) elemental and isotopic abundance ratios.

(2) From the collapsing and increasingly dense cloud, fragments – with several times the mass of our Sun – may separate and become rapidly evolving stars. Such an environment is observed in the Orion nebula; another star-forming region with mainly Solar-like stars is the Auriga nebula. Intermediate-mass and mass-rich evolved stars emit intense stellar winds – followed by a supernova explosion in the case of mass-rich stars – ejecting freshly synthesized elements, including significant amounts of short-lived nuclides like $^{26}$Al, $^{53}$Mn, $^{60}$Fe, $^{129}$I and $^{182}$Hf. Fragments do not collapse simultaneously (e.g. Hartmann, 2001) so short-lived nuclides ejected from one star may contaminate

infant protoplanetary disks, stellar winds from evolved stars may even trigger the collapse of low-mass cloud fragments, or initiate the dissipation of gas in protoplanetary disks.

(3) Radial mixing in protoplanetary disks (see Chapter 4 by Klahr *et al.*) like the early Solar System is effective, as evidenced by the presence of crystalline silicates (that originated close to the early Sun) in comets (Wooden *et al.*, 2000; Wehrstedt and Gail, 2002; Messenger *et al.*, 2003). Hence – with the exception of rare, large interstellar grains – any presolar heterogeneity is erased on the μg level of dust or less, and does not show up in bulk compositions. A considerable fraction of interstellar grains will be processed in the outer parts of the disk by annealing of amorphous grains, grain growth to larger grains, chemical equilibration and/or differentiation via intragrain diffusion, or, in the inner disk, by evaporation and condensation. Hence, large-scale heterogeneity of solids is mainly caused by physical conditions imposed by the central protostar, and grain compositions or – more generally – the distribution of elements between solids and gas is governed by their volatility (Allègre *et al.*, 2001; Palme, 2001). Varying proportions of the major components with different condensation temperatures – metal, enstatite, refractory forsterite and Ca-Al-rich compounds (the latter enriched in $^{16}O$) – will ultimately lead to the observed chondrite groups. For example, modeling the dust component in protoplanetary disks including radial mixing processes (Gail, 2004) predicts different forsterite and enstatite abundances at different radial distances to the Sun. When the gas phase is removed, such large-scale heterogeneities will be preserved in the solid components and (later) in planetary compositions.

(4) Coagulation of dust (Wurm and Blum, Chapter 6; Blum *et al.*, 2000; Henning *et al.*, Chapter 7) and formation of mm- and cm-sized objects like CAIs and chondrules occur very rapidly, within 0.1 Myr (Bizzarro *et al.*, 2004), at least in certain regions of the protoplanetary disk, but in other regions planetesimal formation is not as fast. Apparently, formation of these objects may occur in dust-enriched regions under relatively oxidizing conditions, with dust/gas ratios enhanced by a factor of 1000 compared to Solar or cosmic ratios (= 1:100). Calcium-aluminium-rich inclusions formed by condensation or, as some researchers assume, as evaporation residues. In any case, high temperatures and relatively fast cooling rates are required. On the other hand, refractory forsterite grains present in all types of chondritic meteorites (Pack *et al.*, 2004) reflect reducing conditions as expected from standard models of the Solar Nebula. Refractory material is involved in processes capable of enriching $^{16}O$, possibly gas–solid exchange reactions at high temperatures (Clayton, 1993; Franchi *et al.*, 2001). Hence, the distinct proportions of refractory material in chondritic parent bodies can explain part of their variations in bulk compositions.

(5) A minor proportion of material irradiated close to the Sun by high-energy particle radiation may suffer nuclear reactions producing short-lived radionuclei that lead to isotope anomalies (Lee *et al.*, 1998; Russell *et al.*, 2001). Other solids may suffer low-energy irradiation resulting in Solar-wind ion implantation. Such material may be irradiated either off-disk as single grains (before midplane sedimentation) or in small planetesimals (after gas dissipation). Solar-wind implanted ions are found in meteorites

(Goswami and MacDougall, 1983; Nakamura *et al.*, 1999) and the interiors of terrestrial planets as Solar He and Ne (Trieloff *et al.*, 2000, 2002).

(6) Formation of chondrules and aggregation of chondrules and matrix to larger objects leading to planetesimals of up to 100 km in size occurred rapidly in individual radial zones, as suggested by chemical complementarity of matrix and chondrules (Klerner and Palme, 2000), and the similar $^{26}$Al contents ($= {}^{26}$Al–$^{26}$Mg ages) of chondrules and their chondritic parent bodies: ordinary chondrite parent bodies that were heated – by decay heat of $^{26}$Al – to higher metamorphic temperatures than carbonaceous chondrites (1220 K versus 870 K) should have had initial $^{26}$Al concentrations a factor of two higher, which is supported by measured $^{26}$Al concentrations in LL and CO chondrules (Kita *et al.*, 2000; Kunihiro *et al.*, 2004). The observation that most chondrules are 2–4 Myr younger than CAIs can be easily explained in such a scenario, because the first-formed chondrules were mostly consumed in parent bodies that accreted with high $^{26}$Al. These bodies, however, became molten and differentiated (yielding iron meteorites and basaltic achondrites), and hence could not preserve chondrules. There is no need for a mechanism to produce chondrules exclusively 2–4 Myrs later than CAIs. This model agrees with recent chronological studies that some chondrules (Bizzarro *et al.*, 2004), some iron meteorites (Kleine *et al.*, 2005) and basaltic achondrites (Bizzarro *et al.*, 2005) are nearly as old as CAIs. There are no conflicts with the short dynamic lifetimes of small objects in the Solar Nebula (Weidenschilling, 1977), as most chondrules are formed and immediately consumed in fast-accreting planetesimals, thereby preserving chemical complementarity, while CAIs in chondrites can be treated as independent refractory components, which may well have been redistributed in the Solar System and added (e.g. to CV chondrite regions) or lost (e.g. from E chondrite precursor material) during the 2 Myr before formation of chondrite parent bodies.

(7) While the formation of most meteorite parent bodies occurred rapidly within 4 Myr after CAI formation, cooling extended up to more than 100 Myr in the inner asteroid regions, as evidenced by radiochronometry and cooling rates of highly metamorphosed chondrites or Fe meteorites (Trieloff *et al.*, 2003). Most radiometric ages are cooling ages and not formation ages, thus short formation timescales can still be envisaged.

(8) The formation time interval of asteroid-sized planetesimals of <4 Myr is within the life-time of most protoplanetary disks (Haisch *et al.*, 2001). With such formation timescales, building blocks of terrestrial planets can also be expected during the early evolution of extrasolar systems, i.e. before dissipation. Hence, terrestrial planets can be considered as likely in extrasolar systems. However, their final assembly most probably needs a few tens of Myr, as indicated by core formation ages of Mars and Earth of 13 and 33 Myrs, respectively (Kleine *et al.*, 2002). Late evolutionary stages also include giant impacts like the postulated moon-forming impact on Earth (Benz *et al.*, 1987), and will stochastically produce large amounts of impact debris such as found in many young debris disks.

Future geoscientific investigations will probably help to understand astronomical observations and improve astrophysical modeling of protoplanetary disks, e.g.

details like the abundances and radial distributions of amorphous dust or crystalline forsterite and enstatite, the growth of grains, grain aggregates and planetesimals, and the evolution from gas-dominated disks to debris disks. Within the next decade, combined geoscientific and astrophysical studies may allow us to place our Solar System in context: we may be able to judge if planetary systems like ours are the rule or exception.

## Acknowledgments

We thank the organizers and the participants of the Planets 2004 workshop for the stimulating atmosphere and fruitful discussions at Ringberg Castle. We thank A. P. Boss for constructive review, E. K. Jessberger, H. P. Gail, S. Wolf, C. P. Dullemond, T. Althaus, J. Hopp, and W. Schwarz for helpful comments on the manuscript. We acknowledge support by the Deutsche Forschungsgemeinschaft.

# 6

# Experiments on planetesimal formation

Gerhard Wurm

*Institut für Planetologie*
*Westfälische Wilhelms–Universität Münster, Germany*

Jürgen Blum

*Institut für Geophysik und Extraterrestrische Physik*
*Technische Universität Carolo-Wilhelmina Braunschweig, Germany*

## 6.1 Introduction

Rather few facts can be considered as acceptable to all who are working in the field of planet and planetesimal formation. Starting there, we will explore the possible pathways as suggested by experiments. It is certainly undisputed that the regular mode of planet formation is connected to protoplanetary disks. These disks consist mostly of gas, which makes up about 99% of their mass. The remaining 1% resides in the form of dust and – depending on the temperature – in the form of ice. As terrestrial planets are mostly built from heavier elements it is natural to assume that they are somehow assembled from the dust component in the disk.

Whatever model is placed between the dust and the planets, collisions between the solid bodies are unavoidable. In fact a large part of the process of planet formation can be based on collisions which can and (at least partly) will lead to the formation of larger bodies.

In the following sections we will review experiments that have studied these collisions and eventually put these results in a rough sketch of planetesimal formation. It is sometimes argued that collisions of large bodies might be too energetic to lead to the formation of a still larger body (Youdin and Shu, 2002). As described in this chapter it is true that collisions can lead to erosion rather than growth. However, we will show that this is not necessarily so for all collisions. We must also note that experimental facts are still often ignored for the sake of simplicity in modeling planetesimal formation. With all the results from the experiments during the last one or two decades this simplification is no longer justified. It is quite clear that observations of protoplanetary disks, which rely on the particle distribution resulting from the process of planet formation itself cannot be described properly otherwise.

*Planet Formation: Theory, Observation, and Experiments*, ed. Hubert Klahr and Wolfgang Brandner.
Published by Cambridge University Press. © Cambridge University Press 2006.

In Section 6.2 we consider collisions between two dusty bodies at different collision velocities (energies) without external forces. Section 6.3 deals with the influence of electric charges. Section 6.4 includes the interaction of colliding particles with gas. In Section 6.5 we outline some perspective for future experiments and Section 6.6 will put the experiments in the context of planetesimal formation.

## 6.2 Two-body collisions and the growth of aggregates in dust clouds

If two particles collide there is a number of different outcomes possible. Particles might stick together at low collision energies if sufficient kinetic energy is dissipated and surface forces are strong enough to keep the particles together. Collisions between aggregates of particles might also result in adhesion but, in addition to simple rebound, the possibility of structural changes must be considered. At larger collision energies particles and aggregates can rebound from each other and, one step further, might even be shattered. We do not consider collisions of still higher impact energies which might evaporate part of the colliding bodies since this is of no relevance for the first steps of planet formation, even though such impacts might certainly be of importance for the larger protoplanets. With respect to this we restrict ourselves to collisions relevant for the formation of km-sized objects which are usually termed planetesimals. We note that the reader has to take care not to confuse the different sizes and notions of the term particle as used throughout this chapter. We hope it becomes clear from the context but we would like to sensitize the reader to the problem that people talking about particle collisions in the literature do not necessarily have the same things in mind. Overall the size of the *objects* considered can vary from sub-μm to several km, and km-sized objects can be considered to consist of smaller sub-units again down to the sub-μm size range.

### 6.2.1 Hit-and-stick collisions

A strict classification of collisions is difficult to achieve since details depend on quite a number of parameters, e.g. the morphology of the colliding bodies, the material, the mass, the impact parameters, and the velocity of the impact. In this section we review collision experiments which more or less result in perfect adhesion of the colliding bodies at their contact. A basic question is: under which circumstances two individual solid dust particles stick together or bounce off each other after a collision? This is strongly correlated to the question of adhesion between two dust particles.

### 6.2.1.1 Measurements of adhesion forces

Which forces keep the dust grains together after a (gentle) collision? In the absence of electrostatic charges on the particles (that are very unlikely to be present on the grains in the beginning due to the shielding of ultraviolet and cosmic radiation by an opaque protoplanetary disk), dipole–dipole interactions are considered to be responsible for particle adhesion. For polar materials (e.g. water, ice) these forces are caused by hydrogen bonding; for non-polar materials (e.g. silicates) van der Waals forces are responsible for grain adhesion. Heim *et al.* (1999) experimentally investigated the forces acting between perfectly smooth, spherical $SiO_2$ grains with radii between 0.5 μm and 2.5 μm. They found that the adhesion force is proportional to the (reduced) particle diameter and that the constant of proportionality is close to the bulk value of the surface energy of $SiO_2$. Typical adhesion forces of μm-sized $SiO_2$ particles are $\approx 10^{-7}$ N. Water-ice particles of the same size should have adhesion forces roughly one order of magnitude higher due to the higher surface energy (Dominik and Tielens, 1997).

### 6.2.1.2 Dust–wall collisions

Collisions between single dust particles and flat targets in a broad velocity interval were investigated by Poppe *et al.* (2000a). Particles were typically μm-sized and consisting of a variety of materials, such as silica, enstatite, diamond, and silicon carbide. Due to the mechanism of dust deagglomeration (Poppe *et al.*, 1997), the individual dust grains had velocities of up to $\approx 100\,\mathrm{m\,s^{-1}}$. The targets consisted of polished quartz and silicon. Particle tracks before and after impact were visualized by forward scattering of laser light and were recorded by a high-speed camera attached to a long-distance microscope.

The results of Poppe *et al.* (2000a) showed that spherical particles behave differently from irregular ones. For the $SiO_2$ spheres, a sharp adhesion threshold exists below which particles almost always stick and above which they almost always rebound after impact. This threshold velocity decreases for increasing particle size and is typically $1\,\mathrm{m\,s^{-1}}$ for spheres with 1 μm radius. Irregular grains, however, show a much flatter transition between adhesion and rebound so that they are better described by an adhesion probability. Adhesion probabilities are typically as high as 0.2–0.8 for impact velocities $\lesssim 100\,\mathrm{m\,s^{-1}}$.

### 6.2.1.3 Collisions between dust aggregates

As seen in the experiments by Poppe *et al.* (2000a) μm-sized dust particles colliding at velocities much below $1\,\mathrm{m\,s^{-1}}$ will stick together. If collisions are gentle enough, this instant adhesion at the contact point will also keep dust *aggregates* together after a collision. By this process, larger aggregates can grow. Within the frame of a simple hit-and-stick mechanism different scenarios are possible. If a dust aggregate (cluster) is placed in a reservoir of individual dust particles we get particle–cluster aggregation (PCA). In this case near-spherical morphologies of the growing aggregates result. This is due to the fact that small dust particles can penetrate through the outer rim of the aggregates and fill in the larger gaps. While the resulting aggregates are porous, their mass $M$ still depends on the volume of the aggregate or $M = ar^3$ with $r$ and $a$ being the size of the aggregate and a scaling constant depending on the porosity $P$ of the aggregate. With the bulk density $\rho$ of the material, PCA bodies have $a = 4/3\pi\rho(1 - P)$. Here, the porosity is the ratio of void volume to total volume within the aggregate and $(1 - P)$ is the so-called volume-filling factor. Blum (2004) showed that in the case of PCA the mass of the growing agglomerate can be described as a power law in time (see Eq. 6.5) with a power index determined by the mass dependence of the relative velocity between monomers and agglomerate. Kozasa *et al.* (1992) showed that PCA agglomerates are quasi-spherical in shape and have a volume-filling factor of 0.15, i.e. 15% of the agglomerate volume is occupied by particles. Thus, the porosity of PCA aggregates is $P = 0.85$.

Blum and Schräpler (2004) have experimentally realized the so-called random ballistic deposition (RBD) in which individual monomer grains are ballistically deposited on an agglomerate in a hit-and-stick manner. Although the deposition is unidirectional, the resulting agglomerate structure is morphologically similar to the PCA growth. For RBD, the volume-filling factor is also 0.15 (Watson *et al.*, 1997). In the experiment of Blum and Schräpler (2004), individual monodisperse, spherical $SiO_2$ grains (particle radii 0.75 µm) were, after deagglomeration by a cogwheel (Poppe *et al.*, 1997), dispersed into a streaming rarefied gas and thereafter deposited onto a gas-permeable substrate with velocities of typically $\approx 0.1 \text{ m s}^{-1}$. The deposition velocity was chosen to be so low that adhesion upon first contact (see above) was realized. In that manner, layer upon layer of new particles were deposited on each other. The forming agglomerates had diameters of 25 mm and thicknesses of up to 10 mm (see Fig. 6.1). Analysis of the structure of the agglomerates showed that the volume-filling factor was $0.15 \pm 0.01$, in agreement with the numerical results. Deviation from the sphericity of the monomer grains resulted in a lower volume-filling factor of $\approx 0.10$ (for quasi- monodisperse diamond grains of 1–2 µm diameter), and an additional polydispersity in the monomer sizes further decreased the volume-filling factor to $\approx 0.07$ (for irregular $SiO_2$ grains with diameters between $\approx 0.1$ µm and $\approx 5$ µm) (Blum, 2004).

The macroscopic RBD agglomerates are mechanically stable and can be used for further experiments (e.g. tensile, and compressive strength, impact, or light-scattering measurements).

While the growth via particle–cluster aggregation requires the existence of individual particles (and only a few larger aggregates) and an enhanced collision

Fig. 6.1. Dust aggregates generated by random ballistic deposition. Seen in a, b, and d are images (microscopic in d) of experimentally generated aggregates with a volume-filling factor of 0.15. Seen in c is a Monte Carlo simulation giving the same volume-filling factor. Taken from Blum and Schräpler (2004).

frequency between small and large aggregates, a cloud of individual particles aggregating on its own by a hit-and-stick mechanism will grow in a different manner. A growth model often assumed for this case is based on the fact that, on average, aggregates of the same size collide. This growth mode is called cluster–cluster aggregation (CCA). The coagulation equation is given as (Smoluchowski, 1916)

$$\frac{\mathrm{d}n_i}{\mathrm{d}t} = -n_i \sum_{j=1}^{\infty} K(i, j)n_j + \frac{1}{2} \sum_{j=1}^{i-1} K(i-j, j)n_{i-j}n_j. \tag{6.1}$$

The first term on the right hand side of Eq. (6.1) describes the loss of aggregates with $i$ constituents and number density $n_i$ by aggregation with another aggregate. The second term describes the formation of aggregates with $i$ constituents due to the adhesion of two smaller ones. The collision kernel $K(i, j)$ depends on the collision rate and the adhesion efficiency $\beta$ as $K(i, j) = \beta(i, j)v(i, j)\sigma(i, j)$. Here, $v$ is the collision (relative) velocity between two aggregates and $\sigma(i, j)$ is the collision cross-section. Equation (6.1) can be solved numerically (see Chapter 7 by Henning *et al.*) or analytically in special cases (Wurm and Blum, 1998; Blum, 2004). The cross-section $\sigma(i, j)$ and the collision velocity $v(i, j)$ depend on the morphology of an aggregate. Ad hoc assumptions, Monte Carlo simulations, or experiments are needed for this task.

Numerical simulations and experiments have shown that the mass of CCA aggregates is usually no longer dependent on their volume but follows a power law of the form

$$M = c\, r^{d_f}, \tag{6.2}$$

with $d_f$ and $c$ being the fractal dimension characteristic for a given growth process and a scaling constant.

### 6.2.1.4 Brownian growth

When the dust grains are freshly condensed and are thus still in the (sub-)μm-size range, collisions among the particles are mainly caused by Brownian motion. The mean Brownian collision velocity between two dust grains with masses $m_i$ and $m_j$ is given by

$$v_{\mathrm{Br}} = \sqrt{\frac{8kT}{\pi m}}, \tag{6.3}$$

with $k$, $T$, and $m = \frac{m_i m_j}{m_i + m_j}$ being Boltzmann's constant, the gas temperature, and the reduced mass respectively. For μm-sized grains and gas temperatures around $T = 300$ K, $v_{\mathrm{Br}}$ is typically a few mm s$^{-1}$.

Experiments on the Brownian motion and agglomeration in rarefied gas cannot be performed under laboratory conditions because the gravitational sedimentation velocities exceed the thermal velocities by far. The only attempt to experimentally investigate Brownian agglomeration in the free molecular flow regime is the Cosmic Dust Aggregation experiment CODAG which flew in the microgravity condition of the space shuttle flight STS-95 and onboard the Maser 8 sounding rocket. A detailed technical description of the two experimental setups is given by Blum *et al.* (1999a) and Blum *et al.* (1999b). In both experiments, a cloud of monodisperse, spherical SiO$_2$ grains (particle radii 0.95 μm and 0.50 μm, respectively) was dispersed in a rarefied gas atmosphere and the Brownian agglomeration was observed by long-distance microscopes and high-speed cameras.

The general observation of both experimental runs was that within a few collision timescales $\tau_{\mathrm{coll}}$, chain-like fractal agglomerates with a fractal dimension $d_f \approx 1.4$ (see Fig. 6.2) and a quasi-monodisperse mass distribution form (Blum *et al.*, 2000; Krause and Blum, 2004). Here, the collision timescale is given by

$$\tau_{\mathrm{coll}} = \frac{1}{n\sigma v_{\mathrm{Br}}}, \tag{6.4}$$

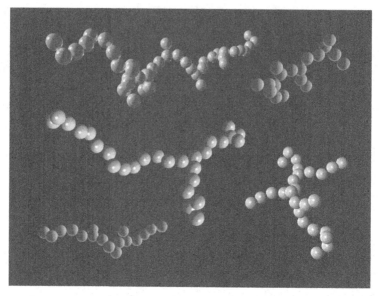

Fig. 6.2. Three-dimensional aggregates grown in the Cosmic Dust Aggregation experiment (CODAG) (from Wurm, 2003).

where $n$ and $\sigma$ denote the particle number density and collision cross-section, respectively. With particle number densities of typically $n = 10^{12}$ m$^{-3}$ and geometrical collision cross-sections (initially) between individual μm-sized particles of $\sigma = 3 \times 10^{-12}$ m$^2$, the experimental collision timescales were of the order of a minute. The temporal evolution of the mean agglomerate mass followed a power law of the form

$$m \propto t^\delta, \tag{6.5}$$

(see Fig. 6.3) with $\delta \approx 2$ (Krause and Blum, 2004), which can be also derived theoretically for monodisperse growth (Blum, 2004).

Growth rates and mass spectra are consistent with the numerical results by Kempf *et al.* (1999), but the fractal dimension of the agglomerates are significantly smaller than derived in the Monte Carlo simulations. Although this has no considerable consequence for the further growth stages, the discrepancy between theoretical and experimental fractal dimensions is so far unexplained. Agglomerate alignment by electrical dipole forces due to grain charging can be experimentally excluded. A hypothesis for the unexpectedly low fractal dimensions in the experiments is the influence of Brownian rotation on the agglomerate structure which was not explicitly treated in the numerical simulations by Kempf *et al.* (1999). The short growth timescales and the low fractal dimensions of the agglomerates imply an

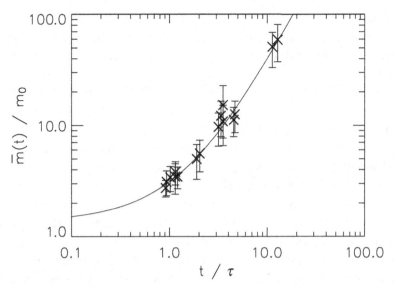

Fig. 6.3. Temporal evolution of mean agglomerate mass due to Brownian motion (taken from Krause and Blum, 2004).

adhesion efficiency of unity and a hit-and-stick behavior of the agglomerates in mutual collisions.

For the agglomerate growth in a protoplanetary disk, the results of the microgravity experiments on Brownian agglomeration imply that within a few years to decades fractal agglomerates consisting of dozens to hundreds of monomer grains form (Blum, 2004, Weidenschilling and Cuzzi, 1993). Due to the decrease in collision velocity with increasing agglomerate mass (see Eq. 6.3), other sources for mutual collisions between agglomerates then become important (see following section).

### 6.2.1.5 Differential sedimentation and turbulent growth

If a particle settles in a protoplanetary disk the sedimentation velocity $v_s$ depends on the gas–grain friction time $\tau_f$ by $v_s = a_g \cdot \tau_f$ with $a_g$ being the vertical component of the star's gravitational acceleration. The gas–grain friction time depends on dust and gas properties (Blum *et al.*, 1996):

$$\tau_f = \frac{m}{A}\frac{\epsilon}{\rho_g v_m}. \tag{6.6}$$

Here, $m$ is the mass, $A$ is the geometrical cross-section, $\epsilon = 0.58$ is a constant which has to be determined experimentally, $\rho_g$ is the gas density and $v_m$ is the mean thermal velocity of the gas molecules. Typical velocities for small dust aggregates are in the mm s$^{-1}$ range though much larger values can be reached at high disk altitudes and further out in the disk. Due to statistical variations in $m/A$ for equal mass agglomerates, collision velocities of typically one order of magnitude smaller than the absolute sedimentation velocities can be reached. Wurm and Blum (1998) observed individual collisions between dust aggregates consisting of 0.95 μm (radii) $SiO_2$ particles due to differential sedimentation at 0.001–0.01 m s$^{-1}$. In the individual collisions a hit-and-stick behavior leading to the formation of larger aggregates could be imaged directly.

As part of each experiment the aggregates had to be generated, which was achieved in a turbulent gas where particles collided to form aggregates with fractal dimensions of $d_f = 1.9$. Analysis of the turbulent growth shows that typical collision velocities during growth were larger than 7 cm s$^{-1}$ (Wurm and Blum, 1998). Thus, though not as unanimous as the direct imaging of the collisions, the turbulent growth suggests hit-and-stick collisions at much higher collision velocities also seen in further experiments described below. In more detail the observed dust aggregates and their size evolution with time (size distribution) can be fitted very well with calculations using Eq. (6.1) (Wurm and Blum, 1998). The fractal dimensions of $d_f = 1.9$ observed for the aggregates are close to the values obtained by an idealized cluster–cluster aggregation (Meakin, 1991). It might be asked if cluster–cluster aggregation in its idealization is possible at all since identical aggregates couple to the gas the same way and have no relative velocities. Only aggregates with different surface-to-mass ratios can collide. However, it can be seen in the experiments on turbulent growth that indeed the mass of the mean colliding aggregate is comparable to the mean aggregate mass at any given time as can be seen in Fig. 6.4. Thus the idea of cluster–cluster aggregation also holds for real growth processes.

Due to the relative velocities between dust aggregates and gas during sedimentation it could be seen in the experiments that the aggregates align (Wurm and Blum, 2000). How much this effect would change the growth timescales and aggregate morphologies is not yet clear. Since the aggregates align preferentially with their longest axes in the direction of sedimentation (gas flow) they can settle slightly faster but at the same time the collision cross-section will be reduced.

It has to be noted, though, that alignment is one of the processes that is important for observations since it leads to polarization effects. The influence of aggregates on light scattering has been studied experimentally on the same kind of aggregates as used in the growth and collision experiments described before (Wurm et al., 2003; Wurm et al., 2004b). Since dust aggregation is a fundamental mechanism present in different astrophysical environments the optical properties of the forming agglomerates should be explored further. The asymmetry parameter which is an average of scattered radiation over scattering angles is, for example, one important quantity. Wurm et al. (2003, 2004b) show that the asymmetry parameter for aggregates

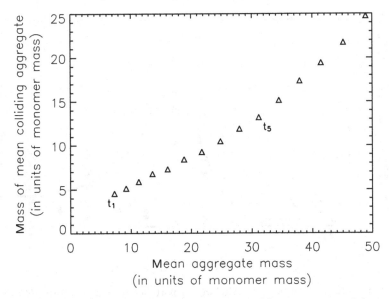

Fig. 6.4. The mass of the mean aggregate colliding with an aggregate of the mean mass (taken from Wurm and Blum, 1998). The individual triangles represent different times in a turbulent growth process. With both masses being similar, the process can be considered a cluster–cluster aggregation.

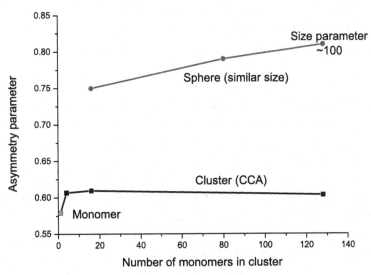

Fig. 6.5. The asymmetry parameter of scattered light for cluster–cluster aggregates and, for comparison, spherical particles (Wurm *et al.*, 2004b).

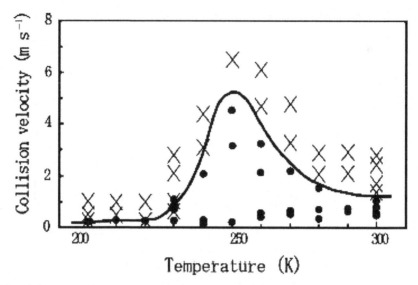

Fig. 6.6. Threshold velocity for sticking of particles coated with organic material (Kouchi *et al.*, 2002). For a small temperature region around 250 K sticking velocities are increased.

that are large compared to the wavelength can decrease as the aggregates grow in contrast to the asymmetry parameter of growing spherical particles which are often taken as model particles (Fig. 6.5). Wurm and Schnaiter (2002) also showed that aggregates might well be suited to explain extinction and polarization features of interstellar dust grains quite naturally.

### 6.2.1.6 Organics

Experiments show (see Section 6.2.2.1; Blum and Wurm, 2000) that the threshold velocities for adhesion of aggregates do not depend too much on the material and on the shape of the individual dust particles as long as refractory materials, e.g. $MgSiO_3$, $SiO_2$, or diamond, are considered. However, organic materials can be much stickier under certain conditions. Kouchi *et al.* (2002) carried out impact experiments on organic-coated copper particles of 1 cm in size at different temperatures and found that for mm-sized organic layers collisions of up to 5 m s$^{-1}$ can lead to immediate adhesion. As seen in Fig. 6.6, this high value is only reached in a very limited temperature range. Nevertheless, at certain times and places this might be an important process for particle growth.

### 6.2.2 Medium/high kinetic-energy collisions

As discussed before, collisions between individual spherical dust grains of 1 μm in size and a flat target lead to rebound for collision velocities above roughly 1 m s$^{-1}$

Table 6.1. *Collisions between cluster–cluster aggregates at different collision energies: (1) first visible restructuring, (2) maximum compression, (3) loss of monomers, (4) catastrophic disruption, adapted from Blum and Wurm (2000).*

| Case | Energy | Experimental velocity (m s$^{-1}$) | Model velocity (m s$^{-1}$) with new data for $E_{roll}$ and $E_{br}$($^+$) | Model velocity (m s$^{-1}$) with old data for $E_{roll}$ and $E_{br}$($^\times$) |
|------|--------|------|------|------|
| (1) | $E_{im} \approx 5 \cdot E_{roll}$ | $0.20^{+0.07}_{-0.05}$(*) | $0.20 \pm 0.04$(*) | $0.047$(*) |
| (2) | $E_{im} \approx 1 \cdot n_k \cdot E_{roll}$ | $0.65^{+0.15}_{-0.10}$ | $0.69 \pm 0.12$ | $0.16$ |
| (3) | $E_{im} \approx 3 \cdot n_k \cdot E_{br}$ | $1.2 \pm 0.2$ | $1.0$ | $0.14$ |
| (4) | $E_{im} > 10 \cdot n_k \cdot E_{br}$ | $1.9 \pm 0.3$ | $1.9$ | $0.26$ |

$^+$ $E_{roll} = 1.7 \cdot 10^{-15}$ J from Heim *et al.*, 1999
$E_{br} = 1.3 \cdot 10^{-15}$ J from Poppe *et al.*, 1999, 2000a
$^\times$ $E_{roll} = 9.5 \cdot 10^{-17}$ J from Dominik and Tielens, 1995
$E_{br} = 2.4 \cdot 10^{-17}$ J from Chokshi *et al.*, 1993 and Dominik and Tielens, 1997
$^*$ valid for $m = 60\, m_0$

(Poppe *et al.*, 2000a). Also larger particles covered by an organic layer will not stick above a few m s$^{-1}$ impact velocity. As aggregates are concerned other processes like restructuring, compaction and fragmentation can occur.

### 6.2.2.1 Aggregate–aggregate collision

Blum and Wurm (2000) studied collisions between fractal dust aggregates and a flat target at collision velocities between a few cm s$^{-1}$ and 30 m s$^{-1}$. As a target they used the 1 μm-thick side of a Si$_3$N$_4$ cantilever as used in atomic force microscopy. The aggregates consisted of up to several tens of SiO$_2$ particles with radii of 0.95 μm and 0.5 μm and of irregular MgSiO$_3$ particles with similar size. In agreement with the results given above, the impacts resulted in adhesion of the dust aggregates at their contact points at low impact velocities. At higher impact velocities more compact dust layers formed varying from slight restructuring to complete compaction on the target. Above $\approx$2 m s$^{-1}$ impact velocity, no growth of dust on the target was observed and the dust aggregates were destroyed during impact. If experimental values for the rolling friction and the break-up energy of two dust grains by Heim *et al.* (1999) and Poppe *et al.* (2000a) are used, the experimental results can well be matched with simulations by Dominik and Tielens (1997). Details can be seen in Table 6.1.

There are different steps that can be identified between perfect adhesion at the first point of contact without restructuring to complete destruction of the aggregates. As seen in Table 6.1 there is a gradual change from slight restructuring through total compression to loss of monomers and complete fragmentation with increasing collision energy.

Experiments suggest a square-root dependence of the velocity threshold for adhesion on the size of the dust grains which build the aggregates. Thus, if all monomer particles were 100 nm in size instead of 1 μm-adhesion at several m s$^{-1}$ would be possible. This does not solve the fundamental problem, though, of how aggregates will stick together at collision velocities of 10 m s$^{-1}$ or higher. Also the threshold velocity for adhesion will not increase continuously towards smaller grain sizes. Different physics applies for very small dust grains of nm size or less. These particles might move significantly with respect to each other due to thermal fluctuations (Jang and Friedlander, 1998). The extreme would be molecules which only stick transiently but can be ejected again after a short time of adsorption.

However, as seen with larger aggregates, the features of, for example, a porous dust aggregate also depend on the size and size distribution of the individual dust grains. Thus it is important to know what size of particles are present in protoplanetary disks.

### 6.2.2.2  Collisions between larger dust aggregates

As shown by Blum and Wurm (2000) (see Section 6.2.2.1), aggregates will get compacted at higher impact velocities or at larger sizes. Impacts between small fractal dust aggregates of several μm in size might not be representative any more. Impacts between macroscopic compacted aggregates will dominate once the restructuring and compaction threshold is reached. Results of the experiments between fractal aggregates cannot be scaled to these conditions. There will, for example, be shielding of the inner parts of these aggregates during an impact and the number of connections between monomer particles is much larger. Wurm *et al.* (2005a) carried out experiments between mm–cm-sized dust projectiles and cm–dm dust targets at impact velocities up to 38 m s$^{-1}$. As dust, μm-sized (SiO$_2$) particles were used. The targets were prepared by sifting the dust sample. This resulted in a granular structure with granule size depending on the mesh size used and high porosities with $P = 0.8$–$0.9$. The projectiles created craters but *no* significant amount of ejecta from the crater faster than about 0.3 m s$^{-1}$ could be detected, which is an upper limit due to the restriction placed by gravity. On the other hand large amounts ($\approx 10$ times the projectile mass) of slow ejecta ($\approx 0.1$ m s$^{-1}$) were observed on the whole surface. The ejection velocities are only 0.5% of the impact velocity. The particles were ejected by elastic waves which were reflected on the target bottom. Due to the small velocities reaccretion by gas drag was possible, as seen in Section 6.4. If ejecta can be observed to reaccrete onto a larger body ($\leq 1$ m) in protoplanetary disks for which the impact velocities are more appropriate remains to be seen. It is possible that no particles are ejected at all and thus net growth would occur. It is also possible that compacted regions within a body reflect the wave back to the surface. It is important to address these questions in the future and preparations are underway. Experiments at high velocities, but for compacted targets, have also been carried out (Wurm *et al.*, 2005b). They show quite different behavior, but also here net growth can be the immediate result of an impact. In fact growth only occurs at collision velocities larger than $\sim 10\,\mathrm{m\,s^{-1}}$.

The RBD dust agglomerates described above are ideal candidates for low-velocity impact experiments as they represent the best analogs of preplanetesimal matter at the start of the runaway growth regime. Recent microgravity drop-tower experiments by Langkowski and Blum (unpublished data) investigated the outcomes of collisions between RBD agglomerates of 25 mm diameter and equal-porosity agglomerates with diameters between $\approx 100\,\mu\mathrm{m}$ and $\approx 3$ mm in the velocity range 0–3 m s$^{-1}$. Impact angles with respect to the perpendicular of the target agglomerate surfaces were randomly distributed. Depending on impact energy and impact angle, mass gain (for low-impact energies and low-impact angles) or mass loss (for high-impact energies and high-impact angles) was observed. In a few cases, the projectile agglomerates were fragmented. Whenever a collision did not result in adhesion, the projectile left the target agglomerate with a considerable energy loss, and the surface of the target agglomerate was cratered. In the case of adhesion, the projectile agglomerates were partially embedded into the target agglomerate. Cratering, partial embedding, and low coefficients of restitution show that the RBD agglomerates were plastically deformed and thus compacted.

### 6.2.2.3 Static measurements of high-energy impacts

The mechanical properties of the macroscopic RBD agglomerates presented in Section 6.2.1.3 were determined by Blum and Schräpler (2004) and Blum (2004). With the knowledge of the compression–density behavior of the samples, predictions of the outcomes in high-velocity impacts can be made. The agglomerate samples were cut into a cylindrical form and were subjected to a compressive force of known strength. With increasing compression, the thickness of the samples decreased and was determined with high precision. By simultaneous measurements of the samples' area perpendicular to the compressive force, the volume-filling factor could be determined. The resulting compression–density curves showed that the agglomerates can resist static compressive pressures of up to a few 100 Pa before the volume-filling factor increases, virtually independently of the material and morphology of the constituent grains (Fig. 6.7).

Tensile strengths of the samples were determined by gluing the surfaces of high-porosity dust samples, using a non-wetting epoxy resin, to two flat, thin glass plates. These glass plates were then attached to an apparatus which allowed the measurement of the tensile forces while the samples were uncompressed until they

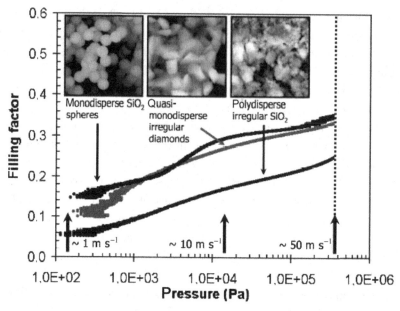

Fig. 6.7. Volume-filling factors of an RBD agglomerate relative to the compression (pressure). For an impact of a body of given density the maximum local compression can be estimated to be $1/2 \, \rho v^2$. The compressions due to impacts at various impact velocities are labeled. This implies that no filling factor larger than 0.35 can be reached at the slow collisions leading to planetesimal formation (Blum, 2004).

broke into two halves. The measured tensile strengths were as low as a few 100 Pa to 1000 Pa, depending on the material, shape, and size distribution of the monomer grains. It is obvious that porosities and tensile strengths of the laboratory samples described in Blum and Schräpler (2004) and Blum (2004) are close to the values of comets (Davidsson and Gutierrez, 2004). As comets are believed to be the sole surviving remnants of the pre-planetesimal phase of the Solar Nebula, the RBD agglomerates are ideal analogs for pre-planetesimal matter.

In conjunction with the static compression measurements, it became clear that any collision between two high-porosity dust agglomerates with velocities $\gtrsim 1 \, \mathrm{m \, s^{-1}}$ results in a compaction of the collision partners. Thus, dust agglomerates which experience no more than 1(10; 100) $\mathrm{m \, s^{-1}}$ should have volume-filling factors in the range 0.07–0.15 (0.15–0.28; 0.20–0.35), depending upon the morphology and size distribution of the monomer grains (Blum and Schräpler, 2004; Blum, 2004).

We note, though, that an impact has other features which cannot be accounted for by the analog of a static compression as described in the previous section, e.g. elastic waves might change the morphology of the body and eject particles.

### 6.2.2.4 Impacts into regolith

Hartmann (1978) observed collisions between a solid impactor of several grams mass at several m s$^{-1}$ into regolith-like material. He found only a small amount of ejecta compared to the projectile mass. The projectile itself seemed to be buried in the regolith. Colwell and Taylor (1999) and Colwell (2003) studied similar kinds of impacts at lower velocities under microgravity. They observed that the cm-sized solid projectile did bounce off the target regolith at velocities above 20 cm s$^{-1}$ while it did stick to the surface at lower impact velocities. The experiments showed that immediate net growth in a collision of this kind was only possible for rather small impact velocities. However, the rebounding projectile was very slow with as little as 1% of the impact velocity (coefficient of restitution $c_r \approx 0.01$) and the ejecta produced were slow, their velocity being typically less than 10% of the impact velocity. Part of the ejecta could be reaccreted by gas drag as described below.

### 6.2.2.5 Ice collisions

Bridges *et al.* (1996), Supulver *et al.* (1995, 1997), and Higa *et al.* (1998) investigated slow collisions between two cm-sized icy bodies. They used pendulums to generate very slow collision velocities in the laboratory. If the icy bodies were coated with a methanol frost layer adhesion at velocities of up to 10 cm s$^{-1}$ was observed (Vance, 2003). Otherwise rebound or, at several m s$^{-1}$ impact velocity, destruction of the ice spheres was the result. These experiments might be viewed in connection with the pendulum experiments with (cold) organic material by Kouchi *et al.* (2002) as described above. However, common (non-organic) icy bodies as expected to be present in the outer regions of protoplanetary disks obviously do not easily stick together if they are compact. This might be different for aggregates of very small ice grains at very low temperatures but this requires further study. It might be worth studying collisions which include mechanisms that on their own seem to promote growth best. Thus bodies of particulate matter consisting of sticky grains colliding with each other might result in a more rapid growth of planetesimals in the outer regions of a protoplanetary disk or somewhere around the snow line. It has to be noted that there is a large difference in size between the ice particles (cm) used in these experiments and the SiO$_2$ dust particles ($\mu$m) mentioned above. This explains why a less sticky dust particle adheres at higher collision velocities than the ice particles.

However, the ice-pendulum experiments provide data for the coefficient of restitution dependence on velocity which finds applications in collisions of planetary ring particles, i.e. around Saturn (Lewis and Stewart, 2000). In the ice experiments the coefficients of restitution were not found to be as low as for the regolith impacts studied by Colwell and Taylor (1999) and Colwell (2003). Nevertheless, for small coefficients of restitution $c_r$ it is sometimes argued that after several impacts the collision velocities might be slow enough for adhesion. This only works if the collisions are in a gas-free environment and thus remember their history. In most cases this is not valid for small bodies in protoplanetary disks. After a collision all particles involved couple to the gas of the disk rather quickly before the next collision occurs. Therefore the next collision will be of the same high velocity as the impact before.

## 6.3 Dust aggregate collisions and electromagnetic forces

Poppe *et al.* (2000b) investigated charge transfer in collisions between µm-sized dust grains and flat targets. They found that for silica spheres of different sizes a relation between the number of collision-induced elementary charges $N_e$ and the collision energy $E_{coll}$ exists of the form

$$N_e = \left( \frac{E_{coll}}{10^{-15} J} \right)^{0.83}. \tag{6.7}$$

Recent laboratory measurements by Poppe and Schräpler (2005) in an extended energy regime showed that the above relation is also valid for a variety of particle and target materials. In these collisions, the impinging dust particles were preferentially negatively charged, even in collisions with targets consisting of the same material.

The potential application of collisional grain charging for planetesimal formation lies in a steady charging of protoplanetesimals due to non-adhesive collisions with smaller dust particles or agglomerates. The cumulative effect of a series of non-adhesive collisions which lead to an accumulation of charges on the larger body might also lead to the build-up of such strong electrical fields at the surface that further collisions with charge separation could lead to an electrostatic trapping of the impinging dust grain or agglomerate. This electrostatic accretion process was introduced by Blum (2004).

More straightforward is the process of aggregation when grains are initially electrically charged. If and how this is possible in a protoplanetary disk is an issue of debate. Since ultraviolet radiation cannot serve as a charging mechanism other ways would need to be considered. It might be possible that after the initial growth of agglomerates from neutral grains, faster collisions lead to a dust cloud of charged particles by the process described by Poppe *et al.* (2000a). However, *if* charges exist, dust particles can aggregate much faster, as has been seen experimentally by Marshall and Cuzzi (2001), Love and Pettit (2004), and described by Ivlev *et al.* (2002). It has to be noted that the aggregates grown by coagulation of charged grains are not more robust against higher velocity impacts as seen, for example, in the experiments by Marshall and Cuzzi (2001) which show that $1 \ m \ s^{-1}$ is also a typical collision velocity before the aggregates get shattered, though it has to be noted that the aggregates consisted of larger (several 100 μm) dust grains.

With a significant fraction of condensable material in protoplanetary disks being Fe the influence of magnetic forces might be considered. The aggregation of magnetic particles was studied by Nuth *et al.* (1994), Dominik and Nübold (2002), and Nübold *et al.* (2003). Due to the strong dipole interactions ferromagnetic particles collide more frequently and preferentially at their edges creating long chain-like or web-like aggregates. As for electrical charges, coagulation rates are enhanced for magnetic particles. Aggregation of this kind has been proposed to explain magnetic features with respect to future measurements of magnetic fields of comets.

## 6.4 Dust aggregate collisions and dust–gas interactions

From the previous sections it is clear that collisions at velocities larger than $\approx 10$ m s$^{-1}$ are still a critical issue and might not lead to the formation of a more massive body. Even choosing the best set of parameters and conditions such bodies will often be disrupted, or might, less spectacularly, be eroded or just bounce off each other in mutual collisions. This is often used against the model of planetesimal formation by collisional growth. Under certain conditions dusty bodies can grow directly in collisions (Wurm *et al.*, 2005b) but this depends on the morphology of the target body. Nevertheless, as, for example, the high velocity impacts into porous dusty bodies show (Wurm *et al.*, 2005a), ejected dust particles can be very slow. If particles are small and slow, the influence of the gaseous environment in which a collision takes place has to be taken into account, especially since the gas is usually moving with up to several tens of m s$^{-1}$ with respect to the larger of the two colliding bodies. Dust particles can thus be reaccreted by the gas flow. This mechanism was proposed and experimentally verified by Wurm *et al.* (2001a, b). Small dust aggregates impinged targets of different widths at different gas pressures and gas-flow conditions. Trajectories of reaccreted dust particles ejected from the surface were imaged (Fig. 6.8) and could be modeled by the known particle–gas interaction. Even at impact velocities above 12 m s$^{-1}$ growth of a dust layer on the target was observed although the threshold velocity for adhesion (see Section 6.2.2.1) was $\approx 1$ m s$^{-1}$. In an environment without gas flow the same aggregates were fragmented at 1–2 m s$^{-1}$ impact velocity (Blum and Wurm, 2000). Thus, reaccretion of dust particles by gas drag can be an efficient mechanism.

The reaccretion of dust grains by a gas flow is restricted to flow conditions in which the streamlines end at the surface or penetrate the surface. Otherwise, as shown by Sekiya and Takeda (2003) impact fragments are transported around the target body by the gas flow. The process of aerodynamic reaccretion as proposed by Wurm *et al.* (2001a, b) is relatively efficient in the regime of free molecular flow

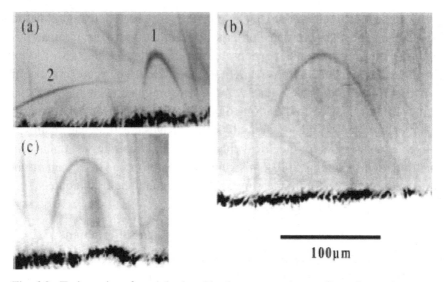

Fig. 6.8. Trajectories of particles bend back to a target by gas flow after an impact. These particles are reaccreted by the target in secondary collisions (Wurm *et al.*, 2001b).

where the size of the body considered is smaller or comparable to the mean free path of the gas molecules. It has to be kept in mind that the growing protoplanetesimals are very porous with porosities of 80% and more as discussed above. Placed in the gas flow of protoplanetary disks part of the gas will flow through these objects and streamlines are again directed towards the surface in a thin layer on the front side facing the gas flow. It has been shown by Wurm *et al.* (2004a) that under typical conditions in a protoplanetary disk, reaccretion of dust particles for a porous body can occur, though the term *typical conditions* still needs to be specified in more detail.

## 6.5 Future experiments

Most experiments reviewed in this paper deal with the collisional interactions between refractory particles, i.e. for those regions in protoplanetary disks where the terrestrial planets form. Most of a nebula's mass, however, is contained in the outer regions in which the giant planets, their icy moons, the Kuiper-Belt objects, and the comets form. The dominant material from which those bodies were originally agglomerated is water ice. Little is known about the adhesion and agglomeration behavior of microscopic water-ice particles. Some experiments were mentioned earlier but the bodies considered were rather solid ice. As with dust particles a more plausible starting point for agglomeration studies might be tiny ice grains. We do not expect tremendous differences between rocky materials and ices at very low temperatures. Hydrogen bonding should lead to slightly enhanced adhesion thresholds and sintering of ice bonds can result in solidified and, hence, impact-resistant bodies (Poppe, 2003). Recently, activity with a view to launching a program for ice particle research under microgravity conditions has increased (Ehrenfreund *et al.*, 2003).

As described in Section 6.3, cumulative collision-induced charging of growing protoplanetesimal bodies can influence the formation of planetesimals (Blum, 2004). Laboratory experiments on such cumulative grain charging are in preparation. These experiments require electrically insulated mounting of the target agglomerate and precise determinations of the accumulating electrical field. In addition to that, observations of particle trajectories are mandatory for the determination of the outcome of individual collisions between small dust particles/agglomerates and dusty targets.

The effect of elastic waves on the ejection of particles will be studied. It is not yet clear what amount of mass can be ejected under different conditions especially in collisions of bodies which are not supported by a tray. This can only be done under microgravity. Impacts of dusty bodies at still higher collision velocities up to $100 \, \mathrm{m \, s^{-1}}$ might also take place in protoplanetary disks depending on the model. It has to be noted that the sound velocity in porous targets can be $50 \, \mathrm{m \, s^{-1}}$ or less (Wurm *et al.*, 2005a). Thus impacts at higher velocities will be mild shocks which can change the outcome of a collision either directly or via shock (elastic) waves. How larger cm-sized porous projectiles will behave in intermediate velocity collisions of a few $\mathrm{m \, s^{-1}}$ has also not yet been studied.

## 6.6 Summary

From these experiments we might sketch the initial growth of larger bodies in a (typical) protoplanetary disk (Blum, 2004). Starting with Brownian motion and continuing during sedimentation, cluster–cluster aggregation will prevail for the first few hundred years at 1 AU. This timescale increases further out to a hundred thousand years at 100 AU distance from the star. After that growth continues but not as cluster–cluster aggregation. The aggregates become non-fractal. A threshold where fragmentation becomes important is reached after 1000 years at 1 AU. This is rather fast and therefore it cannot be expected that "clean" growth be observed in the inner parts of the disk. However, on the outer edges or in less dense regions above the disk, cluster–cluster aggregates might be the dominant type in young systems. If efficient transport mechanisms exist these regions might also have a population of particles from the inner disk which do not have to be cluster–cluster aggregates and might be more compact. The difference in optical properties between morphologically different kinds of particles might be important for the observations or might even give hints on possible transport mechanisms. As far as growth beyond the fragmentation limit is concerned, the experiments show that further growth is possible. At least bodies of several tens of cm can grow quickly by gas-aided growth (Wurm *et al.*, 2001a, b). However, during this growth process fragments will feed the dust reservoir again. Thus small particles and aggregates will be produced which may be morphologically different from cluster–cluster aggregates, e.g. more compact. The experiments suggest that it is pretty likely that as the collision velocities increase growth of still more massive bodies results. Little can be said about the timescales though since different collisions take place during this stage which might be eroding or lead to growth, on average.

Overall the available data set of the outcomes of collision experiments between dust agglomerates is beginning to form a (yet rudimentary) picture of protoplanetesimal dust agglomeration. Still, a lot more data is required to fill in the gaps in our understanding of the formation of planetesimals. Further modeling using the experimental results reviewed here has to be done before it becomes clear which collisions are typical and which are extremes and do not contribute significantly in shaping the distribution of dusty bodies in protoplanetary disks. Although it is premature to conclude about the formation of planetesimals, at least a couple of more or less robust statements about the dust evolution in protoplanetary disks can be given at this time:

- Dust agglomerates in protoplanetary disks grow quickly. Even when the star is still accreting mass and small particles are thus transported inwards with the gas, the timescale for this transport is much longer than the initial timescale of growth. It is inevitable that a significant fraction of at least cm- to dm-sized dust agglomerates exists in any kind of protoplanetary disk.
- The existence of smaller dust particles and agglomerates at all times has to be expected. When macroscopic dusty bodies collide at higher velocities, these collisions will always resupply the reservoir of small grains and agglomerates. It is *not* only coagulation which determines the evolution of dust particles and larger bodies in the disk, fragmentation also has to be considered.

Promising (and logical) types of collisions which might eventually lead to the formation of planetesimals are those between dusty bodies. It might not necessarily be desirable that dust particles are glued together tightly after a collision. While *loose* dust particles can more easily be ejected in an impact, smaller dust aggregates would be ejected which are more susceptible, for example, to gas reaccretion.

Based on the experimental results presented above, we conclude that mechanisms might exist with the potential to explain the formation of planetesimals by collisional growth. High impact velocities of several tens of $m\,s^{-1}$ are not necessarily an obstacle to planetesimal formation.

We are aware of the fact that a couple of alternative formation scenarios for planetesimals have been discussed in the literature. Klahr and Bodenheimer (2006) suggested that particles might be collected in large-scale eddies in protoplanetary disks which arise from baroclinic instabilities. Initial agglomerate sizes of 10–20 cm are required for bodies to get dragged into long-living eddies in which they potentially accumulate into larger objects. As we have shown above, this initial size can be reached by adhesive collisions in the protoplanetary disk. It is yet to be seen how this model evolves further when back-reactions between solid bodies and gas, and collisions among the dust aggregates are considered. Another alternative to direct collisional growth of km-sized dust agglomerates is the *classical* explanation of planetesimal formation by gravitational instability in the dust, originally proposed by Goldreich and Ward (1973). Work on gravitational instability is ongoing but a fundamental problem is reaching the high particle densities required for gravitational collapse. While we will not comment on this concept in general we nevertheless suggest that the experimental results of collision experiments need to be considered here. Youdin and Shu (2002) argue that with chondrule-sized (mm) particles a concentration sufficiently large to trigger the gravitational collapse can be reached within a few million years. As outlined above, dust aggregates evolve beyond chondrule sizes on a much shorter timescale. At least the formation of 10 cm particles can easily be explained and will occur (Wurm *et al.*, 2001a, b; Blum, 2004). The overall distribution of particles sizes, however, depends on the interplay between growth and fragmentation.

## Acknowledgments

We thank Hubert Klahr and Wolfgang Brandner for the organization of this "event." It was a pleasure to enjoy the relaxed atmosphere of the workshop. We are also grateful to Mario Trieloff for his constructive review of our manuscript.

# 7

# Dust coagulation in protoplanetary disks

Thomas Henning, Cornelis P. Dullemond, Sebastian Wolf

*Max-Planck-Institut für Astronomie, Heidelberg, Germany*

Carsten Dominik

*Sterrenkundig Instituut "Anton Pannekoek", Amsterdam, the Netherlands*

## 7.1 Introduction

According to the core-accretion model for planet formation, the building blocks of planets are formed by the coagulation of dust grains, growing from the initial sub-micron sizes inherited from the interstellar medium to the 100 kilometer sizes of full-grown planetesimals. This is a growth process over 12 orders of magnitude in size and 36 orders of magnitude in mass. The physics of dust is of crucial importance for the study of planet formation. It also plays a major role in the structure and evolution of protoplanetary disks, since the dust carries most of the opacity of the dust–gas mixture in these disks and provides the surface for chemical reactions. Moreover, infrared and (sub)millimeter observations of dust continuum emission from these disks can be used as a powerful probe for the disk structure and mineralogical composition. A deep understanding of the physics of dust and the coagulation of grains is therefore of paramount importance for the study of the formation of planets and the circumstellar disks in which they are formed.

The study of grain coagulation and the formation of planetesimals has a long history. At the start of the twentieth century an equation for coagulation of colloidal particles was formulated by Smoluchowski (1916), though not related to astrophysical applications. The continuous form of that equation was later used to study the size distribution of fog particles in the Earth's atmosphere (Shumann, 1940). The applications of these ideas to the formation of planets started at the end of the 60s, with many of the fundamental ideas already being discussed in the book by Safronov (1969). Since that time the topic of grain coagulation in the "Protosolar Nebula" has gained much attention, and considerable progress has been made by authors like J. G. Wetherill, S. Weidenschilling, J. N. Cuzzi, and many more.

*Planet Formation: Theory, Observation, and Experiments*, ed. Hubert Klahr and Wolfgang Brandner.
Published by Cambridge University Press. © Cambridge University Press 2006.

However, until not long ago most of such studies were devoted to understanding the formation of our *own* Solar System. Theories were therefore mostly compared with observations of the Solar System planets, comets and asteroids, data from meteorites, and data from space probes. But the enormous increase in sensitivity and spatial resolution of infrared and (sub)millimeter telescopes in recent years has opened a new window on grain growth: the coagulation of grains in the dusty protoplanetary disks surrounding nearby newly formed stars. In this review we would therefore like to focus on recent developments in this direction, while referring the reader to earlier reviews (e.g. Beckwith *et al.*, 2000; Weidenschilling and Cuzzi, 1993) for more background on previous developments.

This review is divided into three parts. In Section 7.2 we review evidence of grain growth in protoplanetary disks obtained by observations from the optical to the millimeter. In Section 7.3 we discuss the importance of interpreting such observations with detailed radiative-transfer modeling. And finally, in Section 7.4 we discuss detailed theoretical modeling of the coagulation process itself, and how these studies can be related to the aforementioned observations.

## 7.2 Observational evidence for grain growth

Scattered light images, mid-infrared dust spectroscopy, and millimeter interferometry together with a detailed analysis of spectral energy distributions provide accumulating evidence of grain growth in protoplanetary disks. These dust grains grow to sizes larger than the typical 0.1 µm size of interstellar dust. For interpreting such data correctly, one needs to take into account that radiation at different wavelengths traces distinct regions in a disk (see Fig. 7.1).

Near-infrared data mostly probe the inner 0.1 AU of disks around Solar-type stars. Mid-infrared spectroscopy characterizes the disk atmosphere in the 1–5 AU domain. Millimeter continuum observations probe the colder disk material at larger radial distances from the star. Furthermore, as a rule of thumb, radiation of a certain wavelength can only trace particles of sizes comparable to this wavelength. All compact particles much larger in size show a similar gray extinction behaviour. In addition, the absorptivity decreases for particles if they grow to such dimensions (see, for example, Krügel and Siebenmorgen, 1994).

The first evidence for grain growth comes from statistical investigations of the presence of dust emission as a function of age. Near-infrared observations of nearby star-forming regions suggest that the percentage of stars with 2–4 µm excesses is more than 80% for $\sim 1$ Myr stars and diminishes to $\sim 50\%$ by an age of 3 Myr (Haisch *et al.*, 2001). By ages of about 10 Myr, the inner dust disk traced by continuum infrared radiation disappears (Mamajek *et al.*, 2002). Far-infrared observations seem to suggest disk evolution on similar timescales. An extensive

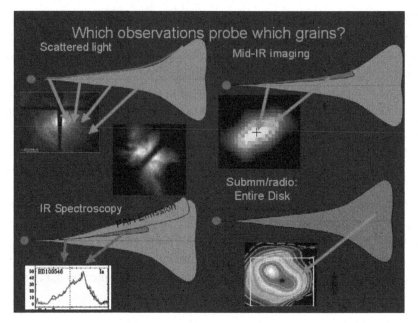

Fig. 7.1. Radiation from a disk: where is which radiation produced?

millimeter study by Carpenter *et al.* (2005) demonstrated a decrease in the mass of cold small grains by stellar ages of 10–30 Myr. All these observations point to grain growth with the consequence of decreased opacities and dust emission.

A more direct way to probe growth of grains to sizes of several microns is dust spectroscopy in solid-state features. The most prominent band, also accessible by ground-based observations, is caused by Si–O stretching vibrations in silicates around 10 µm. If grains grow to sizes of a few microns the feature gets a typical flat-top structure and decreases in strength before it completely vanishes for even larger sizes. This phenomenon (see Fig. 7.2) has been observed for Herbig Ae/Be stars (van Boekel *et al.*, 2003, see Fig. 7.2; see also Bouwman *et al.*, 2000; Meeus *et al.*, 2001) and T Tauri stars (Przygodda *et al.*, 2003). Van Boekel *et al.* (2004) could demonstrate grain growth in the inner regions of disks by mid-infrared inter-ferometry coupled with spectral resolution in the silicate band. Here we should note that the band-to-continuum ratio is not only a function of the size of the particles, but is also strongly influenced by the optical depth, and by the abundance of other materials contributing to continuum emission. This means that disk inclination, flaring, and mass-accretion rates will influence this ratio and the appearance of the band (see, for example, Men'shchikov and Henning, 1997).

The analysis of sub-millimeter and millimeter radiation provides an independent line of evidence for grain growth up to centimeter sizes. In the case of optically thin emission and the Rayleigh-Jeans part of the spectrum, we can expect that the

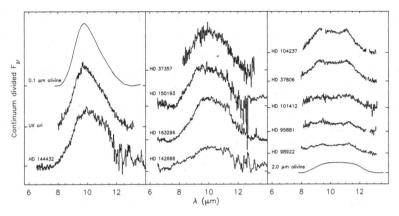

Fig. 7.2. Mid-infrared spectra of Herbig Ae/Be stars in the wavelength range of the 10 μm silicate feature (from van Boekel *et al.*, 2003).

wavelength dependence of the flux is given by $S \propto \lambda^{-\alpha}$ with $\alpha = 2 + \beta$. Here the dust absorption coefficient is given by $\kappa \propto \lambda^{-\beta}$. For the diffuse interstellar medium it is commonly assumed that $\beta$ is about two (e.g. Draine and Lee, 1984; Henning *et al.*, 1995). Earlier studies of dust in disks mainly relied on interpreting the submillimeter–millimeter spectral energy distributions (see, for example, Beckwith and Sargent, 1991; Mannings and Emerson, 1994). These studies led consistently to lower values of the $\beta$ index with values around 0.6, indicating the presence of larger grains. The problem with this approach is that one cannot exclude contributions from optically thick emission, which would lead to lower $\beta$ values than the right numbers. This problem can be solved as soon as interferometrically resolved images at millimeter wavelengths become available. The first such observations were performed for the T Tauri star DO Tauri, using the Owens Valley Radio Observatory (OVRO) and the Very Large Array (VLA) for wavelengths between 1.3 mm and 7 mm (Koerner *et al.*, 1995). These authors derived a $\beta$ value of 0.39±0.23. The error is usually dominated by the quality of the flux calibration for the different wavelengths and telescopes and the estimate of the flux contribution from free–free emission. A similar uncertainty is found for $\beta$ values derived in other interferometric studies. Dutrey *et al.* (1996) found dust emissivity values between 0.5 and 1.0 for T Tauri stars in the Taurus–Auriga region. Measurements for the 10 Myr-old pre-main sequence stars TW Hya (Calvet *et al.*, 2002) and CQ Tau (Testi *et al.*, 2003) revealed $\beta$ values of 0.6, again pointing to larger particles. A larger sample of Herbig Ae stars was investigated by Natta *et al.* (2004) which led to similar results. Rodmann *et al.* (2006) did a comprehensive study of T Tauri stars and obtained $\beta$ values between 0.8 (large grains) and 2.4 (ISM-like).

Here one should note that $\beta$ is an averaged quantity which is not only influenced by the grain size distribution, but also the chemical and physical structure of the particles and their temperature (see, for example, Henning *et al.*, 1995; Krügel and Siebenmorgen, 1994; Semenov *et al.*, 2003; Boudet *et al.*, 2005). Therefore, it is not uniquely possible to derive a size distribution, although power-law size distributions with maximum sizes of a few centimeters are consistent with the observations (Miyake and Nakagawa, 1993; D'Alessio *et al.*, 2001).

A completely independent view on the dust properties in disks is offered by analyzing the size and optical depth structure of silhouette disks such as those found in Orion (McCaughrean and O'Dell, 1996). These disks appear as dark shadows in narrow-line filters centered on nebula emission lines from the background molecular cloud surface. Choosing different emission lines allows a direct study of the integrated optical depth of the disk along the line of sight. Throop *et al.* (2001) found that the size of the disk shadow was identical in the H$\alpha$ and Pa$\alpha$ lines at 656 and 1870 nm, respectively, concluding that the dust opacity was gray in this wavelength region and that the grains should be larger than 5 $\mu$m. However, the dust opacity law is not the only parameter determining this measurement. The radial profile of the dust surface density is also important. If the disk has a sharp outer edge, a population of small grains making the disk still optically thick at 1870 nm is an alternative explanation. Shuping *et al.* (2003) extended this experiment with ground-based measurements of the Br$\alpha$ line at 4.05 $\mu$m, using the Keck telescope with adaptive optics, and found a marginally smaller disk at this wavelength, indicating that the grains should be larger than 1.9 $\mu$m, but not much larger than 4 $\mu$m. However, given the spatial variations of the background brightness and the complex and time-variable adaptive optics point spread function (PSF), this result may not yet be fully conclusive.

## 7.3 Radiative transfer analysis

The analysis of the radiation scattered and re-emitted by the dust component in circumstellar disks allows us to derive the opacity structure of the disk, and thus to constrain both the dust density distribution and optical properties of the dust grains. Besides the chemical composition and internal structure of the dust grains (crystalline/amorphous), their optical properties are determined by their geometrical shape, macroscopic structure (e.g. porous/compact, spherical/spheroidal) and size parameter ($\propto$ grain size/wavelength). Therefore, the interpretation of the optical properties of the dust in principle allows us to derive conclusions about grain sizes and thus about the coagulation process.

Due to the high complexity of the radiation transport process even in a simply structured disk model with rotation symmetry, only numerical simulations provide the necessary basis for the interpretation of observed spectral energy distributions (SEDs), intensity and polarization maps. In order to take into account the viewing-angle effect, the minimum requirement for simulating circumstellar disks are two-dimensional models, which have been developed since the late 80s (e.g. Dent, 1988). However, since then the numerical description of physical processes considered and the applied algorithms have been improved significantly (see, for example, Henning, 2001, for a review).

Given the small number of spatially well-resolved circumstellar disks, a proper interpretation of their SEDs is of decisive importance for studying the disk evolution. Unfortunately, there are many adjustable parameters in the disk models, and several different parameter combinations usually produce acceptable fits to the same SEDs (Thamm *et al.*, 1994). For this reason it is usually necessary to use additional constraints, such as done by Wood *et al.* (2002) who concluded from models for the SED of the classical T Tauri star, HH 30 IRS, that dust grains have grown to larger than 50 μm within its circumstellar disk, taking into account the known disk inclination and geometry.

D'Alessio *et al.* (2001) presented detailed models of irradiated T Tauri disks including dust-grain growth with power-law size distributions. These models assume complete mixing between dust and gas and solve for the vertical disk structure self-consistently including the heating effects of stellar irradiation as well as local viscous heating. For a given total dust mass, grain growth is found to decrease the vertical height of the surface where the optical depth to the stellar radiation becomes unity, while increasing the disk emission at (sub)millimeter wavelengths.

Another approach for investigating the grain size in the circumstellar environment and disk surface has been suggested by Fischer *et al.* (1996). The characteristic decrease of the polarization degree of the scattered near-infrared radiation with increasing grain size can be used to constrain grain sizes in the optically thin regions. For example, polarization measurements allowed the verification of the model of the prototype low-mass star HL Tau by Men'shchikov *et al.* (1999). Based on the SED of this object, the authors concluded that very large particles causing gray extinction are abundant in the dense torus of this object, while wavelength-dependent extinction points to submicron-sized grains in the circumstellar envelope, which is in rough agreement with the measured polarization degrees.

As outlined in Section 7.4, the evolution of the dust phase in circumstellar disks is expected to depend on the radial and vertical position in the disk. Based on the above selected studies, we conclude that multi-wavelength observations tracing different regions and relevant physical processes with high spatial resolution are mandatory in order to rule out ambiguities. As an example, we refer to the model of the circumstellar environment of the Butterfly star in Taurus by Wolf *et al.* (2003) which is based on high-resolution continuum observations at near-infrared and millimeter wavelengths. On the one hand, the millimeter observations were sensitive to the long-wavelength radiation being re-emitted from the dust in the central parts close to the midplane of the circumstellar disk. Furthermore, the resolved millimeter images allowed discrimination between different disk models with similar far-infrared/millimeter SEDs and therefore the disk geometry could be disentangled much more precisely. On the other hand, the near-infrared observations traced the envelope structure and dust properties in the envelope and the disk surface. The authors found that the grains in the envelope could not be distinguished from those of the interstellar medium, while coagulation had already resulted in grain sizes up to ~100 μm in the circumstellar disk.

## 7.4 Theoretical models of dust coagulation

### 7.4.1 Important processes

Dust aggregate growth is a result of the complex interaction between gas and dust in a protoplanetary disk. This interaction determines: (1) the relative velocity between grains and (2) the spatial distribution of grains in the disk, both critical parameters for grain growth. The relative velocities between grains set the collision rates and determine the outcome of the collision, which may be sticking, bouncing, and the modification (restructuring), erosion or destruction of the aggregates involved. Collisions of particles lead to sticking and growth if the relative velocities are smaller than about $1\,\text{m s}^{-1}$ (Poppe *et al.*, 2000). Larger collision velocities can cause destruction of aggregates and reinsertion of small grains into the disk (Dominik and Tielens, 1997; Blum and Wurm, 2000). For a detailed discussion of the conditions under which sticking occurs, we refer to the review by Wurm and Blum (Chapter 6).

The interaction between particles and the gas is governed by the stopping time which, for small dust particles, is given by

$$t_s = \frac{3}{4c_s\rho_{\text{gas}}}\frac{m}{\sigma} \tag{7.1}$$

where $m$ and $\sigma$ are mass and average projected cross-section of the particle, and $c_s$ and $\rho_{\text{gas}}$ are the speed of sound and the mass density of the gas. The limit of small dust particles, for which the above formula is valid, is called the "Epstein regime:" in this regime the mean free path of the gas particles is larger than the size of the dust grain. This is typically valid for compact particles smaller than about a centimeter at 1 AU in a protoplanetary disk.

The $m/\sigma$ ratio depends not only on the aggregate mass, but also on its structure. For relatively compact aggregates this will typically be larger than for very porous grains. The $m/\sigma$ ratio of the aggregates is strongly affected by which process is responsible for the aggregation. Conversely, the $m/\sigma$ ratio will determine which

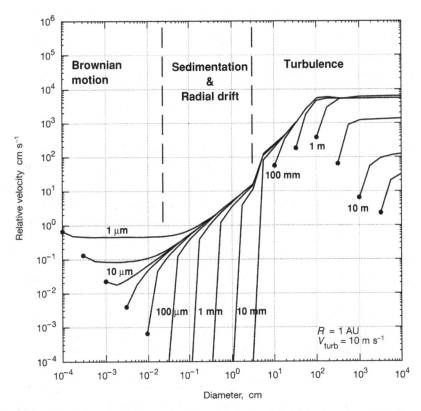

Fig. 7.3. Relative velocities between compact and spherical grains of various sizes. The dominant origins of relative velocities are shown as well (courtesy, S. Weidenschilling).

process dominates the production of relative velocities necessary for the aggregation process. In Fig. 7.3 typical relative velocities between grains are shown. The processes that cause these relative velocities are discussed below.

### Brownian motion growth

For the smallest particles with typical ISM grain sizes between 0.01 and 0.1 µm, the stopping time is shorter than the relevant timescales of the gas motion in a disk, implying almost perfect coupling of the particle to the gas. This means that these particles stay well mixed with the gas except for the very tenuous layers at the disk surface. They also follow any gas motions and therefore are easily mixed throughout the disk by turbulence (see Chapter 4 by Klahr *et al.*). For these small particles, growth is caused by low-velocity collisions ($\sim$cm s$^{-1}$) due to Brownian motion. Because the smallest grains have the largest Brownian motion velocities, they will be the first to start to aggregate. Aggregates formed by this process are of open structure, with fractal dimension below two (Blum *et al.*, 2000; Kempf *et al.*, 1999) and therefore stopping times very similar to those of individual grains. To get into a different growth regime, the particles have to continue to grow until a spread in the surface-to-mass ratio is realized. At that point, a number of different processes start to become important.

## Turbulent growth

Grains with relatively small surface-to-mass ratios (stopping time equal to or longer than the eddy turn-over time) start decoupling from turbulent eddies (Völk *et al.*, 1980). Because different grains couple to different eddies, a velocity dispersion between grains develops which causes collisions preferentially between grains of *different* friction time, thus between large or compact particles and small particles. These processes are likely to produce more compact particles with fractal dimensions of close to three, because small particles penetrate deep into large fluffy aggregates, filling the volume to a constant density (Ball and Witten, 1984).

## Settling and the formation of a dust sublayer

As aggregates grow larger and more compact, their sedimentation velocity toward the midplane increases. For compact grains of 0.1 μm at 1 AU in a disk of surface density 1000 g cm$^{-2}$ the sedimentation down to the midplane takes about $10^7$ yr, while for a compact 1 mm grain, it takes only 1000 yr. Even if we assign the large grains a porosity of 99%, the settling timescale of $10^5$ yr is still much faster than that of a small grain. Therefore, the large grains will settle increasingly faster and sweep up smaller grains. Also this process will lead to non-fractal compact grains if the size difference between sweeping and swept-up grains is large. In a single drop of a particle to the midplane, this process can create grains with masses equivalent to compact millimeter to centimeter-sized grains (Safronov, 1969; Weidenschilling, 1980). Turbulent motions can prolong this phase as particles are mixed up in the disk and are allowed to fall again.

Depending on the strength of turbulent mixing, the equilibrium distribution of dust in the disk produces a thin dust subdisk, the thickness of which is a function of $m/\sigma$. Increasingly large grains settle to an increasingly thin dust disk (see Fig. 7.4). Dubrulle *et al.* (1995) have derived analytical expressions for the thickness of this dust subdisk. If the dust mass density in the subdisk exceeds the gas mass density (i.e. if the scale height of the dust distribution is smaller than the scale height of the gas by a factor in excess of the initial gas-to-dust ratio), the dust will dominate the dynamics of the midplane layer and the gas will be dragged along with the dust. This effect reduces the headwind on individual particles which reduces the radial drift (e.g. Nakagawa *et al.*, 1986) and diminishes the relative velocities between particles of different size.

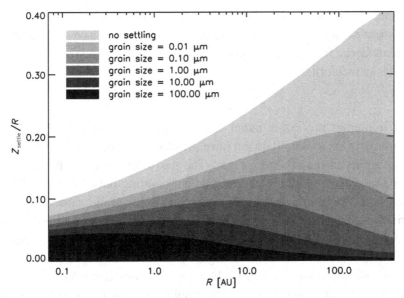

Fig. 7.4. Sedimentation of small grains: the figure shows how deep grains of a certain size sediment into the disk after equilibrium between sedimentation and turbulent mixing has set in for a turbulent $\alpha$-parameter of $\alpha = 10^{-4}$ (Dullemond and Dominik, 2004).

This effect is also directly related to the growth of particles to meter sizes and beyond, which still poses significant problems. In this size range random (non-Keplerian) velocities tend to exceed 1 to 10 m s$^{-1}$. Theoretical and laboratory experiments have shown that collisions between such aggregates lead to fragmentation instead of growth (Blum and Münch, 1993; Dominik and Tielens, 1997; Blum and Wurm, 2000). And if bigger bodies (meter-sized or larger) are hit by smaller ones, the impacts tend to erode the bigger bodies instead of adding matter to them (see review by Wurm and Blum, Chapter 6).

## Trapping

Non-linear effects of turbulence can also affect the distribution of solid particles. The trapping of such particles between eddies and in vortices has been studied in a number of papers (Squires and Eaton, 1991; Cuzzi *et al.*, 1993; Klahr and Henning, 1997; Cuzzi *et al.*, 2001; Tanga *et al.*, 1996; Barge and Sommeria, 1995; Hodgson and Brandenburg, 1998; Johansen *et al.*, 2004). Trapping of meter-sized boulders in anti-cyclonic vortices can have the interesting consequence that at the center a dense midplane concentration of dust can gather that does not experience turbulence due to shear, and may become gravitationally unstable (see Klahr and Bodenheimer, 2005, and Chapter 4 by Klahr *et al.*). Vortices and turbulent eddies alike trap particles of a relatively narrow size range. This means that if a rapid growth mechanism operates in such locations, the objects it creates should consist of constituent particles of nearly the same size. According to Cuzzi *et al.* (2001) there is indeed evidence for such a size-selective concentration in the size distribution of chondrules in meteorites.

## *Radial drift*

Another important process is radial drift of dust grains and larger bodies. Because dust particles as a fluid do not have pressure, such particles would move on Keplerian orbits in the absence of gas. The gas in the disk, on the other hand, has pressure support, with a pressure gradient pointing outward in the disk. The gas itself is therefore orbiting at sub-Keplerian speeds. Solid particles moving at Keplerian speeds therefore experience a head wind slowing the particles down and causing radial drift (Whipple, 1972; Weidenschilling, 1977). This effect becomes strongest for about meter-sized objects, for which the stopping time becomes comparable to the orbital time. Much larger particles ($\sim$km) fully decouple from the gas and are not affected at all. The implications of this process for planet formation are significant. Radial drift can lead to a redistribution of dust in the disk, draining the outer disk of solid material and enhancing the surface density in the inner disk (Youdin and Chiang, 2004). At locations in the disk where certain materials evaporate (ice, silicates, carbonaceous grains), evaporation and outward mixing may cause local enhancements of dust and vapor surface density (Morfill and Völk, 1984; Cuzzi and Zahnle, 2004). Radial migration of particles also sets important constraints on the growth of particles. As meter-sized objects drift so quickly, they would disappear from the orbit of the Earth into the Sun within a mere 100 years. Therefore, either the growth through this phase must be extremely rapid, or the midplane physics must change in order to decrease the effects of radial migration.

## *Radial turbulent mixing and meridional flow*

Throughout the disk, complex mechanisms operate governing the flow of dust grains of various sizes in protoplanetary disks. An example is the meridional flow in accretion disks first discussed by Urpin (1984): inward gas flow in an accretion disk takes place predominantly in the surface layers while the midplane layers tend to move outward. Particles located near the midplane will then be dragged outward even if this midplane dust has not yet concentrated in a massive enough midplane layer to dominate the gas. Turbulent mixing also transports grains, both in the vertical and in radial directions (e.g. Morfill and Völk, 1984). Johansen and Klahr (2005) showed recently that in magneto-turbulent disks the strength of this mixing is roughly equal to the strength of angular momentum transport (50% larger in the radial direction and 30% lower in the vertical direction). A comprehensive study of this and many other processes that move dust particles in various directions was published by Takeuchi and Lin (2002, 2003). Paardekooper and Mellema (2004) found that a growing planet may already open a gap in the dust disk before being massive enough to create a gap in the gas disk.

The radial transport of grains (and gas) can have important consequences for the chemical/mineralogical composition of dust particles. The primordial silicate dust grains from the interstellar medium are amorphous. The temperature in all but the very inner regions of the disk is too low to anneal these grains to a crystalline form (e.g. Fabian *et al.*, 2000). But dust grains that initially reside in the hot inner regions of the protoplanetary disk will become crystalline due to annealing, and later get diffusively mixed outward (Morfill and Völk, 1984; Gail, 2001; Wehrstedt and Gail, 2002; Gail, 2002; Bockelée-Morvan *et al.*, 2002; Cuzzi *et al.*, 2003). Also, high temperature condensates like calcium-aluminium-rich inclusions (CAIs) are more likely to form in the inner nebula and will then be mixed out (Cuzzi *et al.*, 2003). The dust grains from which the planetesimals and comets are formed will then contain crystalline silicates that were not formed *in situ*. This is comparable to the presence of CAIs in meteorites (Cuzzi *et al.*, 2003). Infrared observations of comets show that a significant fraction of their dust is of crystalline form (e.g. Hanner *et al.*, 1999; Wooden *et al.*, 1999). Moreover, the crystallinity of dust in protoplanetary disks appears to be a clearly decreasing function of radius, at least for disks around Herbig Ae/Be stars (van Boekel *et al.*, 2004). These observations lend support to the radial transport scenario. Whether this transport is due to turbulent mixing, meridional flow or surface outflow remains to be answered.

### Gravitational instability of the dust layer

The process of dust settling and radial transport has significance in the discussion whether the dust subdisk can become gravitationally unstable (Goldreich and Ward, 1973). Weidenschilling (1977) was the first to note that the minute amounts of turbulence caused by the shear between the dust layer and the gas is enough to puff up the dust layer sufficiently to inhibit this gravitational instability (see also Cuzzi *et al.*, 1993). But Youdin and Shu (2002) have recently claimed that if the dust-to-gas mass ratio is higher than the 1/100 usually assumed, then the instability may still be operating. Such an enhanced dust-to-gas ratio could be the result of dust–gas separation (e.g. gas evaporation from the disk while leaving the dust behind). It may also be caused by radial drift depleting the outer disk and enhancing the dust surface densities in the inner disk (Youdin and Chiang, 2004), but it requires relatively small grains to drift over large distances without growing. The thickness of the dust layer (which is essential for its gravitational stability) depends, however, critically on the mass of the grains and coupling strength of the dust particles to the turbulence (Dubrulle *et al.*, 1995). A detailed recomputation of this turbulence coupling strength was recently published by Schräpler and Henning (2004).

It is still a matter of debate whether a gravitationally unstable midplane layer can be formed or not. The different models all face different problems which we do not want to cover here in detail. In the following we will focus on models with a stable midplane.

### 7.4.2 Global models of grain sedimentation and aggregation in protoplanetary disks

To model the process of grain growth in a comprehensive way, all the processes mentioned above have to be combined into a single model which calculates the motion of dust grains and the coagulation process at all locations in the protoplanetary disk. Various models have been made over the years.

Some models have focused on the sedimentation process only, but computed that for the whole disk in a time-dependent way and emphasizing observational features. Miyake and Nakagawa (1995) showed that if the dust sediments in a protoplanetary disk, the far-infrared part of the spectral energy distribution (SED) is strongly affected. The SED becomes steeper toward long wavelengths. Dullemond and Dominik (2004) computed the dust sedimentation process coupled to vertical turbulence, and applied a multi-dimensional radiative transfer code to compute the SED and images at different times. They found that, for weak enough turbulence, the disk can become self-shadowed in the outer regions. This is because the dust in the outer regions can sediment closer to the midplane than in the inner regions (see Fig. 7.4). Similar to Miyake and Nakagawa (1995), this has the effect of steepening the SED (see Fig. 7.5), and it makes the outer disk regions virtually invisible in scattered light, an effect indeed observed in some objects (Grady *et al.*, 2004).

The modeling of the dust growth on a global scale is quite challenging. The aggregation (coagulation) of dust grains has presumably already started before or during disk formation, when matter is still located in the (collapsing) protostellar core (Ossenkopf, 1993; Weidenschilling and Ruzmaikina, 1994; Suttner and Yorke, 1999). However, it is likely that such loosely bound aggregates, presumably held together in part by the icy mantels around the elementary grains, do not survive the passage through the stand-off shock near the disk surface as the envelope matter accretes onto the disk (Schmitt *et al.*, 1997). Most models of grain coagulation in protostellar/protoplanetary disks in fact assume a pristine grain size distribution to start with. The global modeling of dust growth in these disks was pioneered by Weidenschilling (1980) and by Nakagawa *et al.* (1981). These models include dust

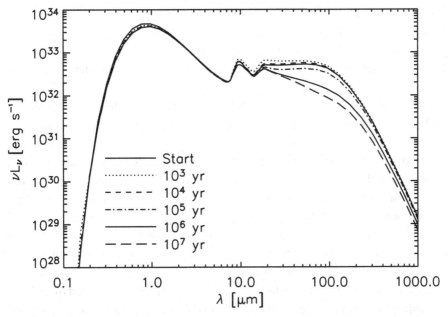

Fig. 7.5. Effect of settling on the SED (from Dullemond and Dominik, 2004) for a low value of the turbulent $\alpha$ parameter ($\alpha = 10^{-5}$) and for a Mathis, Rumpl, Nordsiek (MRN) dust grain size distribution. After about $10^6$ yr the equilibrium state is reached in which the far-infrared emission is clearly suppressed.

sedimentation both for the vertical drift as well as a source of differential velocities for the aggregation. The coagulation equation (Smoluchowski equation) for the grain mass distribution function $f(m)$ is

$$\left. \frac{\partial f(m)}{\partial t} \right|_{coag} = \int_0^{m/2} f(m')f(m - m')\sigma(m', m - m')$$

$$\times \Delta v(m', m - m')dm'$$

$$- \int_0^{\infty} f(m')f(m)\sigma(m', m)\Delta v(m', m)dm', \qquad (7.2)$$

where $m$ is the mass of the grain, $\sigma(m_1, m_2)$ is the coagulation kernel and $\Delta v(m_1, m_2)$ is the average relative velocity between grains of mass $m_1$ and $m_2$. The first term in the equation is the gain term when two grains stick and form an aggregate of mass $m$. The second term is the loss term when a grain of mass $m$ sticks to another grain to form an aggregate with larger mass. It should be noted that the distribution function $f(m)$ is also a function of $z$ (height above the midplane) and $r$ (radial coordinate of the disk), even though this is not explicitly stated in the equation. The Smoluchowski equation is only valid when the number of particles is so large that one can treat them in a statistical manner. In some circumstances a run-away growth might favor the growth of a few (or even a single) bigger bodies over the growth of all the other bodies. The Smoluchowski equation then breaks down, making an explicit multi-particle approach necessary. Typically this may happen when the particles reach planetesimal size, and is not likely for smaller particles.

In principle, aside from the mass and the coordinates $r$ and $z$, one should also include a coordinate $p$ in the model: the "porosity" of the aggregates (the factor by which $\sigma/m$ exceeds that of a compact grain). Another way of formulating this is to introduce two "fractal dimensions:" $D_m$ and $D_s$ (mass and surface fractal dimensions of the aggregate with $m \propto r^{D_m}$ and $\sigma \propto r^{D_s}$). The reason for these additional dimensions is that aggregates of the same mass can still have a different shape and internal structure, depending on where and how they were formed. The "porosity" (or better: $\sigma/m$ ratio) of such an aggregate does not only affect the drift velocity, but also the binding strength: compact aggregates are more difficult to destroy. However, the computational demand for a model that also includes this additional coordinate is quite high (Kempf *et al.*, 1999), and so far most models used in the astrophysical literature *assume* a porosity as a function of aggregate mass: cluster–cluster aggregates (CCAs) have a fractal structure while particle–cluster aggregates (PCAs) are somewhat more compact.

Grain coagulation in protoplanetary disks turns out to be very rapid. Estimates of Safronov (1969) show that growth up to about a millimeter happens on timescales of a few thousand years at 1 AU (see also Weidenschilling, 1980). If grains would grow without fragmentation taking place, all small opacity-carrying grains would be locked up in bigger aggregates within a time much smaller than the disk lifetime (e.g. Weidenschilling, 1984; Schmitt *et al.*, 1997; Dullemond and Dominik, 2005). The rapid coagulation can strongly reduce the optical depth of the disk (Weiden-schilling, 1984). This may quench any possibly existing convection in these disks, since energy can then escape more easily from the disk. Until the discovery of the magneto-rotational instability (MRI, Balbus and Hawley, 1991), convection-driven turbulence was believed to be the main origin of "alpha-viscosity" in protoplanetary accretion disks, and therefore the coagulation process could stop accretion (Mizuno *et al.*, 1988). In fact Klahr *et al.* (1999) demonstrated that convection cannot drive turbulence in disks. Nowadays it seems ever more convincing that MRI is the main cause of turbulence in gravitationally stable disks. If true, the effect of coagulation stopping the accretion no longer plays a significant role. But another (opposite) effect may come to replace it: the coagulation of grains can reduce the total effec-tive grain surface that can remove free electrons from the gas (Sano *et al.*, 2000; Semenov *et al.*, 2004). The gas will therefore become more conducting and will couple more strongly to the magnetic fields, making the MRI more effective. On the other hand, the reduced optical depth may cause the disk to be cooler (at least in the early phases, where viscous dissipation of gravitational energy dominates the energy balance of the disk), which could make parts of the disk too cool to have a sufficient amount of free electrons for MRI to operate. Even if the MRI (or another cause of "viscosity") is active nonetheless, by the nature of "$\alpha$-disks" a cooler disk accretes more slowly than a hot one. In these various ways, dust coagulation strongly affects the structure *and* evolution of the disk (e.g. Schmitt *et al.*, 1997).

The rapid depletion of small grains can be counteracted by aggregate fragmentation (Weidenschilling, 1980). Small grains can also be replenished by the addition of new envelope material onto the disk in the early phases of disk evolution (Mizuno *et al.*, 1988; Mizuno, 1989).

The models described above solve the coagulation–sedimentation equation in the vertical direction. If one solves a set of these one-dimensional vertical coagulation problems at different radii independently, one obtains the dust evolution of the entire disk. However, in reality these vertical slices are not independent: radial drift and turbulent radial mixing can strongly affect the grain distribution, and transport grains from one radius to another. A simple way to model this is to ignore for the moment the vertical coagulation and concentrate fully on the radial motions. Models of this kind were made, for example, by Mizuno (1989), Sterzik and Morfill (1994), and Schmitt *et al.* (1997). Recently a model appeared by Kornet *et al.* (2005) in which the entire growth from dust to planetesimals was modeled in this way, and conclusions were drawn on the relation between the presence of planets and the metallicity of the system. But clearly, ultimately a two-dimensional axisymmetric (so to say 2.5-D) or a fully three-dimensional model is required to include all effects of radial and vertical drift. Suttner and Yorke (2001) have presented such a 2.5-D model, based on radiation-hydrodynamical simulations of a forming protoplanetary disk, though with a still somewhat simplified coagulation kernel and only for the earliest stages of disk evolution.

In the light of the enormous increase in observational data on grain growth, it is important that detailed models of coagulation are coupled to radiative transfer codes to obtain model predictions for observations. Two recent papers address this issue: Tanaka *et al.* (2005), and Dullemond and Dominik (2005). They calculate the time-dependent coagulation in a 1+1-D fashion (a series of independent one-dimensional slices at different radii), and for different time intervals they compute the spectral energy distributions. These papers predict that an inner opacity hole should form in these disks (see also Schmitt *et al.*, 1997), which may well be similar to the holes found in T Tauri disks (Calvet *et al.*, 2002; Forrest *et al.*, 2004) and Herbig Ae/Be stars (Bouwman *et al.*, 2003). Dullemond and Dominik (2005) show that, in particular when turbulence is included, the timescale for the removal of small grains is so short that statistically speaking almost all T Tauri-star disks should have become optically thin. Since T Tauri-star disks are typically optically thick, it is inferred that aggregate destruction should play a role in these disks to replenish the small grain population.

## 7.5 Summary

Grain evolution in protoplanetary disks is a complex process, including gas dynamics, coupling between gas and grains, and sticking properties of solid particles at the relevant collisional velocities. Disk models have just started to integrate the processes of grain diffusion, sedimentation, and coagulation and now make predictions for the time-dependent spectral energy distributions. The coagulation of particles will strongly influence the thermal and dynamical structure of disks, closing the cycle between gas dynamics and grain evolution. The complete modeling of this cycle remains a challenging numerical problem in star and planet formation.

Observational evidence supports the idea of grain growth, covering the size range from a few microns to centimeters. However, grain growth seems to happen on timescales longer than usually predicted by models. Observations at centimeter wavelengths with the VLA and millimeter interferometric studies with higher sensitivity with the Atacama Large Millimeter Array (ALMA) will add to our empirical knowledge of grain growth.

# 8

# The accretion of giant-planet cores

Edward W. Thommes

*Canadian Institute for Theoretical Astrophysics, Toronto, Canada*

Martin J. Duncan

*Queen's University, Kingston, Canada*

## 8.1 Introduction

The count of extrasolar giant planets detected by radial velocity measurements is now well over a hundred, accounting for about 5% of F, G and K main-sequence stars in the Solar neighborhood; about 10% of the planets are in multiple systems[1]. It thus seems an inescapable conclusion that giant planet formation is a ubiquitous and robust process. There is also strong observational evidence for a correlation between the occurrence rate of (detectable) planets and the metallicity of the parent star (Gonzalez, 1997; Fischer and Valenti, 2003). There are two possible explanations for this phenomenon: first, the planet formation process may tend to "pollute" the parent star with higher-metallicity material, as giant planets (Laughlin and Adams, 1997) or planetesimals (Murray *et al.*, 2001) migrate in and are engulfed. If this is the case, higher-mass stars, which have thinner convective envelopes in which to preserve the pollution, ought to display a systematically higher metallicity. However, no such trend has been observed so far (Wilden *et al.*, 2002; Dotter and Chaboyer, 2002; Quillen, 2002; Fischer and Valenti, 2003). Furthermore, Fischer and Valenti (2005) found no sign of various other potential accretion signatures, such as dilution of metallicity in subgiants with planets. The other explanation is that higher metallicity – and thus a higher fraction of solids in the protoplanetary disk – increases the chances of forming a giant planet. Such a direct reliance on solids would offer strong support

---

[1] California and Carnegie Planet Search: www.exoplanets.org; IAU Working Group on Extrasolar Planets: http://www.ciw.edu/boss/IAU/div3/wgesp/planets.shtml

*Planet Formation: Theory, Observation, and Experiments*, ed. Hubert Klahr and Wolfgang Brandner.
Published by Cambridge University Press. © Cambridge University Press 2006.

for the nucleated instability model of giant-planet formation, wherein a large protoplanet forms first, then accretes a massive gas envelope. The alternative model of giant-planet formation, in which parts of the gas disk itself become gravitationally unstable and collapse to form planets (Boss, 1997; Mayer *et al.*, 2002; see also the review by Boss, Chapter 12), does not benefit in any obvious way from the presence of extra solids. More recently a third model of giant-planet formation has been proposed, which may be viewed as intermediate between the two traditional rivals of core accretion and disk instability. In this picture, a giant vortex forms in the gas disk and concentrates solids at its center (Klahr and Bodenheimer, 2003, 2006). This model has a number of attractive features, among them a short accretion time for gas giants ($\lesssim 10^6$ yr) and the ability to function even in low-mass disks. It is not yet clear whether the required vortices can actually form in a protoplanetary disk, but future observing programs are planned to search for them around young stars. A detailed review is given by Klahr *et al.* in Chapter 4; in the rest of this article, we will focus on the "pure" core accretion model, in a simple non-turbulent Keplerian gas disk.

Observational evidence notwithstanding, the nucleated instability model has an Achilles' heel, namely the very first step. The accretion of a massive atmosphere requires a solid core $\sim 10 \, M_\oplus$ in mass (Mizuno, 1980; Pollack *et al.*, 1996; see also the review by Hubickyj in Chapter 10). Assembling such a large body, it turns out, offers some serious challenges to the theory of planet formation as it currently stands. The difficulties are threefold: first, the accretion process has to be efficient enough to concentrate such a large mass in (at least) one single body; second, everything has to happen fast enough that when the putative core is ready, there is still enough gas – of order $10^2 \, M_\oplus$ – left in the nearby part of the disk to furnish its envelope. Observations of T Tauri stars (Haisch *et al.*, 2001; see also the reviews by Richling *et al.* in Chapter 3, and Bouwman *et al.* in Chapter 2) show that gas disks, at least insofar as they are traced by the presence of dust in the inner AU as well as accretion onto the star, have lifetimes of $\sim$1–10 Myr, with even a single Jupiter mass of gas unlikely to be found in the disk much later than that. The final problem concerns migration due to planet–disk tidal interactions, which threatens to drop core-sized bodies into the central star faster than they can accrete (e.g. Ward, 1997; see also the review by Masset & Kley in Chapter 14).

In this review, we begin with an overview of the post-runaway accretion regime (called "oligarchic" growth) in which the vast majority of a giant-planet core's growth very likely occurs (Section 8.2.1). We then consider various ways in which accretion rate and efficiency might be elevated above that given by this standard model. We examine the particularly problematic issue of Uranus and Neptune's formation in Section 8.4. In Section 8.5, we comment briefly on the problems posed to core formation by the migration resulting from planet–disk tidal interactions. We summarize and conclude the chapter in Section 8.6.

## 8.2 Estimating the growth rate

The issue of formation time is of central importance in the study of planet accretion, but it is in a sense most critical in the formation of gas-giant cores. In the simplest view, the time to form the terrestrial planets of our Solar System has just one truly hard upper limit, namely the 4.5 Gyr age of the system itself (in reality more stringent limits exist, in particular cosmochemical evidence seems to restrict the Earth's formation to the first $\sim 10-30$ Myr of the Solar System's existence; Yin *et al.*, 2002). In contrast, gas-giant planets have no choice but to form while their parent protoplanetary disk is still young enough to contain a significant amount of gas, i.e. in $\lesssim 10^7$ yr (Section 8.1).

Terrestrial planet formation can be divided into three qualitatively different phases; however, in the formation of giant-planet cores, the third of these is very likely absent. In the first stage, planetesimals grow by runaway accretion (wherein the largest bodies grow the fastest) to produce bodies of order 100 km in size (Wetherill and Stewart, 1989). In the middle stage, accretion changes from runaway to self-regulating, as the largest bodies become big enough to gravitationally "stir their own soup" of planetesimals (Ida and Makino, 1993; Kokubo and Ida, 1998, 2000; Thommes *et al.*, 2003, hereafter TDL03). This stage ends at a given location in the disk when the local "oligarchy" of largest bodies reach their isolation mass (Eq. 8.16 below), meaning that they have consumed all planetesimals within their gravitational reach. In the terrestrial-planet region, typical disk models produce isolation masses of only about Mars mass, thus a third phase must take place in which these bodies' orbits cross and they collide to form Earth- and Venus-mass bodies. This takes $\sim 10^8$ yr in the standard gas-free model (Chambers and Wetherill, 1998; Chambers, 2001), or only a few tens of Myr if one takes into account the dynamical shakeup caused by sweeping secular resonances of the giant planets as the gas disk dissipates (Lin *et al.*, in preparation). The former timescale is much too long for gas giants, and the latter mechanism requires pre-existing giant planets. However, as we will see below (Fig. 8.1), assuming a somewhat enhanced disk mass, an isolation mass of $\sim 10^1$ $M_\oplus$ is produced in what corresponds to the Jupiter–Saturn region. Thus, oligarchic growth should in principle suffice to produce giant-planet cores.

### 8.2.1 Oligarchic growth

The first step in calculating accretion rates is to adopt a model for both the planetesimal and the gas distribution of a protoplanetary disk. It is convenient to scale

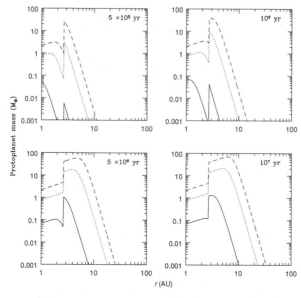

Fig. 8.1. The mass of the locally largest protoplanet as a function of stellocentric radius and time. The simple version of the oligarchic growth model (Eq. 8.15), without the effect of planetesimal orbital decay by gas drag, is used to compute the curves; it gives a qualitatively correct picture of oligarchic growth but overestimates the final mass. The central star has a mass of 1 $M_\odot$, protoplanets and planetesimals have densities of 1.5 g cm$^{-2}$, planetesimals have a radius of 10 km, and an orbital spacing of $b = 10r_H$ between adjacent protoplanets is assumed. Gas and solids surface densities are scaled relative to the minimum-mass Solar Nebula model (Eqs. 8.1 and 8.3). The calculation is performed for: i) $1 \times \Sigma_{gas}^{min}$ and $1 \times \Sigma_{solid}^{min}$ (solid curve); ii) $5 \times \Sigma_{gas}^{min}$ and $5 \times \Sigma_{solid}^{min}$ (dotted curve); and iii) $10 \times \Sigma_{gas}^{min}$ and $10 \times \Sigma_{solid}^{min}$ (dashed curve).

these relative to the "minimum-mass" model of Hayashi (1981):

$$\Sigma_{gas}^{min} = 1700 \left(\frac{r}{1\,\text{AU}}\right)^{-3/2} \text{g cm}^{-2} \qquad (8.1)$$

$$\rho_{gas}^{min} \approx \Sigma_{gas}/2H, \quad H = 0.0472 \left(\frac{r}{1\,\text{AU}}\right)^{5/4} \text{AU} \qquad (8.2)$$

and

$$\Sigma_{solid}^{min} = 7.1 F_{SN} \left(\frac{r}{1\,\text{AU}}\right)^{-3/2} \text{g cm}^{-2}, \qquad (8.3)$$

where

$$F_{SN} = \begin{cases} 1, & r < r_{SN} \\ 4.2, & r > r_{SN} \end{cases} \qquad (8.4)$$

is the "snow line" solids enhancement factor: beyond $r_{SN}$ (= 2.7 AU in the Hayashi model), water freezes out, thus adding to the surface density of solids.

Ida and Makino (1993) derived the condition for the end of runaway growth as

$$2\Sigma_M M > \Sigma_m m, \tag{8.5}$$

where $m$ is the characteristic planetesimal mass, $\Sigma_m$ the surface density of planetesimals, $M$ the characteristic mass of the local population of largest protoplanets, $\Sigma_M$ the (effective) surface density of these protoplanets, and $\Sigma_M + \Sigma_m = \Sigma_{solid}$.

This condition yields an end to runaway at a protoplanet mass of only $\sim 10^{-6} - 10^{-5} M_\oplus$ for $\Sigma_{solid} = 1 - 10 \Sigma_{solid}^{min}$ (TDL03). The rest of the accretion takes place in the "oligarchic" regime (Kokubo and Ida, 1998). Herein, the mode of growth is still the sweep-up of planetesimals by much larger protoplanets, but it is now the gravitational influence of the protoplanets which determines the random velocities of the planetesimals, rather than self-stirring of the planetesimals.

We will assume that the RMS planetesimal random velocity $v_m$ remains in the dispersion, rather than shear-dominated, regime (however, see Section 8.3.2 below); that is,

$$v_m \gtrsim \Omega r_H, \tag{8.6}$$

where

$$r_H = \left(\frac{M}{3M_*}\right)^{1/3} r \tag{8.7}$$

is the Hill (or Roche) radius of a protoplanet of mass $M$. This amounts to saying that planetesimal random velocities are larger than the Keplerian shear across a protoplanet's Hill radius. Furthermore, as long as $\Sigma_M << \Sigma_m$, dynamical friction from the planetesimals will keep the protoplanets on very nearly circular orbits, thus $v_m$ is also the RMS relative velocity between protoplanets and planetesimals. The mass accretion rate of a protoplanet is then well described by the "particle-in-a-box" approximation (Safronov, 1969; Wetherill, 1980; Ida and Nakazawa, 1989)[2]:

$$\frac{dM}{dt} \approx F \frac{\Sigma_m}{2H} \pi R_M^2 \left(1 + \frac{v_{esc}^2}{v_m^2}\right) v_m, \tag{8.8}$$

where $H$ is the disk scale height, $R_M$ the protoplanet radius, $v_{esc}$ the escape velocity from the protoplanet's surface, and $F \approx 3$ is a correction factor (Greenzweig and Lissauer, 1992). We assume that RMS planetesimal eccentricities $e_m$ and inclinations $i_m$ are related to each other and $v_m$ by $v_m \approx e_m r \Omega$, $i_m \approx e_m/2$, and approximate the scale height of the planetesimal disk as $H \approx i_m r \approx e_m r/2$ (details

---

[2] This equation in TDL03 (their Eq. 6) has a typo; it is missing a factor of 1/2.

are given in Kokubo and Ida, 1996). With these approximations, the protoplanet accretion rate becomes

$$\frac{dM}{dt} \propto \frac{\Sigma_m M^{4/3}}{e_m^2 r^{1/2}}. \tag{8.9}$$

By definition, the planetesimals' random velocities in the oligarchic regime are dominated by stirring due to the protoplanets. One can estimate the resulting equilibrium RMS eccentricity by equating the gravitational viscous stirring timescale due to the protoplanet, with the random-velocity damping timescale due to aerodynamic drag by the nebular gas (Ida and Makino, 1993; Kokubo and Ida, 2000); $e_m^{eq} \propto M^{1/3} \rho_{gas}^{-1/5}$. One obtains (for details see TDL03)

$$\frac{dM}{dt} \approx A \Sigma_m M^{2/3}, \tag{8.10}$$

where

$$A = 3.9 \frac{b^{2/5} C_D^{2/5} G^{1/2} M_*^{1/6} \rho_{gas}^{2/5}}{\rho_m^{4/15} \rho_M^{1/3} r^{1/10} m^{2/15}}. \tag{8.11}$$

$C_D$ is a dimensionless drag coefficient $\sim 1$ for km-sized or larger planetesimals, and $b$ is the spacing between adjacent protoplanets in units of their Hill radii. An equilibrium between mutual gravitational scattering on the one hand and recircularization by dynamical friction on the other keeps $b \sim 5-10$ (Kokubo and Ida, 1998).

The qualitative difference between runaway and oligarchic growth stems from the fact that the former has $\dot{M} \propto M^{4/3}$, while for the latter, because $e_m$ depends on $M$, $\dot{M} \propto M^{2/3}$ (Ida and Makino, 1993). The mass ratio of two (nearby) protoplanets changes as

$$\frac{d}{dt}\left(\frac{M_1}{M_2}\right) = \frac{M_1}{M_2}\left(\frac{\dot{M}_1}{M_1} - \frac{\dot{M}_2}{M_2}\right), \tag{8.12}$$

thus, supposing $M_1 > M_2$, in the runaway regime we have $d/dt(M_1/M_2) = (M_1/M_2)(M_1^{1/3} - M_2^{1/3}) > 1$, while in the oligarchic regime we have $d/dt(M_1/M_2) = (M_1/M_2)(M_1^{-1/3} - M_2^{-1/3}) < 1$. In other words, in runaway growth the mass ratio of two nearby protoplanets diverges from unity, while in oligarchic growth it approaches unity.

Equation (8.10) was used by TDL03 as a starting point to get a global picture of how oligarchic growth converts a planetesimal disk into protoplanets. In the simpler of two approaches, they assumed that planetesimals undergo no net radial migration. The relationship between protoplanet mass and planetesimal surface density is then

$$\Sigma_m(M) = \Sigma_m(0) - \frac{M}{2\pi r \Delta r} = \Sigma_m(0) - \frac{3^{1/3} M_*^{1/3} M^{2/3}}{2\pi b r^2}, \qquad (8.13)$$

where we have used $\Delta r = b r_H$. Using Eqs. (8.10) and (8.13),

$$\frac{dM}{dt} \approx A M^{2/3} (\Sigma_m(0) - B M^{2/3}), \qquad (8.14)$$

where $A$ is given by Eq. (8.11) and

$$B = \frac{3^{1/3} M_*^{1/3}}{2\pi b r^2}.$$

Solving this differential equation, one obtains

$$M \approx \left(\frac{\Sigma_m(0)}{B}\right)^{3/2} \tanh^3 \left[\frac{AB^{1/2}\Sigma_m(0)^{1/2}}{3} t + \tanh^{-1}\left(\frac{B^{1/2}M(0)^{1/3}}{\Sigma_m(0)^{1/2}}\right)\right]. \qquad (8.15)$$

This describes an outward-expanding front of growth, which reaches a given radius of the disk on a timescale $3A^{-1}B^{-1/2}\Sigma_m(0)^{-1/2}$; this is the time to reach the isolation mass

$$M_{\text{iso}} = \left(\frac{\Sigma_m(0)}{B}\right)^{3/2}. \qquad (8.16)$$

Examples are shown in Fig. 8.1. After about a million years (in the cases with disks enhanced above minimum mass), there exists an annulus within the disk where protoplanets have grown to a mass $\gtrsim 10\ M_\oplus$. On the inside, where growth has finished, this annulus is bounded by the final (i.e. isolation) mass. On the outside, it is bounded by how far out the accretion front has progressed. The effect of the snow line at 2.7 AU is clearly visible (though in a real disk it will inevitably involve a smoother transition in surface density than the step function we use here).

Although Eq. (8.15) gives the right qualitative picture of oligarchic growth, it constitutes a significant simplification because it neglects the role of planetesimal radial migration. This is an important effect because aerodynamic gas drag extracts energy from planetesimal orbits as it damps their random velocities (Adachi *et al.*, 1976). With the balance between damping and gravitational stirring by the proto-planets maintaining a non-zero equilibrium planetesimal random velocity, there is a continuous net orbital decay of planetesimals. The surface density of planetesimals thus changes at a given radius not just because planetesimals are swept up by protoplanets, but also because of this migration:

$$\frac{d\Sigma_m}{dt} = \left.\frac{d\Sigma_m}{dt}\right|_{\text{accr}} + \left.\frac{d\Sigma_m}{dt}\right|_{\text{migr}}, \qquad (8.17)$$

where $\mathrm{d}\Sigma_m/\mathrm{d}t|_{\mathrm{accr}}$ is the time derivative of Eq. (8.13), and, taking $\dot{r}_m$ as the radial migration speed of planetesimals, continuity gives us

$$\frac{\mathrm{d}\Sigma_m}{\mathrm{d}t}\bigg|_{\mathrm{migr}} = -\frac{1}{r}\frac{\partial}{\partial r}(r\Sigma_m \dot{r}_m). \qquad (8.18)$$

$M = M(r, t)$ was taken by TDL03, turning Eqs. (8.10) and (8.17) into a pair of coupled partial differential equations, which they solve numerically. They find that gas drag acts as a two-edged sword in the accretion of massive bodies: on the one hand, increasing the strength of gas drag (by increasing the gas density, decreasing the planetesimal size, or a combination thereof) damps random velocities more strongly and speeds the accretion rate (see Eq. 8.9). On the other hand, as also found by Inaba and Wetherill (2001), stronger gas drag also increases the rate at which $\Sigma_m$ is depleted by planetesimal orbital decay; this causes growth to stall earlier. In other words, stronger gas drag increases the speed of protoplanet growth at the cost of decreasing its efficiency.

In the growth rate estimate of TDL03, assuming a characteristic planetesimal size of 10 km, the production of $\sim 10\,M_\oplus$ protoplanets in what corresponds to the Jupiter–Saturn region requires a disk with, at least in that region, $\sim 10\Sigma_{\mathrm{solid}}^{\mathrm{min}}$ (assuming that $\Sigma_{\mathrm{gas}}^{\mathrm{min}}$ is multiplied by the same factor). The time to reach $\sim 10\,M_\oplus$ at $\sim 5$ AU is then $\sim 1$ Myr. Decreasing the characteristic planetesimal size decreases the mass at which growth stalls. N-body simulations were also performed by TDL03 which confirmed this effect.

The calculations of Inaba and Wetherill (2001) were performed using a statistical rather than an N-body code, however they added a fragmentation model, which led to considerably less optimistic results than those of TDL03. Fragmentation reprocesses a large fraction of the planetesimals to much smaller size, decreasing the efficiency of accretion so that the largest protoplanet produced has a mass of less than $2\,M_\oplus$. Fragmentation models are of course fraught with uncertainty, but the dynamical regime we are considering – wherein the random velocities of the planetesimals are determined by stirring from much larger bodies – makes it likely that fragmentation will play a role, since planetesimals will collide with relative velocities much larger than their surface escape velocities. The above result is thus worrisome for the core accretion model.

## 8.3 Possibilities for boosting accretion speed and efficiency

The relatively simple model laid out above is capable of producing giant-planet core-sized bodies quickly in the Jupiter–Saturn region. However, the (at least locally) rather massive disk required, as well as the danger posed by fragmentation, suggest that we may not yet have the full picture. Below, we look at recent work which suggests several promising avenues for "plugging the holes" in the oligarchic growth model for giant-planet core accretion.

### 8.3.1 The role of protoplanet atmospheres

The model of Pollack *et al.* (1996) for the accretion of giant-planet cores also models the slow accretion of a gas atmosphere onto a growing giant protoplanet, prior to the final runaway gas accretion phase. This atmosphere acts to enhance the capture cross-section for additional planetesimals, but because they assume planetesimal sizes of 1–100 km, the tenuous atmosphere present during the initial solids-dominated accretion phase does not play a very large role in raising the accretion rate (they use an accretion model which underestimates stirring of planetesimals, thus core growth still only takes $< 10^6$ yr). However, as a planetesimal's size is decreased, the strength of gas drag it feels is increased. So, too, is its capture radius with respect to a protoplanet possessing a gas atmosphere. Inaba and Ikoma (2003), hereafter II03, studied the capture of planetesimals in the atmosphere of a growing core. They performed numerical simulations, then constructed analytic approximations which are in good agreement. They showed that, as long as random velocities of the planetesimals are small compared to the escape velocity from the core's surface, the largest radius of planetesimal, $r_m$, which can be captured at a radius $R_c$ in the core atmosphere, is given by

$$r_m = \frac{3}{2} \frac{\rho(R_c)}{\rho_m} r_H, \tag{8.19}$$

where $\rho(R_c)$ is the gas density at $R_c$. For a given planetesimal size and envelope model (which determines $\rho(r)$ and thus $R_c$), the accretion rate onto the core is then calculated using an envelope-enhanced core radius of $R_c$, rather than the core's physical radius $R_M$. This can lead to a substantial increase in accretion rate; using a simple estimate, II03 obtained a value of less than $10^5$ yr for the growth of a $10 \, M_\oplus$ body by accretion of 100 m planetesimals distributed with a surface density of $\Sigma_{solid} = 10 \, \text{g cm}^{-2}$. However, this estimate does not take into account radial migration of planetesimals. Inaba *et al.* (2003), hereafter IWI03, put all the pieces together and found that the gas-enhanced capture cross-section of growing cores in fact *rescues* the core-accretion model from the perils of fragmentation. The important point is that fragmentation now has a positive as well as a negative consequence: smaller planetesimals are lost more rapidly by migration, but are also accreted more readily by cores with atmospheres. Simply put, the two effects largely cancel each other out, and the results of IWI03 are not dissimilar from those of TDL03: for a disk of about ten times the solids and gas density of the minimum-mass model, a $\sim 20 \, M_\oplus$ body forms at 5 AU in $\sim 3 \times 10^6$ yr (with a reduced grain opacity and hence a lower critical core mass, IWI03 found that a disk of about five times minimum mass suffices). It should be pointed out that this model, and to a lesser extent that of TDL03, fall somewhat short of producing a 10 $M_\oplus$ core at Saturn's orbital radius.

### 8.3.2 Accretion in the shear-dominated regime

We stated previously the assumption of dispersion- rather than shear-dominated planetesimal random velocities (Eq. 8.6). However, if gas drag damps planetesimal random velocities strongly enough, things can get pushed back into the shear-dominated regime. Rafikov (2004), hereafter R04, studied this situation, finding that shear-dominated oligarchic growth proceeds in a qualitatively different way, and can be much more rapid, than dispersion-dominated growth. The important distinction is that damping of random velocities is so rapid that between consecutive close encounters with a growing protoplanet, a planetesimal loses almost all of its eccentricity and inclination. Thus at each close encounter, the relative velocity between planetesimal and protoplanet is approximately just that due to the Keplerian shear in the disk, and significant eccentricities and inclinations are possessed by the planetesimal only immediately after a scattering event. This is the opposite of the dispersion-dominated case, wherein the non-circular (or random) velocity $v$ of a planetesimal at any time is in general large compared to the kick $\Delta v$ it picks up in a single close encounter. Furthermore, individual scatterings are more efficient at increasing eccentricity than inclination:

$$\Delta v_r \sim \Omega r_H, \ \Delta v_z \sim v_z \text{ if } v << \Omega r_H,$$

thus $i << e$, unlike the dispersion-dominated regime in which $i \sim e$. This makes the planetesimal disk very thin, which increases the accretion rate onto a protoplanet (Eq. 8.8). In the most extreme case, if

$$v_z \lesssim 0.07 \Omega r_H \left( \frac{r}{1 \text{ AU}} \right), \tag{8.20}$$

the entire vertical column of the planetesimal disk is within the protoplanet's capture radius, thus making accretion a two-dimensional process. The protoplanet accretion rate is then

$$\frac{dM}{dt}_{2d} \approx \Omega p^{1/2} \Sigma_{\text{solid}} r_H^2 = 0.45 G^{1/2} \frac{\Sigma_{\text{solid}} M^{2/3}}{\rho_M^{1/6}}, \tag{8.21}$$

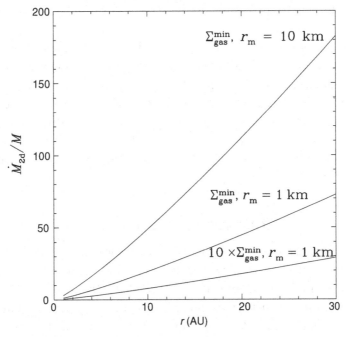

Fig. 8.2. The ratio (Eq. 8.22) of the two-dimensional accretion rate, applicable in the extreme case of shear-dominated accretion wherein the entire vertical column of planetesimals is accreted (Eq. 8.21), to the standard oligarchic accretion rate (Eq. 8.10).

where $p \equiv R_M/r_H$. One can compare this to the standard oligarchic growth rate of Eq. (8.10):

$$\frac{\dot{M}_{2d}}{\dot{M}} \approx 0.1 \frac{\rho_M^{1/6} \rho_m^{4/15} r^{1/10} m^{2/15}}{b^{2/5} C_D^{2/5} M_*^{1/6} \rho_{gas}^{2/5}}. \tag{8.22}$$

Fig. 8.2 plots comparisons of the two accretion rates as a function of stellocentric distance, for several different values of gas surface density and planetesimal size. As can be seen, the two-dimensional enhancement to the growth rate can exceed an order of magnitude at 5 AU (or two orders of magnitude at 30 AU), as compared to oligarchic growth with a minimum-mass gas disk and 10 km-sized planetesimals.

Relative velocities are shear-dominated for planetesimals having sizes below $\sim 100$ m–1 km; the transition size increases with distance from the star. In reality, it is likely only a fraction of the total planetesimal population finds itself in this regime. However, R04 showed that only 1% of the total mass in planetesimals needs to be shear-dominated in order for the accretion rate of embedded protoplanets to be dominated by this part of the population. This is analogous to the findings of Wetherill and Stewart (1993), who demonstrated the important contribution of a planetesimal "fragmentation tail" to the growth rate of the largest bodies in the runaway regime.

Fast accretion would then seem to require simply that a population of $< 1$ km bodies constituting $\gtrsim 1\%$ of the total mass is maintained by fragmentation. A complete picture, however, will also require taking into account the role of planetesimal radial migration. Just as a shear-dominated planetesimal receives discrete kicks in random velocity which are then damped, so too will it orbitally decay in discrete jumps, which will tend to remove planetesimals from a protoplanet's feeding zone (Section 8.2.1). Between these two effects, it needs to be determined what the net consequence for accretion is.

### 8.3.3 *Local enhancement of solids*

The recurring theme thus far has been that rapid inward migration of planetesimals acts to frustrate planet growth. However, under the right circumstances, the opposite can also be true. In the model of Cuzzi and Zahnle (2004), hereafter CZ04, things actually work *best* if planetesimals are of order meters rather than kilometers in size. The resulting rapid radial migration serves as a means to produce a high local concentration of condensible material, specifically ice. This happens because when ice-rich planetesimals from the outer disk arrive at the snow line (Section 8.2.1), their complement of water begins to evaporate. The inward flux of ice-rich planetesimals, the water evaporation rate and the diffusion rate of the vapor plume results were modeled by CZ04, and they showed that a steady state may not exist until a very large local enhancement in water, between 1 and 2 orders of magnitude, has occurred. Since this vapor plume straddles the snow line, water will recondense onto solid bodies at its outer edge. In this way, the protoplanetary disk receives a large enhancement in ice over a relatively small radial range ($\Delta r \lesssim 1$ AU). The picture CZ04 proposed is similar to the earlier "cold-finger" model of Stevenson and Lunine (1988), the main difference being that the latter invoked diffusively redistributed water vapor from the part of the nebular *interior* to the snow line as the source of solids enhancement at the snow line. This immediately places an upper limit on the available water. Bringing the water in from the outer disk instead makes much more water potentially available. Assuming the characteristic planetesimal size really is meters rather than kilometers, CZ04 showed that unless turbulent transport is very large in the nebula, water delivery by inward planetesimal migration is both faster and more robust than delivery by outward diffusion of water vapor. Since accretion rate is $\propto \Sigma_{\text{solid}}$, and isolation mass is $\propto \Sigma_{\text{solid}}^{3/2}$, an increase in the surface density of solids makes it possible to grow bigger bodies faster.

It was pointed out by CZ04 that a pivotal parameter their model rests on is the actual fraction of solids in meter-sized planetesimals. Their base-line calculation assumes a quite substantial fraction of $\sim 10\%$. They took the view, in contrast to the above models, that such small bodies are largely primordial rather than products of fragmentation. This assumption is based on earlier work which found that particle growth faces a number of obstacles, most notably the problem of fragmenting in mutual collisions, beyond the meter-size regime. If true, this has important implications for the models described in Section 8.3.1 and Section 8.3.2; we will return to this issue in Section 8.6.

## 8.4 Ice giants: the problem of Uranus and Neptune

We have so far concentrated on the issue of producing solid bodies to serve as the cores of gas-giant planets. Leaving aside for the moment the possibility of significant planet migration, this makes our zone of interest the inner $\sim 10$ AU of a protoplanetary disk, since the furthest-orbiting gas-giant planet which we know of is Saturn at 9.5 AU. This includes all detected extrasolar planets, among which the largest orbit is that of 55 Cancri d at 5.9 AU (Marcy *et al.*, 2002), though this upper limit is subject to a strong two-fold selection effect: closer orbits produce larger radial velocity signatures, plus it takes about one orbital period of the planet to reliably detect the signature. However, in our own Solar System we have the "ice giants" (thus called because they are basically of cometary composition), Uranus and Neptune, at 19.2 AU and 30.1 AU, respectively. A glance back at Fig. 8.1 immediately reveals a problem, though: even after $10^7$ yr of accretion, protoplanets have only grown to about Mars mass or less at the orbit of Uranus. This is a problem because Uranus and Neptune, their ice-giant status notwithstanding, do each have several $M_\oplus$ of H and He in their atmospheres. The most natural way to account for this is if these planets finished their accretion in $\sim 10^7$ yr, before the gas nebula was completely depleted. To what extent can the mechanisms described in Section 8.3 help out? A protoplanet must reach one to several $M_\oplus$ in order for its gas atmosphere to start playing an appreciable role in enlarging its capture radius (Section 8.3.1). However, assuming the primordial size of planetesimals is $\gtrsim 1$ km, fragmentation must become effective before the protoplanet can reap the benefits of its extended reach. A similar argument applies to the model of Section 8.3.2; the planetesimal population has to form a $\sim 1\%$ fragmentation tail before shear-regime accretion becomes dominant. This introduces a sort of Catch 22: until a protoplanet grows that is big enough to induce significant fragmentation, strong gas drag cannot begin to boost the accretion rate. So the fact that an Earth-mass body cannot grow in $10^7$ yr unless one invokes a disk far more massive than the minimum-mass model casts doubt on whether, during the lifetime of the gas nebula, these putative growth acceleration mechanisms can even begin to become effective. The situation is different if, as assumed by CZ04, a significant fraction of planetesimals is primordially very small. However, in the extreme of meter-sized bodies, the rapid migration of planetesimals due to gas drag, while potentially of great help in enhancing the surface density of solids at the snow line at $\sim 1-5$ AU, instead depletes the available solids at larger radii and thus becomes a nuisance again.

Goldreich *et al.* (2004) explore one approach to accounting for the formation of Uranus and Neptune. Also invoking collisional fragmentation, they show that a sufficiently small planetesimal size allows collisions among planetesimals to take over the role of damping by gas drag and keep random velocities low. For the extreme case of centimeter-sized planetesimals, they show that Uranus and Neptune can accrete *in situ* in $\lesssim 1$ Myr. However, as the authors pointed out, such a small planetesimal size, while beneficial in reducing the growth timescale of the ice giants, brings with it a new problem. The high optical depth of planetesimals makes the final "clean-up" of the leftovers by gravitational scattering (which also populates the Oort cloud) problematic, because a planetesimal's ejection timescale is far longer than its collision time. Remnant planetesimals in the outer Solar System are thus in danger of overstaying their welcome. This problem, which to a lesser extent plagues even kilometer-sized planetesimals, was first pointed out by Stern and Weissman (2001) (though subsequent work by Charnoz and Morbidelli (2003) challenged their findings). Another issue not explored by the authors is whether the large amount of energy dissipation in the disk will cause significant radial rearrangement of planetesimal mass. In particular, collisional grinding down of the planetesimal disk needs to be delayed until after the dispersal of the gas disk, otherwise gas drag is likely to remove a lot of mass to smaller radii (Section 8.2.1). In any case, from an observational point of view, a very nice feature of such a mode of accretion is that it likely involves the generation of a large amount of readily visible dust. Future observations of debris disks, with upcoming instruments such as ALMA, will furnish us with useful constraints on this model in the relatively near future.

Thommes *et al.* (1999, 2002), hereafter TDL99, TDL02, developed a model for the origin of the ice giants which sidesteps the problem of accretion at large heliocentric distances. They proposed that *all* of our giant planets – gas and ice alike – originated in the same region of the protoplanetary disk, $\sim 5-10$ AU, in other words the present-day Jupiter–Saturn region. Even with just a moderate increase above minimum mass in the surface density, oligarchic growth predicts (assuming an inter-protoplanet spacing of $\sim 10 r_H$ is respected until the end of oligarchic growth) that three to five bodies of mass $\sim 10 M_\oplus$ will form in the Jupiter–Saturn region. Numerical simulations were performed by TDL99 and TDL02 in which one of these bodies had its mass increased to that of Jupiter over a $10^5$ yr timescale, in keeping with the final runaway gas accretion phase in the Pollack *et al.* (1996) model. They found that this rapidly destabilizes the orbits of the adjacent oligarchs, causing them to acquire large eccentricities and cross the outer part of the disk where, it is assumed, a largely pristine population of planetesimals still exists. Dynamical friction then acts to recircularize the orbits of these bodies, though leaving them with a far larger than original orbital spacing. The typical end result ("end" meaning after a few Myr, when rapid orbital evolution has ceased) is similar to the architecture of the present-day outer Solar System, with the initially neighboring oligarchs now having semimajor axes comparable to those of Saturn, Uranus and Neptune. All that remains is for the next-innermost protoplanet to become Saturn by acquiring its own (less) massive gas envelope, and for the remaining planetesimals among the orbits of the giant planets to be cleaned up. In this picture, then, Uranus and Neptune are simply "failed" gas giants that lost the race to accrete a massive atmosphere and were unceremoniously scattered into the outer Solar System. Saturn may represent an intermediate case, which still managed to retain a substantial atmosphere; since Saturn's core is somewhat more difficult to produce *in situ* than Jupiter's (Section 8.2.1), an origin involving outward scattering will tend to be helpful for this planet, too. A limitation of the model is that it does not account for interactions between the protoplanets and a gas disk, in particular the rapid migration which may result (Section 8.5); including these effects will be needed to complete the picture. As it stands the model is most appropriate to the case of a tenuous, or even truncated (Shu *et al.*, 1993), gas disk.

## 8.5 Migration and survival

The topic of tidal planet–disk interactions is technically beyond the scope of this chapter, especially as it is covered in depth by the review by Masset and Kley (see Chapter 14). Nevertheless, because tidal-torque migration is inextricably linked with giant-planet formation, we will touch briefly on the issue here. Planet–disk interaction is typically broken down into two qualitatively different regimes. In the first, the planet launches density waves at resonances, but is too small to significantly perturb the azimuthally averaged surface density profile of the disk. This has come to be called the "Type I" regime (Ward, 1997). However, as the planet mass increases, so does the strength of the planet–disk torques, until an annular gap opens about the planet's orbit; to the extent that the gap is clean, the planet is then locked into the disk's viscous evolution, in what is called the "Type II" regime. In typical disk models, a gap does not begin to open until the planet mass is of gas-giant rather than core mass (several tens of $M_\oplus$), so it is the Type I regime which is relevant to core formation.

A body orbiting in a gas disk experiences a repulsive torque from both inner and outer Lindblad resonances (Goldreich and Tremaine, 1980); for typical model disks, the outer torques are stronger – the main source of this asymmetry is the sub-Keplerian (due to pressure support) rotation of the disk – and inward migration results. The timescale for a body to migrate all the way in to the star is

$$t_{\text{Type I}} \sim 10^4 - 10^5 \left(\frac{M}{10^1\,\mathrm{M}_\oplus}\right)^{-1} \left(\frac{\Sigma_{\text{gas}}}{10^2\mathrm{g\,cm}^{-2}}\right)^{-1} \left(\frac{r}{\mathrm{AU}}\right)^{-1/2} \left(\frac{H/r}{0.07}\right)^2 \,\mathrm{yr}$$

(8.23)

(Ward, 1997; Papaloizou and Larwood, 2000). We recall from Section 8.2.1 that the timescale for dispersion-dominated oligarchic growth of a core-sized body at $\sim 5\,\mathrm{AU}$ is $\sim 10^6$ yr. This may be reduced by one or more of the potentially accretion-enhancing mechanisms of Sections 8.3.1, 8.3.2 and 8.3.3, though shaving off up to two orders of magnitude may not be a realistic expectation. But if the growth timescale of a giant-planet core does exceed the time it has before it plunges into the central star, how can any ever form? This is perhaps the single most troubling issue in the core-accretion model of gas-giant formation. Several survival mechanisms have been proposed. To begin with, Eq. (8.23) is obtained from a two-dimensional calculation of disk torques; Tanaka *et al.* (2002) calculated the differential torques three-dimensionally and found a reduction by a factor of $\sim 2$–3 in the migration rate. Also, the gas disk surface density profile may not be smooth as is usually assumed. Laughlin *et al.* (2004) and Nelson and Papaloizou (2004) showed that density fluctuations due to magnetohydrodynamic (MHD) turbulence can give Type I migration the character of a random walk in radius (it remains to be seen, though, whether this random walk truly eliminates the net orbital decay, or is simply superimposed on it). Menou and Goodman (2004) combine calculations of Type I migration with a more detailed disk model, and find that at locations of opacity transition in a disk, variations in surface density can increase the Type I migration timescale of a $\sim 10\,\mathrm{M}_\oplus$ body to more than $10^6$ yr, thus creating local migration bottlenecks. Another possibility is that once planets reach Type II mass and lock themselves into the (generally slower) inward accretion of disk material, they act as a barrier for faster-migrating Type I bodies approaching from exterior orbits. Thommes (2005) showed that the usual outcome when a Type I body catches up with a gap-opening body is that it becomes locked in a first-order mean-motion resonance with the latter. This suggests that if a *first* gap-opening planet can somehow be formed, subsequent giant-planet migration may proceed more readily. In summary, then, although Type I migration seems a severe problem when looking at the simple case of a core in a smooth disk, when one considers the bigger picture, various potential mechanisms of survival do present themselves.

## 8.6 Discussion and conclusions

The formation of gas-giant planets by gas accretion onto a solid core has some very attractive features as a model. Most notably, it offers a natural explanation for two observational facts: (1) in our own system, Jupiter and Saturn are significantly enhanced in solids relative to Solar abundance, and (2) there is a robust correlation between metallicity and extrasolar planet occurrence rate, suggesting that solids play an important role in gas-giant formation around other stars as well. We have attempted to give an overview of how, in our current understanding, the formation of a giant planet's solid core would proceed. One thing that seems fairly certain is that the dominant mode of core growth is the devouring of a large number of planetesimals by a small population of much larger protoplanets; though this starts as a runaway process, it very early on becomes self-regulating as growing cores gravitationally stir their own food supply. Under the assumption that the characteristic planetesimal size remains in the kilometer regime, analytic estimates and numerical simulations of this process suggest that giant-planet cores can be formed on a $10^6$ yr timescale – fast enough to still grab a massive atmosphere of gas from the dissipating nebula – provided that solids and gas are enhanced, at least locally, by roughly an order of magnitude above the standard minimum-mass Solar Nebular model (Hayashi, 1981). The picture gets more complicated if one relaxes either assumption, and it may then become necessary to appeal to additional effects in order to keep accretion fast and efficient enough. Several such mechanisms have been proposed; central to all of them is the role played by gas drag on the planetesimals.

Smaller planetesimals experience stronger gas drag, which is both good and bad: it makes accretion faster by keeping random velocities low, but it also increases the rate at which planetesimals orbitally decay out of a protoplanet's feeding zone. One possible route around this problem is to simply make accretion *so* fast that the accompanying inefficiency is no longer fatal. The work of Inaba and Ikoma (2003) and Inaba *et al.* (2003), as well as that of Rafikov (2004), falls under this general theme. In the former case, gas atmospheres accumulating on protoplanets provide an additional way to make small planetesimals accretionally advantageous (stronger atmospheric drag = larger capture radius). In the latter, small planetesimals' random velocities are simply damped so much that the character of oligarchic growth qualitatively changes and, in the most extreme case, accretion actually becomes two-dimensional. The work of Cuzzi and Zahnle (2004) suggests a different view altogether. They have shown that the rapid orbital decay of small planetesimals, rather than killing accretion efficiency, can actually provide a large local enhancement in solid material – just the thing needed to grow large bodies rapidly, without needing to appeal to an uncomfortably large mass of the *whole* disk.

What is most exciting to note in looking at these three different models is that they are complementary; rather than having to choose among them, one can readily envision all operating at the same time in a protoplanetary disk. In fact, the somewhat provocative idea of a substantial (and perhaps primordial) fraction by mass of the planetesimals being meter-sized, as invoked by CZ04, would immediately lend the other two effects a higher importance, too. Clearly, what is needed now is for all of these pieces to be put together into one picture. As such, we may be just one step away from a coherent, robust, end-to-end model of giant-planet core formation. A model which readily and rapidly produces core-sized solid bodies will sit comfortably with the observational data: (1) from radial velocity surveys, gas giants are ubiquitous, and (2) from observations of young stars, gas disks only last for a few million years.

In the model of TDL99,02, the ice giants Uranus and Neptune are just "failed" versions of Jupiter and Saturn; in fact Saturn itself may be somewhat of an intermediate case. If this picture is correct, then a way to prolifically build gas-giant cores will benefit ice giants too. We thus arrive at a picture where all giant-planet formation takes place in the presence of gas, in the first few million years of a star's lifetime. If this is *not* the case, Uranus and Neptune must have formed in a fundamentally different manner, likely with planetesimal collisions taking on the role of dissipating random velocities in lieu of gas (Goldreich *et al.*, 2004). However, the high optical depth of planetesimals required for this makes their ultimate removal problematic. Future high-resolution observations of dust evolution in debris disks are likely to give us a better idea of how the post-gas evolution plays out. In addition, important clues are likely to come from the discoveries of new extrasolar planets. In particular, the recent radial velocity detections of very low projected mass planets ($M \sin i \lesssim 20\,M_\oplus$) in $\mu$ Arae (Santos *et al.*, 2004a), Gliese 436 (Butler *et al.*, 2004), and $\rho$ Cancri (McArthur *et al.*, 2004) may constitute our first glimpse of extrasolar core-sized bodies, though it is doubtful whether, on their close-in orbits, they retain enough water to qualify as ice giants.

The issue of core survival in the face of migration by nebula tides remains an open problem. However, different variations on the theme of migration bottlenecks (Menou and Goodman, 2004; Thommes, 2005) may help here too. Alternatively, more realistic models of disk MHD may do away altogether with the idea of monotonic orbital decay by tidal torques (Nelson and Papaloizou, 2004; Laughlin *et al.*, 2004).

# 9

# Planetary transits:
# a first direct vision of extrasolar planets

Alain Lecavelier des Etangs and Alfred Vidal-Madjar

*Institut d'Astrophysique de Paris-CNRS, Paris, France*

## 9.1 Introduction

The observation of a transit in our own Solar System is a long-lasting experience. Historical events related to transits can be traced back to Ptolemy who mentioned in his "Almagest" that the lack of detections of transits was not in contradiction with Mercury and Venus being closer to the Earth than the Sun (in the geocentric system) simply because they could be either too small to be detected or their orbital plane could be slightly tilted to the Solar one (Gerbaldi, personal communication). In 1607 Johannes Kepler thought he had directly observed a predicted Mercury transit but in fact only followed sun spots. He did, however, predict the next transits of Venus and Mercury to take place in 1631 following the extremely accurate observations of the planets by Tycho Brahe. The first transit to be observed was the Mercury transit in 1631 with the best observations leading Pierre Gassendi to evaluate its diameter to be less than 20 arcsec, much smaller than ever thought before. All the following transit observations led to new ephemerides and estimates of the size of the Solar System, but not as accurate as expected because of the difficulty of locating in time the entrance and exit of the planetary disk over the Solar one. First pictures of the Venus transit were made as early as in 1874 (Fig. 9.1). The transit of Mercury was also observed with the Solar and Heliospheric Observatory (SOHO) spacecraft from the L5 Lagrange point of the Earth.

Finally the transit of the Earth in front of the Sun can be "indirectly" observed by measuring the reflected moonlight during total lunar eclipses. This shows that absorption spectroscopy observations of transit can efficiently probe the planet's atmosphere, including the detection of the Earth's ozone layer even in the visible range (Chalonge *et al.*, 1942).

*Planet Formation: Theory, Observation, and Experiments*, ed. Hubert Klahr and Wolfgang Brandner.
Published by Cambridge University Press. © Cambridge University Press 2006.

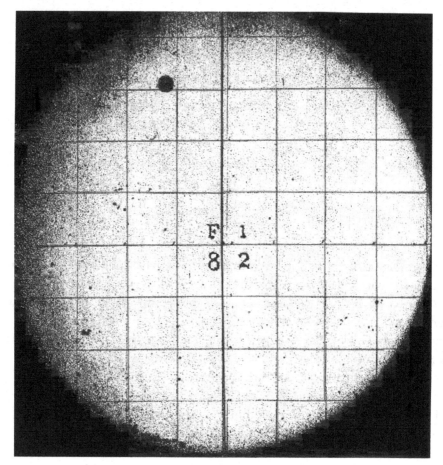

Fig. 9.1. Picture of the transit of Venus taken in 1874 (Janiczek, 1983).

These first studies within our own Solar System show the great variety of impacts that transit observations can have on our understanding of a system. These impacts range from dynamical constraints to specific characteristics of the transiting objects including size and shape as well as atmospheric content and composition.

Very early, it was also recognized that the transit of stars in front of other stars, in binary or multiple systems, was an extremely powerful tool to investigate the quality of stellar models, including stellar evolutionary models. In effect, from these observations coupled with radial velocity measurements it was possible to have access to both the stellar masses as well as to their diameters. These were for a long time the only possible direct test of stellar models. But in addition, these observations allowed many more surprising discoveries like the presence of additional invisible low stellar mass members in the systems via the slow evolution of the timing, the presence of mass exchange between the different stars, and the analysis of extended structures above the stellar surfaces. All these potentialities also apply for planetary transit which lead to similar results related to mass, size, environment, and atmospheres, as we will see below.

To illustrate these capabilities in the case of extrasolar systems, a few particular and important examples can be given regarding various aspects. When seen under a favorable orientation, the transit technique allows a detailed view of the nature, extent and evolution of circumstellar dust orbiting some stars. As a recent example, occultations of the star KH15D revealed detailed information on the geometry of this puzzling system (Kearns and Herbst, 1998; Hamilton *et al.*, 2001; Chiang and Murray-Clay, 2004; Winn *et al.*, 2004). In other systems, transits of gaseous material have been interpreted as comet-like bodies in debris disks, giving in the 80s the first hint of solid bodies orbiting in extrasolar systems (Section 9.5.1). The transit observation of the planetary mass object orbiting around HD209458 was the first demonstration that radial velocity pulsations were indeed due to planets. Having access to its size and to its mass from previous radial velocity detection, it was easy to estimate its density and to conclude that the discovered object was a gaseous planet (Charbonneau *et al.*, 2000; Henry *et al.*, 2000; Mazeh *et al.*, 2000). Finally the most recent development made with transits is the direct detection of different species in the extrasolar planets' atmosphere, ranging from the lower atmospheric levels just above the limb to the upper atmospheric layers revealing an unexpected extension and thus a peculiar escape situation (Charbonneau *et al.*, 2002; Vidal-Madjar *et al.*, 2003, 2004).

## 9.2 Probability and frequency of transits

In the case of stellar–stellar or planetary–stellar transits one can estimate the probability to observe transits as a function of three parameters, the main star radius $R_*$, the transiting object radius $R_t$, and the distance between both objects at the time of transits $a$. The probability to observe a transit assuming random orientation of the orbit is (Sackett, 1999):

$$P = \frac{R_* + R_t}{a}$$

(see Bodenheimer *et al.*, 2003 for the case of an eccentric orbit). In the case of planetary transits, this probability becomes close to $R_*/a$ showing that it is more likely to observe transiting planets at small orbital distances. In the case of the Solar System observed by a random observer at infinity, the probability to observe a transit of Mercury is of the order of 1% while it drops down to 0.1% for Jupiter.

It is thus quite unlikely to detect transits and pre-selecting favorable inclination targets (for example, stars with high rotational inclination) could be a way to favor transit discoveries. However due to the precision of such evaluations and to the probable spread of the planets' orbital plane inclinations relative to the star (in the Solar System the spread of planetary orbits is of the order of 7°) a gain of no more than a factor of three to five in detection probability is expected (Sackett, 1999). Other ways to favor transit detection are to select already transiting binary star systems by assuming that planets could be present in the orbital plane of the stellar system (Schneider and Chevreton, 1990) or by looking for systems with disks seen edge-on. However in both cases the gain in probability remains of the order of a factor of three to five.

In a planetary system, comets can also be transiting objects. If the number of comets is high enough, some trajectories can be star grazing and their inclination broadly spread away from the orbit plane of the main planets; the probability of detecting such transits is then certainly much higher. As an example in the Solar System several Sun-grazing comets are discovered every year, many of these being potentially transiting. Although it is difficult to estimate the increase in transit probability in the case of extrasolar comets, it seems that, in some circumstances, comets can be easier to detect, as with the ones discovered around the star $\beta$ Pictoris more than ten years before the discovery of the first extrasolar planet (Section 9.5.1).

Since there is no clear way to improve the probability of detecting planetary transits, searches have to cope with probabilities in the range of 0.1 to 1%, except in the case of the "very hot Jupiter" which can be detected with a much higher probability, up to 20% for the shortest orbital periods (Bouchy et al., 2004). The consequent strategies are to look for a large number of targets, either with robotic telescopes fast-scanning wide fields to survey photometric variations of bright stars, or to look to fainter stars in a given field of view. Both strategies recently led to the discovery of new transiting planets (Section 9.4).

Finally, the probability of transit is relatively high for an already identified "hot Jupiter" (for example, identified with radial velocity), with a probability of about 10%. Follow-up observations are crucial and can lead to potentially important discoveries of transits. Such a follow-up allowed the first detection of a transiting planet, HD209458b, which is, up to now, the planet transiting the brightest star identified, allowing fruitful spectroscopic observations (Section 9.5.2). It is fun to see that, with a 10% probability of transit, this HD 209458 transit was discovered in the fall of 1999 when about 10 hot Jupiters were known, and that we have now identified about 20 hot Jupiters. As the sample of known hot Jupiters increases, the probability of the discovery of a second similar case increases.

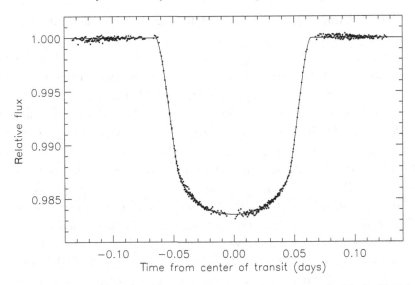

Fig. 9.2. Plot of the HD 209458b transit light curve as measured with the Hubble Space Telescope (Brown *et al.*, 2001).

### 9.3 Basics of transits

### *9.3.1 Photometric transits*

If one assumes that the shapes of both the occulting body and of the star are known, it is possible from simple geometrical calculations to infer what must be the duration of the transit event and the shape of the photometric light curve. In the case of the Solar System seen by an edge-on observer, the maximum duration of the planets' transit ranges from 0.35 days for Mercury to 3.5 days in the case of Pluto. The maximum transit duration $T_{max}$ (when crossing over the center of the stellar disk), in front of a star of radius $R_*$ and mass $M_*$, for a planet orbiting with a period $P$ at a distance $a$ is:

$$T_{max} = \frac{P \times R_*}{\pi \times a} \approx 13.0 \left(\frac{a}{1\,\text{AU}}\right)^{1/2} \left(\frac{M_*}{M_\odot}\right)^{-1/2} \left(\frac{R_*}{R_\odot}\right) \text{ hours,}$$

assuming a planet radius much smaller than $R_*$.

To calculate the theoretical transit light curve, one must take into account the non-homogeneity of the stellar flux over the different parts of the stellar disk. It is in general affected by the limb darkening (at least in the visible spectral range) which induces, for the transit profile, a curved bottom shape, as can be seen in the very high quality transit profile of HD 209458b obtained by Brown *et al.*, (2001) with the Hubble Space Telescope (Fig. 9.2). This observed profile can be very well fitted by the theoretical light curve taking into account the limb darkening of the Solar-type star HD209458.

The shape of a transit curve at a wavelength other than a broad band in the visible directly depends upon the appearance of the stellar disk at this wavelength. When moving away from the black-body emission, either at short wavelength as in the far ultraviolet or in narrow spectral ranges as at the bottom of strong photospheric lines, the stellar appearance can present limb brightening and/or be much more inhomogeneous than in the visible. The variations at the bottom of the transit light curve are expected to be of the order of the stellar flux inhomogeneities averaged over a size equal to the occulting planetary disk. As an example, in the far-ultraviolet at Lyman $\alpha$, chromospheric inhomogeneities are well known and observed in the case of the Sun (Prinz, 1974), but when averaged over the planetary disk size, they cannot produce fluctuations larger than a few %. In the case of the Solar-type star HD 209458 where a 15% absorption is observed in Lyman $\alpha$ during the transit, these fluctuations are marginal. However these fluctuations are more important for smaller transiting objects, the transit becoming more difficult to detect not only because the transit depth decreases but also because the intrinsic fluctuations at the bottom of the transit curve increase.

The detailed photometry of the transit light curve gives detailed information on the occulting body shape. As an example one can assume that the planetary disk is not circular but oblate (Seager and Hui, 2002; Barnes and Fortney, 2003). This shape can be the signature of a rapid rotation, as in the case of Jupiter, or of upper atmospheric winds. In the same way, signatures of satellites or ring systems can be searched for as additional absorptions before or after the planetary transits (Barnes and Fortney, 2004). All departures from the standard light curve are thus the signature of many different possibilities concerning the planet structure and environment.

In addition to the shape of the transit light curve, the exact timing of the successive events also carries important information. An accurate timing of the transit ingress and egress could reveal the presence of additional planets gravitationally perturbing the observed planet, or the presence of orbiting satellites too small to be directly detected.

Finally, the shape of the transit light curve can also be used to discriminate between a comet and a planet. It has been shown that most of the comets' transit light curves have an asymmetrical triangular shape (Lecavelier des Etangs *et al.*, 1999). This may be an important issue because cometary transits are more probable than planetary ones. A large number of the transits will be discovered by the upcoming space instruments like CoRoT (Convection, rotation and planetary transits) and Kepler able to reach $10^{-4}$ photometric accuracy; most of the observed photometric transits may be due to comets instead of planets (Lecavelier des Etangs, 1999).

### 9.3.2 Spectroscopic transits

Transits are usefully observed with spectroscopy. At the simplest level, assuming a circular shape for the occulting body, one can consider the radius of the occulting body as a function of the wavelength. This can lead to the observation of atmospheres. As an example, the Earth's atmosphere being opaque below 3000 Å, the occulting body looks larger by about 60 km when an Earth transit is observed at shorter wavelengths. It is, of course, more difficult to observe a transit in spectroscopy than in photometry, because the number of photons in the limited spectral domain is lower and the required accuracy is higher. The increased fraction of the stellar disk occulted by a typical atmosphere is of the order of $10^{-4}$ to $10^{-8}$, depending on the atmospheric scale height (related to the planet total density) and the searched component of the atmosphere (Brown *et al.*, 2001). This requires large space instruments to be used.

## 9.4 Observed photometric transits

### 9.4.1 β Pictoris

Ten years ago, it was thought that the best places to detect transits of extrasolar planets were the systems seen edge-on, because this increases the transit probability (Schneider and Chevreton, 1990). In this frame, the $\beta$ Pic system was among the best candidates: first, planets are believed to orbit in the system because they are needed to explain both the large number of comets (Section 9.5.1) and the morphological properties of the circumstellar dust disks (see, for example, Mouillet *et al.*, 1997); second, the disk is seen almost exactly edge-on (Kalas and Jewitt, 1995). In addition, photometric measurements of $\beta$ Pic were done at La Silla by photometrists of the Geneva observatory, from 1975 to 1981, when they detected significant and unexplained variations.

These 1981 variations can be described by an increase in the star brightness over about ten days and symmetrical return to the normal level. Moreover, exactly at the middle and top of this brightness increase, a sharp decrease in the star brightness was measured over a few hours (Lecavelier des Etangs *et al.*, 1994). The analysis of these variations showed that: (1) these variations are real; (2) they are consistent with the transit of a body in front of $\beta$ Pictoris. Two possible interpretations of this phenomenon arise: either the transit of a massive planet, possibly with rings (Lecavelier des Etangs *et al.*, 1995, 1997), or the transit of a giant comet (Lamers *et al.*, 1997). Hence this likely represents the first historical record and interpretation of an extrasolar transit. But to discriminate between the two possible interpretations, new observations of the same phenomenon are required. Despite dense surveys, similar variations have never been observed again (Nitschelm *et al.*, 2000; Lecavelier des Etangs *et al.*, 2005), leaving this photometric transit unconfirmed.

### 9.4.2 HD 209458b

The first confirmed planetary transit is the transit of the extrasolar planet HD 209458b in front of its parent star. This hot-Jupiter planet has been discovered by radial velocity searches (Mazeh *et al.*, 2000). It has a mass of $0.699\pm0.007$ $M_J$ and a period of about 3.52 days (Naef *et al.*, 2004). The discovery of the transit (Charbonneau *et al.*, 2000; Henry *et al.*, 2000) was a strong confirmation that the radial velocity oscillations of the stars were indeed due to planets. This transit gave the first estimate of the radius of an extrasolar planet ($\approx 1.42^{+0.10}_{-0.13}R_J$, Cody and Sasselov, 2002) confirming it to be a gas giant. A posteriori, Robichon and Arenou (2000) and Castellano *et al.* (2000) found photometric measurements of this transit in the Hipparcos data collected from the beginning of 1990 to the end of 1993. Together with more recent measurements, this allowed them to obtain a very accurate estimate of the period which is now estimated to be $P = 3.5247542\pm0.0000004$ (Wittenmyer *et al.*, 2004), that is with an uncertainty of less than 4s yr$^{-1}$.

Because HD 209458b is transiting a bright star ($m_v = 7.6$), this allows detailed observations of the transit light curve. First, the limb darkening effect is seen in detail in the data collected with the Hubble Space Telescope (Brown *et al.*, 2001; Fig. 9.2). In addition, a careful analysis of the light curve and timing gives constraints on potential satellites or rings: the upper limits on satellite radius and mass are 1.2 $R_\oplus$ and 3 $M_\oplus$, respectively; opaque rings, if present, must be smaller than 1.8 planetary radii in radial extent (Brown *et al.*, 2001).

Finally, the measurements of the planet radius give a value which appears to be larger than expected (Bodenheimer *et al.*, 2003). This measured radius is still unexplained and challenging in regards to other subsequently observed transits by the Optical Gravitational Lensing Experiment (OGLE) (see Sections 9.4.3 and 9.4.6).

### 9.4.3 OGLE planets

Transiting planets can also be searched for in deep surveys of selected fields. Deep surveys of several fields toward the Galactic center and the Carina region of the Galactic disk by OGLE have identified 137 transiting candidates (Udalski *et al.*, 2002a, b, c, 2003). Because a stellar grazing transit, an M dwarf in a binary system, or a stellar transit in a triple system produce about the same light curve (Konacki *et al.*, 2003a; Torres *et al.*, 2004a), each identified transiting candidate must be confirmed by a discriminating observation. Now, radial velocity measurements have been successful in confirming five OGLE candidates: OGLE-TR-56 (Konacki *et al.*, 2003b; Torres *et al.*, 2004b), OGLE-TR-113 (Bouchy *et al.*, 2004; Konacki *et al.*, 2004), OGLE-TR-132 (Bouchy *et al.*, 2004), OGLE-TR-111 (Pont *et al.*, 2004), and OGLE-TR-10 (Bouchy *et al.*, 2005; Konacki *et al.*, 2005).

The new point raised by this survey is the discovery of very short period planets. If we except the recent discoveries of OGLE-TR-111 and OGLE-TR-10, the first three OGLE planets have orbital periods ranging from 1.2 to 1.7 days, which is shorter than any planet discovered before by almost a factor of two. Even the last two planets have short orbital periods of 3.1 and 4.0 days. The apparent contradiction of the efficient discovery of these "very hot Jupiters" relative to the radial velocity method can be explained by comparing the efficiency of both methods and the presumed frequency of the planets as a function of the orbital period (Bouchy *et al.*, 2004; Pont *et al.*, 2004). In short, the OGLE program efficiently probes a large number of targets ($\geq 150\,000$ stars) but is not sensitive to periods larger than about 3 days, therefore OGLE finds the very few "very hot Jupiters;" the radial velocity searches have been limited to about 2000 stars, therefore these searches have found the more numerous planets with longer orbital periods (Gaudi *et al.*, 2005).

### 9.4.4 TrES-1

A strategy different from OGLE has been used by many other searches for planets via transit photometry. This strategy consists of the survey of wider fields to look at brighter stars with smaller telescopes. Among these surveys, the Hungarian-made Automated Telescope Network (HATnet) uses 11 cm robotic telescopes and has chosen fields overlapping with the planned Kepler mission (Hartman *et al.*, 2004).

The first (and presently the only) positive detection with this kind of approach has been recently announced with the discovery of the planet GSC 02652-01324b nick-named 'TrES-1' (TrES for Trans-atlantic Exoplanet Survey) with the combination of three telescopes: STARE (Stellar Astrophysics and Research on Exoplanets) in the Canary Islands, PSST (Planet Search Survey Telescope) in Arizona, and Sleuth in California (Alonso *et al.*, 2004). This planet orbits a relatively bright K0V star ($m_v = 11.8$). It has a mass of $0.075 \pm 0.07$ $M_J$, a radius of $1.08^{+0.18}_{-0.04}$ $R_J$ and an orbital period of 3.0 days.

This discovery opens a new field of discoveries with wide surveys, which are of extreme importance because transits in front of bright stars offer unique capabilities to scrutinize planetary atmospheres in detail (Section 9.5.2). With one transiting hot Jupiter per one thousand stars in the sky, there are several dozens of hot Jupiters transiting a K, G or F main-sequence star brighter than $V = 10$.

### 9.4.5 Missing photometric transits

Several surveys of open and globular clusters have been made to search for extrasolar planets' photometric transit signatures (e.g. Burke *et al.*, 2004). For instance, a

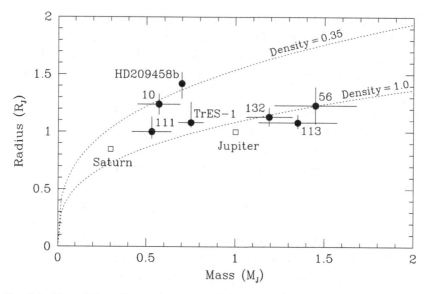

Fig. 9.3. Plot of the radius versus mass of the extrasolar planets for which these two quantities have been measured. The mass is given by radial velocity and the radius is given by transit observations. The labels give the object names, numbers alone are for OGLE-TR-xxx. Courtesy of F. Pont.

photometric survey of about 34 000 main-sequence stars in the globular cluster 47 Tucanae has been made with the wide-field camera of the Hubble Space Telescope (Gilliland *et al.*, 2000). Assuming the same frequency of hot Jupiters in the Solar Neighborhood as in 47 Tuc, about 17 planets should have been detected. The absence of close-in planets in 47 Tuc is assumed to be due to either the low metallicity and/or the crowding of the central part of this globular cluster.

As there are many current programs dedicated to this approach, results are awaited in the coming years.

### 9.4.6 The planet radius problem

One important parameter constrained by the photometric measurements of transit is the planet size. We have now in hand seven measurements of planet size (e.g. Moutou *et al.*, 2004) together with their mass estimates. A plot of the planet radius versus mass (Fig. 9.3) shows that, except for HD 209458b, the measured sizes are in agreement with theoretical prediction (Bodenheimer *et al.*, 2003; Burrows *et al.*, 2004). The origin of the HD 209458b radius is still a matter of debate; this could be due to an over-interpretation of the measured radius and uncertainties (Burrows *et al.*, 2003), an internal heating source (e.g. Chabrier *et al.*, 2004), or tidal heating, possibly requiring the presence of an additional planetary companion

(Bodenheimer *et al.*, 2001; Gu *et al.*, 2003, 2004). New observations are needed to discriminate between these different possibilities. In particular, it will be crucial to know if HD 209458b is exceptional or not, and to see if the "radius problem" correlates with the planetary mass, stellar type or age, the presence of a close-by companion, or another parameter.

## 9.5 Observed spectroscopic transits

### 9.5.1 β Pictoris

As for the photometric transits, the presence of a circumstellar disk seen edge-on put β Pictoris in the top list for the detection of orbiting material through absorption spectroscopy. The first high-resolution spectra obtained soon after the discovery of the dust disk were aimed at observing the gas disk (Hobbs *et al.*, 1985; Vidal-Madjar *et al.*, 1986). The surprise came from the recognition that the observed spectra present rapid variations in short timescales (Ferlet *et al.*, 1987). These variations have been observed and analyzed in great detail over many aspects like composition, ionization, velocity fields, lines widths, etc. (see review in Vidal-Madjar *et al.*, 1998). The main result is the interpretation of these spectroscopic signatures as the transits of evaporating (extrasolar) comets (Ferlet *et al.*, 1987; Beust *et al.*, 1990). Interestingly, we have now a large number of detected comets' transits to draw statistical conclusions and detailed explanations on the dynamical origin of these comets. The most detailed and predictive model proposes that these comets are large ($\geq 10$ km) asteroidal bodies trapped in 4:1 resonance with a massive planet at $\sim 10$ AU (Beust and Morbidelli, 1996, 2000). As an example, this model can explain the observed relationship between the radial velocity and line widths as being due to a precession of the periastron angle as a function of the eccentricity inside the resonance (Beust and Morbidelli, 2000).

### 9.5.2 HD209458b

The discovery of the first transiting extrasolar planets offered unprecedented capabilities to scrutinize the planetary environment, in particular through spectroscopy because HD 209458 is a bright star. The spectroscopic signature of the transit has been observed in the radial velocity measurements as distortions of the stellar line profile during the transit; these distortions are due to the small size of the planet occulting a fraction of the rotationally Doppler-shifted stellar disk (Queloz *et al.*, 2000). This allows the confirmation that the planetary orbit is in the same direction as the stellar rotation. An upper limit of the angle between the orbital plane and the stellar equatorial plane can also be determined, and is found to be less than $30°$ in the case of HD 209458.

But the most promising technique related to the spectroscopic observation of planetary transits is certainly the possibility to detect atmospheric signatures through additional absorption in spectroscopic lines of atoms and ions present in the atmosphere. This technique takes advantage of the numerous stellar photons going through the atmosphere making atmospheric characterization much easier than by detecting the few photons coming from the planet. The first detection of an extrasolar planet atmosphere was made rapidly after the discovery of the transit of HD 209458b. Using very high-precision spectrophotometric observations obtained with the Hubble Space Telescope, Charbonneau *et al.* (2002) detected an additional absorption of $(2.32\pm0.57)\times10^{-4}$ in the Na I doublet at 589 nm. This signature of the atmosphere is lower than predicted (Brown, 2001). This could be interpreted in terms of atomic Na depletion into molecules or grains, or more likely photoionization or clouds in the atmosphere (Charbonneau *et al.*, 2002), this last hypothesis being consistent with the non-detection of CO (Brown *et al.*, 2002; Deming *et al.*, 2005).

In the case of HD 209458b, the detection of an atmospheric signature in the spectroscopic transit has also been made for three other species, namely H I, C II and O I (Vidal-Madjar *et al.*, 2003, 2004). In these last cases, the absorptions have been detected against stellar emission lines in the far-ultraviolet (Lyman-$\alpha$ at 121.6 nm for H I, 130 nm for O I, and 133 nm for C II). The first surprise came from the observation that the absorption of the stellar H I Lyman-$\alpha$ line is about 15±4%, which corresponds to a hydrogen cloud in the upper atmosphere more extended than even the Roche lobe of the planet (a cloud with the size of the Roche lobe would produce a 10% absorption depth). H I in this extended exosphere is seen with radial velocity exceeding $100\,\mathrm{km\,s^{-1}}$; this velocity is explained by the radiation pressure on the hydrogen atoms (Vidal-Madjar and Lecavelier des Etangs, 2004). The cloud's geometrical extension beyond the Roche lobe and the velocity exceeding the escape velocity both show that hydrogen is escaping the planet. The planet is losing mass, justifying the proposal of the nickname "Osiris" (Vidal-Madjar and Lecavelier des Etangs, 2004). A numerical simulation shows that the transit signature as observed in Lyman-$\alpha$ can be produced by a mass-loss rate of at least $\sim 10^{10}\mathrm{g\,s^{-1}}$. This evaporation rate is in agreement with theoretical calculation of the upper atmosphere structure, taking into account the tidal forces and the heating of the upper atmosphere by ultraviolet and extreme-ultraviolet (Lecavelier des Etangs *et al.*, 2004). But another surprise was the observation that atomic oxygen and ionized carbon are also present in the upper atmosphere and also produce about 10% absorption of the C II and O I stellar lines. The presence of these heavier elements high in the upper atmosphere requires that the atmosphere is not under the regime of Jeans escape but is hydrodynamically escaping in a "blow-off" state (Vidal-Madjar *et al.*, 2004). From this observation and through the evaluation of the total density at the level of

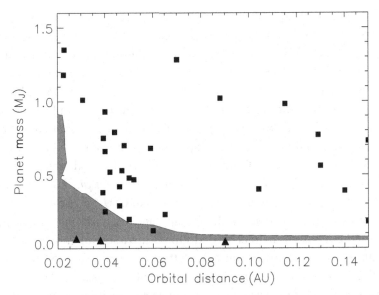

Fig. 9.4. Plot of the domain where the planet lifetime against evaporation is shorter than $10^9$ years as a function of the initial mass and orbital period. In the grey region, the time to evaporate the full hydrogen envelope of a planet with a 15 $M_\oplus$ solid core is shorter than $10^9$ yr. This calculation uses the planet's radii given by Guillot (2005) as a function of time, mass and effective temperature. The time variations of the far-ultraviolet/extreme-ultraviolet stellar flux is also taken into account, but it is found to have a negligible effect in contrast to the assumed planet radii. The squares show the positions of the known planets. The three planets with mass between 14 and 21 $M_\oplus$ have been plotted with triangles.

the Roche lobe ($\sim 10^6$ cm$^{-3}$), the escape rate can also be independently estimated to be of the order of $\sim 10^{10}$g s$^{-1}$ in agreement with the estimate obtained with the H I observation.

### 9.5.3 Evaporation of hot Jupiters

The transit observation of HD 209458b allowed the realization that a hot Jupiter can evaporate, and under certain circumstances can be significantly modified by this evaporation. In addition to an understanding of the observed escape rate from HD 209458b, the escape rate of hot Jupiters as a function of their mass and orbital period was evaluated by Lecavelier des Etangs *et al.* (2004, Fig. 9.4). From this evaluation, it turns out that even "very hot Jupiters" can be stable against evaporation if they are massive enough, like OGLE-TR-56, OGLE-TR-113 or OGLE-TR-132. On the other hand, planets with an orbital period shorter than about three days and less massive than about $\sim$0.5 $M_J$ should be the subject of intense evaporation. If this takes place, most of the atmosphere must disappear on a short timescale leaving the remaining central core of about $\sim$15 M$_{\oplus}$ of heavy material, with a shallow atmosphere or no atmosphere at all. The state of the surface of these putative bodies is still to be determined and could be similar to the lava surface of Io with its 1500 K temperature. The emergence of the inside core of former and evaporated hot Jupiters may constitute a new class of planets which we proposed to call the "chthonian" planets in reference to the Greek god of the Earth: Khthôn ("chthonian" is used to name the Greek deities who come from the hot infernal underworld).

This suggestion of the existence of evaporated planet remnants has been recently emphasized by the discovery of three planets with short periods and low mass from $m \sin i \sim$14 to $\sim$21 M$_{\oplus}$. In one case ($\mu$ Arae, Santos *et al.*, 2004a), the orbital period of 9.55$\pm$0.03 days and the orbital distance of 0.09 AU seem unfavorable for an evaporation scenario. However, the initial planetary mass and radius, and early evolution of the X-ray, ultraviolet and extreme-ultraviolet flux from the star are still to be evaluated to draw a firm conclusion. In the two other cases (55 CnC, McArthur *et al.*, 2004; GJ 436, Butler *et al.*, 2004), the orbital periods are 2.81 and 2.64 days. These two last planets fall exactly at the mass-period position predicted for most of the evaporation-modified ("chthonian") planets (Lecavelier des Etangs *et al.*, 2004). In conclusion, the nature, origin and history of these newly discovered planets is still to be clarified, but they appear to be possibly the result of the evaporation mechanism which has been uncovered thanks to spectroscopic observations of the transit of HD 209458b. If evaporated planet remnants are numerous, they will also be uncovered by future space programs dedicated to the search for planetary transits (Section 9.5.4). We have to wait for the coming deep-transit surveys with dedicated spatial missions to understand the properties, distribution and evolution of these low-mass planets which we are starting to discover.

### *9.5.4 The search for transits with space observatories*

In the near future, several space missions will be dedicated to the search for extrasolar planets by transits, and in particular to the search for Earth-size planets. CoRoT will be the first high-precision photometric satellite to be launched in 2006 with the aim of detecting extrasolar planets by the transit method (Bordé *et al.*, 2003). CoRoT will be followed by the Kepler NASA mission scheduled to be launched in 2007. The Kepler mission will determine the distribution of Earth-size planets (0.5 to 10 M$_{\oplus}$) in the habitable zones of Solar-like stars. This mission will monitor more than 100 000 dwarf stars simultaneously for at least 4 yr (Duren *et al.*, 2004). After these missions dedicated to the search for transiting extrasolar planets, the Global Astrometric Interferometer for Astrophysics (GAIA), the European Space Agency astrometric mission to be launched around 2011, will also detect a large number of transiting planets (Robichon, 2002).

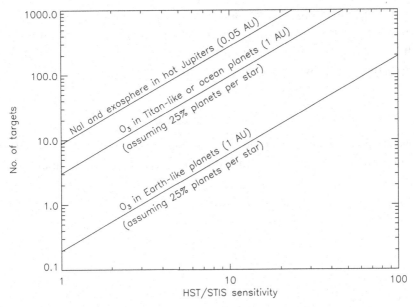

Fig. 9.5. Estimate of the number of transiting planets orbiting K, G and F main sequence stars which can be observed as a function of the telescope sensitivity. For hot Jupiters, the fraction of planets orbiting at ~10 stellar radii has been assumed to be 1% hot Jupiters per star. For Titan-like satellites, Earth-like and ocean planets, the fraction of planets orbiting in the habitable zone at ~1 AU has been assumed to be 25% planets per star. With these numbers, we can estimate the telescope sensitivity needed to detect atmospheric signatures of sodium and exosphere in the first case, or ozone in the case of small planets. The telescope sensitivity is obtained by scaling the result of *real* observations and given in units of the sensitivity of the Hubble Space Telescope with the Space Telescope Imaging Spectrograph.

Algorithms are being developed and tested for future space missions (e.g. Aigrain and Favata, 2002; Aigrain and Irwin, 2004). The best strategy can certainly be refined, by, for instance, looking at M stars in the searches for habitable-zone planets (Gould *et al.*, 2003). Even reflected light from close-in extrasolar giant planets will likely be feasible with the Kepler photometer (Jenkins and Doyle, 2003).

### 9.6 Conclusion

As we have seen above, planetary transits offer a first direct vision of extrasolar planets. They will give unprecedented capabilities to probe the extrasolar planets' characteristics and their environment. As a last demonstration of these capabilities, we have plotted in Fig. 9.5 the number of expected transits in the sky as a function of the instrument sensitivity needed to detect important species like ozone which could be considered as a potential biomarker.

In the past, transit observation has been crucial for Solar System exploration. In the present, it is providing a powerful tool to characterize extrasolar planets and their atmospheres. In the future, it will allow important progress and likely numerous discoveries which are now impossible to predict.

## Acknowledgments

We warmly thank T. Brown for the kind permission to reproduce the HD 209458b light-curve in Fig. 9.2, and F. Pont for providing Fig. 9.3 and fruitful discussions.

# 10

## The core accretion–gas capture model
## for gas-giant planet formation

Olenka Hubickyj

*UCO/Lick Observatory, University of California*
*and Space Science Division, NASA-Ames Research Center*

### 10.1 Introduction

At the time of this workshop, there are now more than 150 detected extrasolar planets discovered and 13 confirmed multiple planet systems with more candidate planets and systems being evaluated (G. Marcy in Chapter 11 of this book). There is a growing number of ground-based observations of young circumstellar and protoplanetary disks as well as volumes of data from Spitzer and Cassini. There are two scenarios for gas-giant planet formation that are sufficiently sophisticated to provide results and predictions. Clearly, the ingredients are present for planetary scientists to develop a comprehensive or, at least, a cohesive model for the formation of the gas-giant planets in the Solar System and in extrasolar systems and, to initiate resolution of the age old question: how do planets form?

The subject of this chapter is the core accretion–gas capture model, the more generally accepted scenario for gas-giant planet formation (see also Chapter 8 by Thommes and Duncan for the core accretion itself). Please note that this model has had many names over the decades of its development: planetesimal hypothesis, nucleated instability model, and core instability model. However, it seems best to call it by the more descriptive label: the **C**ore **A**ccretion–**G**as **C**apture model, the CAGC or, the short version, the core accretion model. Computer models simulating the core accretion model (Perri and Cameron, 1974; Mizuno *et al.*, 1978; Mizuno, 1980; Bodenheimer and Pollack, 1986; Pollack *et al.*, 1996, hereafter referred to as Paper 1) have evolved over the years and have contributed much to the general knowledge of gas-giant planet formation.

Another much discussed planet-formation scenario is the gas-instability model (alternatively referred to as the gas-giant protoplanet model: GGPP model), which

*Planet Formation: Theory, Observation, and Experiments*, ed. Hubert Klahr and Wolfgang Brandner.
Published by Cambridge University Press. © Cambridge University Press 2006.

is described and discussed by A. Boss in Chapter 12 in this book. The gas-instability model forms planets initiated by a gravitational instability of the gas in the Solar Nebula which rapidly forms a gravitationally bound subcondensation known as a giant gaseous protoplanet (Kuiper, 1951; Cameron, 1978; DeCampli and Cameron, 1979; Boss, 1998b, 2000; Mayer *et al.*, 2002; Pickett *et al.*, 2003; Rice *et al.*, 2003a, b; Boss, 2003; Mayer *et al.*, 2004).

Gammie (2001) investigated the stability of cooling, gaseous disks and reported conditions under which fragmentation would occur. Not only is it necessary for the disk to be gravitationally unstable, but it must be able to cool efficiently. He found that the cooling time must be less than half an orbital period. Detailed radiation-transfer calculations in a disk by Mejía *et al.* (2005) and Cai *et al.* (2004) indicate that the cooling time is too long to allow fragmentation to take place in the inner regions of disks. Boss (2004), on the other hand, showed that convection currents can bring energy out from the interior of a disk fast enough so that the cooling time requirement can be met outside about 8 AU. A recent study by Rafikov (2005) re-examined the conditions for planet formation by the GGPP mechanism. Overall, he stated that isothermal simulations should not be used to study planet formation in real protoplanetary disks and thermal conditions in the disk must be considered carefully. He determined that the cooling timescale ought to be comparable to the local disk orbital period. Interpreted into disk parameters, for planets to form at about 10 AU, the gas temperature of the disk should be $> 10^3$ K with a mass of 0.7 $M_\oplus$ and a luminosity of 40 $L_\odot$. These parameters are outside the range of observationally deduced disk properties.

The evolution of a gas-giant protoplanet in the context of the core-accretion scenario is viewed to occur in the following sequence (as described in Bodenheimer *et al.*, 2000, hereafter referred to as BHL00):

(1) Dust particles in the Solar Nebula form planetesimals that accrete into a solid core surrounded by a low-mass gaseous envelope. Initially, solid runaway accretion occurs, and the gas accretion rate is much slower than that of solids. As the solid material in the feeding zone is depleted, the solid accretion rate is reduced.

(2) The gas accretion rate steadily increases and eventually exceeds the solid accretion rate. The protoplanet continues to grow as the gas accretes at a relatively constant rate. The mass of the solid core also increases but at a slower rate and eventually the core and envelope masses become equal.

(3) Runaway gas accretion occurs and the protoplanet grows rapidly. The evolution (1) through (3) is referred to as the *nebular* stage, because the outer boundary of the proto-planetary envelope is in contact with the Solar Nebula and the density and temperature at this interface are given nebular values.

(4) The gas accretion rate reaches a limiting value defined by the rate at which the nebula can transport gas to the vicinity of the planet. After this point, the equilibrium region of

the protoplanet contracts inside the effective accretion radius, and gas accretes hydro-dynamically onto this equilibrium region. This part of the evolution is considered to be the *transition* stage.

(5) Accretion is stopped by either the opening of a gap in the disk as a consequence of accretion and the tidal effect of the planet, or by dissipation of the nebula. Once accretion stops, the planet enters the *isolation* stage.

(6) The planet contracts and cools to the present state at constant mass.

The discussion in this chapter will start with an overview of the historical development of the core accretion–gas capture model, highlighting the achievements by the various investigators. This will be followed in Section 10.3 by a description of the fundamental observations that any formation model needs to explain. The computer-simulation technique based on the CAGC model is described in Section 10.4 followed by the presentation of the most recent studies based on the CAGC model in Section 10.5. Section 10.6 presents a summary of the most up-to-date results and conclusions of the CAGC simulation.

## 10.2 The development of the CAGC model

Planets have been discovered and studied since antiquity, but it wasn't until the Age of Reason that methodical thought was given to the formation of the Sun and planets. A historical review of planet formation is presented by P. Bodenheimer (Chapter 1 in this book) and a brief historical overview of the CAGC model is offered in this section.

Early Solar-System formation theories were based on a hypothesis stating that planets formed from mass thrown off the Sun after it had condensed into its current state. The proponents of this treatise were Descartes (1644), Kant (1755), Laplace (1796), and W. Herschel (1811). Buffon (1749) considered a "building up" formation process rather than a condensing mechanism proposing that a comet, passing close to the Sun, pulled matter off the Sun which then accreted into planets.

The origin of planets from Solar material was the dominant model until the 1840s. By that time geologists and Darwinian evolutionists were trying to determine the age of the Earth (Brush, 1990) by using the laws of thermodynamics that had been recently formulated. Geologist T. C. Chamberlin (1899), using the kinetic theory of gases, argued against the belief that planets formed from a hot gaseous environment, the prevalent view at the time. He argued that if Earth had been hot enough to vaporize iron, vaporized molecules of lighter elements would have acquired velocities high enough to escape and Earth would have lost its atmosphere. He wrote that Earth might have formed by accretion of cold particles that he called "planetesimals." In collaboration with Moulton (1905), he proposed that tidal action of a near encounter with another star released Solar material that condensed into planetesimals which accreted by mutual gravitational attraction to form planets.

The hot gas condensation hypothesis continued to be the popular explanation for forming planets until Jeffreys (1917, 1918) and Jeans (1919) showed that gravitational forces were not strong enough to initiate the condensation process for planets with the amount of mass believed to be in the Solar System at that time. The "tidal" model proposed by Chamberlin and Moulton became the generally accepted formation scenario until Russell (1935) demonstrated that the tidal interaction would not put material into orbit with the required angular momentum. Additionally, using Eddington's stellar-structure theory that was based on radiative energy transfer rather than convection, Russell argued that the temperatures of the Sun's gases were about a million degrees which is hot enough for the velocities to be high enough for the atoms in the gas to escape into interstellar space, precluding any kind of condensation.

A return to the nebular hypothesis occurred when von Weizsäcker (1944) and Kuiper (1951) proposed that the planets in the Solar System formed from a slowly rotating cloud of dust and gas, and as it contracted, the cloud started to rotate faster in its outer parts and eddies were created. These eddies were small near the center of the cloud and larger at greater distances from the center. Urey (1951) supported the nebular hypothesis until he began studying terrestrial planets, after which he rejected the idea in favor of assuming numerous small objects like asteroids and Lunar-sized objects were the building blocks of the planets.

The earliest quantitative work was undertaken by Safronov during the 1960s. He developed a model based on the work of Shmidt (1944) who postulated that the Sun captured material from interstellar space called a "protoplanetary cloud." Safronov (1969) created an analytical formulation for the accumulation of solid particles from the protoplanetary cloud into planets. The Safronov accretion model and the burgeoning capabilities of computers prompted extensive work on planet formation. Wetherill was one of the earliest researchers who adopted Safronov's planetary-accretion model for a computer simulation of Earth and terrestrial-planet formation.

In tandem to Safronov's work, a series of papers by Kusaka *et al.* (1970), Hayashi (1981), and Nakagawa *et al.* (1981) investigated the growth of solid particles in the Solar disk. Mizuno *et al.* (1978) included the effect of the gaseous nebula on the buildup of planetesimals into planets, and then Mizuno (1980) extended this accretion model to the formation of Jupiter and Saturn. Their work was based on the computation of a series of protoplanetary models of increasing core mass with a gaseous envelope in hydrostatic equilibrium that extends out to the protoplanet's tidal radius. Mizuno determined that there is a maximum core mass, called the "critical" core mass, $M_{crit}$, for which a static solution for the envelope with a core mass greater than $M_{crit}$ was not possible. This value was determined to be

$M_{\text{crit}} \approx 10 \, M_\oplus$. They also found that $M_{\text{crit}}$ was insensitive to the distance from the Sun. The success of the study by Mizuno and his collaborators at Kyoto University marked a clear advantage of the accretion model over the gaseous-condensation model.

Bodenheimer and Pollack (1986) computed the first evolutionary calculation of gas-giant planets based on the core-accretion model. These models are based on an adapted stellar evolution code with constant accretion rates. They found that the critical core mass was most sensitive to the rate of planetesimal accretion, namely, that $M_{\text{crit}}$ decreased as the planetesimal accretion rate was reduced. They corroborated Mizuno's results that $M_{\text{crit}}$ was not dependent on Solar-Nebula boundary conditions but they found that $M_{\text{crit}}$ was less sensitive to micron-sized grains in the envelope than was determined in Mizuno's calculation.

Wuchterl (1991a, b) analyzed the core-accretion model with a radiation-hydrodynamics code rather than the quasi-static one used by previous investigators. He found that once the envelope mass became comparable to the core mass, a dynamical instability develops that results in the ejection of much of the envelope. Discussion of these results is continued in Section 10.4.

The CAGC model has now become a sophisticated model with computer simulations that explain many features of the gas-giant planets. Interior model calculations of the gas-giant planets (e.g. Hubbard *et al.*, 1999; Saumon and Guillot, 2004) are an important component to the general investigation of gas-giant planet formation. Based on actual observations of the giant planets in the Solar System (e.g. gravitational moments) substantial information about the presence of a solid core and the size and composition of it can be extracted when structural model parameters are matched with observed values (Marley *et al.*, 1999). The discovery of the first planetary companion to 51 Peg (Marcy and Butler, 1995; Mayor and Queloz, 1995) introduced a challenge to the CAGC model. Since then the original sample of the four gas giants in our Solar System has increased substantially and the characteristics of these extrasolar planets exhibit a diversity (Bodenheimer and Lin, 2002) that planet scientists are presently in the process of categorizing and modeling.

## 10.3 Observational requirements for planet-forming models

Any theoretical model that explains gas-giant planet formation should explain these basic characteristics:

(1) The observed bulk composition characteristics of Jupiter, Saturn, Uranus, and Neptune are: (a) the similarity of the total heavy element contents of the four giants, (b) the very massive $H_2$ and He envelopes of Jupiter and Saturn and much less massive gaseous envelopes of Uranus and Neptune, and (c) the enhancement of metals over Solar abundance in the atmospheres of all four giant planets (e.g. Pollack and Bodenheimer, 1989;

Owen *et al.*, 1999). Interior models of Jupiter (Saumon and Guillot, 2004) indicate that the total solid mass ranges from 8–39 $M_\oplus$, of which 0–11 $M_\oplus$ is concentrated in the core. Saturn models indicate a total heavy element mass of 13–28 $M_\oplus$, with a core mass of 9–22 $M_\oplus$. Uranus and Neptune models indicate heavy element masses ranging from 10–15 $M_\oplus$ and a gaseous mass of 2–4 $M_\oplus$ (Pollack and Bodenheimer, 1989).

(2) Giant planets need to form quickly. Observed dust disks around young stellar objects indicate disk ages of <10 Myr (Cassen and Woolum, 1999; Haisch *et al.*, 2001; Lada, 2003; Chen and Kamp, 2004; Metchev *et al.*, 2004); initial Spitzer results (Bouwman *et al.* in Chapter 2 in this book) are consistent with this conclusion.

(3) The extrasolar planets exhibit a wide range of eccentricities and semimajor axes. In a few cases there are long-period, low-eccentricity planets whose orbits are comparable to that of Jupiter.

Historically, both formation models (CAGC and GGPP) have addressed these observations and each has had some advantage over the other. The bulk composition and the metal enhancement in the envelopes of gas-giant planets in the Solar System are a direct consequence of the CAGC process: early accretion creates a solid core that can acquire a gaseous envelope which is eventually massive enough that the incoming planetesimals interacting with the gas either break up or ablate in the envelope, thereby depositing solids. Though it was suggested by Boss (1998b, 2000) that ice and rock cores *should* be able to form inside Jupiter after the occurrence of gravitational instability, there are no completed computations demonstrating this.

The disk instabilities are a dynamical effect, and if the planets formed by the process of GGPP formation, they would do so very rapidly on timescales of at most a few tens of orbits. The giant gaseous protoplanets computed by Boss (1998b) are formed in $\sim 10^3$ yr, much less than the 1 to 50 Myr range reported in Paper 1. This short formation timescale has been a significant advantage of the GGPP model over CAGC until recently. Hubickyj *et al.* (2005) (hereafter referred to as HBL05) have demonstrated that gas giants can form on timescales well within the observational limit and in much shorter times than the standard case from Paper 1.

Recent observational studies of extrasolar planets have shown that planets are discovered much more frequently around metal-rich stars (Santos *et al.*, 2000; Fischer and Valenti, 2003). This observation is explained easily within the context of the CAGC model as demonstrated by Kornet *et al.* (2005). Their computation of protoplanetary disks with different metallicities indicates that planets form easily in more metal-rich disks. This result is underscored by the theoretical studies in Paper 1 which showed that the formation time of Jupiter decreases with increasing surface density of planetesimals in the Solar Nebula. The sample of extrasolar planets has significantly increased since the discovery of the companion to 51 Peg. An extensive discussion of the observational properties of extrasolar planets is given by Bodenheimer and Lin (2002) including a review of the trends in masses, orbital periods around their central star, eccentricity distributions, and multiple-planet systems. Work continues to determine how the CAGC model can explain these attributes.

## 10.4 The CAGC computer model

As of this Ringberg Workshop active modeling of CAGC formation of gas-giant planets is being rendered by four groups: the collaborators at NASA–Ames Research Center and University of California at Santa Cruz (referred to as ARC/UCSC group), the "Kyoto" group, the group in Bern, and G. Wuchterl. The first three groups use a similar technique based on a modified stellar structure evolution code, and Wuchterl uses a fully hydrodynamical computer code to model the evolving protoplanet. The discussion of the computer code technique will concentrate on the one used by the ARC/UCSC group, and variations on this work by others will be noted and described.

The core accretion–gas capture model is currently the best candidate model for the formation scenario of gas-giant planets. Simulations based on the CAGC model have been successful in explaining many features of the giant planets in the Solar System (Paper 1), and have shown that *in situ* formation of extrasolar planets is possible (BHL00). Motivated by the interior models of Jupiter and Saturn by Guillot *et al.* (1997) and Saumon and Guillot (2004) who call for low solid-mass cores for Jupiter and Saturn, the ARC/UCSC group studied the effects on Jupiter's core mass and evolution timescale of varying the parameters shown in Paper 1 to affect planet formation (Ikoma *et al.*, 2000; HBL05). Migration is more than likely a viable aspect of gas-giant planet formation in explaining the wide range of eccentricities and semimajor axes deduced from the observations of extrasolar planets. Alibert *et al.* (2005) incorporated migration into their core-accretion computer simulation and examined the effects on giant-planet formation. The method and results of these last two papers will be described in this and the next section. A brief summary of the hydrodynamic method (Wuchterl *et al.*, 2000) will also be presented.

The ARC/UCSC code consists of three main components:

(1) The calculation of the rate of solid accretion onto the protoplanet with an updated version of the classical theory of planetary growth (Safronov, 1969) to calculate the rate of growth of the solid core. The gravitational enhancement factor, which is the ratio of the total effective accretion cross-section to the geometric cross-section, is an analytical expression that was derived to fit the data from the numerical calculations of Greenzweig and Lissauer (1992) consisting of a large number of three-body (Sun, protoplanet, and planetesimal) orbital interaction simulations.

(2) The calculation of the interaction of the accreted planetesimals with the gas in the envelope (Podolak *et al.*, 1988) which determines whether the planetesimals reach the

core, or are dissolved in the envelope, or a combination of the two. Calculations of trajectories of planetesimals through the envelope result in the radius in the envelope at which the planetesimal is captured (required to compute the accretion rate of the planetesimals), and the energy deposition profile in the envelope (required for the structure computation).

(3) The calculation of the gas accretion rate and evolution of the protoplanet under the assumption that the planet is spherical and that the standard equations of stellar structure apply. The conventional stellar structure equations of conservation of mass and energy, hydrostatic equilibrium, and radiative or convective energy transport are used. The energy generation rate is the result of the accretion of planetesimals and the quasi-static contraction of the envelope.

The following assumptions were applied:

(1) The growing protoplanet is a lone embryo that is surrounded by a disk with an initially uniform surface density in the region of the protoplanet consisting of planetesimals with the same mass and radius.

(2) The protoplanet's feeding zone is assumed to be an annulus extending to a radial distance of about 4 Hill-sphere radii on either side of its orbit (Kary and Lissauer, 1994) which grows as the planet gains mass. Planetesimals are spread uniformly over the zone and do not migrate into or out of the feeding zone.

(3) The equation of state is non-ideal and the tables used are based on the calculations of Saumon *et al.* (1995), interpolated to a near-protosolar composition of $X = 0.74$, $Y = 0.243$, $Z = 0.017$. The opacity tables are derived from the calculations of Pollack *et al.* (1985) and Alexander and Ferguson (1994).

(4) The capture criterion includes planetesimals that deposit 50% or more of their mass into the envelope during their trajectory.

(5) Once the mass and energy profiles in the envelope have been determined, the planetesimals are assumed to sink to the core, liberating additional energy in the process.

(6) The rate of planetesimal accretion near gas runaway is limited to its value at crossover in order to account for the depletion of the planetesimal disk by accretion onto neighboring embryos.

The inner and outer boundary conditions are set at the bottom and top of the envelope, respectively. The core is assumed to have a uniform density and to be composed of a combination of ice, CHON, and rock, depending on the conditions in the nebula in which the planet forms. The outer boundary condition of the protoplanet is applied in three ways, depending on the evolutionary stage of the planet (as noted in Section 10.1 and described in full in BHL00).

The calculations start at $t = 10^4$ yr with a core mass of 0.1 $M_\oplus$ and an envelope mass of $10^{-9}$ $M_\oplus$. These initial values were chosen for computational convenience and the final results are insensitive to initial conditions. Near the end of the evolution of the protoplanet, the gas accretion rate is limited by the ability of the Solar

Nebula to supply gas at the required rate. When this limiting rate is reached, the planet contracts inside its accretion radius (evolution enters the transition stage). The supply of gas to the planet is eventually assumed to be exhausted as a result of tidal truncation of the nebula, removal of the gas by effects of the star, and/or the accretion of all nearby gas by the planet. The planet's mass levels off to the limiting value defined by the object that is being modeled. Since this process is not yet modeled in the ARC/UCSC code, the gas-accretion rate onto the planet is assumed to reduce smoothly to zero as the limiting mass value is approached. The planet then evolves through the *isolated* stage, during which it remains at constant mass.

Fig. 10.1 illustrates the typical nature of the simulations. The mass, luminosity, solid- and gas-accretion rates, and radii are plotted as functions of time. The solid core is accreted during Phase 1. This phase ends when the feeding zone is depleted of solids, thus the protoplanet reaches its isolation mass. Phase 2 is characterized by a steady rate of both solid and gas accretion but with the gas rate being slightly greater. The duration of Phase 2 ends at the *crossover* point, when the gas mass is equal to the solid mass. It is evident that Phase 2 determines the overall timescale for the protoplanet to form, since this time is much longer than the times for the solid core to accrete and for the gas runaway to occur. Phase 3 is characterized by the gas runaway which lasts until both the gas- and solid-accretion rates turn off. The protoplanet then cools and contracts to its presently observed state.

The Bern group's CAGC code (Alibert *et al.*, 2005) is similar to the ARC/UCSC code *except* for the inclusion of the components that compute the evolution of the protoplanetary disk and the migration of the growing protoplanet. The disk structure and its evolution are based on the method of Papaloizou and Terquem (1999) which is in the framework of the $\alpha$–disk formulation of Shakura and Sunyaev (1973). Migration occurs when there is a dynamical tidal interaction of the growing protoplanet with the disk which leads to two phenomena: inward migration and gap formation (Lin and Papaloizou, 1979; Ward, 1997; Tanaka *et al.*, 2002). For low-mass planets, the tidal interaction is a linear function of mass and the migration is type I (i.e. inward migration with no gap opening). Higher-mass planets open a gap, leading to a reduction of the inward migration, and this is referred to as type II migration. The rest of the procedure used by the Bern group is similar to that used by the ARC/UCSC group: the solid-accretion rate uses the gravitational enhancement factor based on that of Greenzweig and Lissauer (1992); the interaction between the infalling planetesimal and the atmosphere of the growing planet is based on the work of Podolak *et al.* (1988); and, the standard planetary structure and evolution scheme is applied. The rate for type I migration is a free parameter in the Alibert *et al.* (2005) calculations, therefore, the starting location of the embryo is adjusted for each choice of the migration rate, in order for the

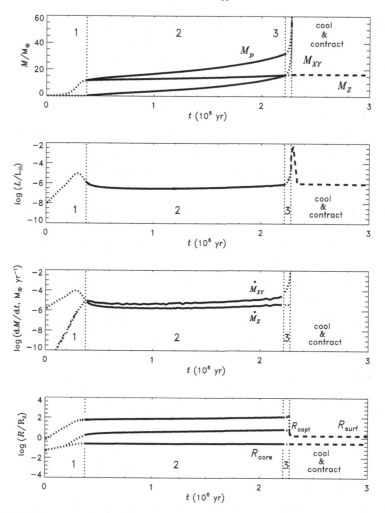

Fig. 10.1. The mass (units of $M_\oplus$), the luminosity (units of $L_\odot$), the accretion rates (units of $M_\oplus$ yr$^{-1}$), and the radii (units of $R_J$) are plotted as a function of time (units of Myr) for the baseline case 10L$\infty$. The three phases of evolution and the cooling and contraction of the protoplanet are marked. *Dotted line:* phase 1, *solid line:* phase 2, *second dotted line:* phase 3, *dashed line:* contraction and cooling.

protoplanet to reach the crossover mass at 5.5 AU. The profile of the initial disk surface density, $\Sigma$, is the same for each of the migration rates, namely a power law $\Sigma \propto r^{-2}$, which was chosen to correspond to case J2 (i.e. $\sigma_{init,Z} = 7.5$ g cm$^{-2}$) of Paper 1.

An analysis of the gas flow in the envelope was undertaken by Wuchterl (1991a, 1991b, 1993) who developed a hydrodynamic code to study the flow velocity of the gas by solving an equation of motion for the envelope gas in the framework of convective radiation–fluid dynamics. This allows the study of the collapse of the envelope, of the accretion with finite Mach number, and of linear, adiabatic and non-linear, non-adiabatic pulsational stability of the envelope. Furthermore, the treatment of convective energy transfer has been improved by calculations using a time-dependent mixing length theory of convection (Wuchterl, 1995, 1999) in hydrodynamics. Wuchterl starts his calculations at critical mass, $M_{crit}$, and finds that instead of collapsing as Mizuno surmised from his models, these envelopes begin to pulsate, resulting in mass loss. A new quasi-equilibrium state is achieved after the protoplanet loses a large fraction of its envelope mass. A major result of the hydrodynamical studies is that the protoplanet may pulsate and develop pulsation-driven mass loss. This leaves a planet with a low-mass envelope and properties similar to Uranus and Neptune but the model does not account for Jupiter and Saturn. His results are contrary to the linear stability analysis of Tajima and Nakagawa (1997), who find the envelope models in Bodenheimer and Pollack (1986) to be dynamically stable. This issue with the quasi-static models is still unresolved.

The Kyoto group (called as such more for historical than location reasons) continues to work on all aspects of planet formation. Kokubo and Ida (1998, 2002) studied the formation of planets and planet cores by oligarchic growth. Ikoma *et al.* (2000, 2001) examined the effect of opacity of the grains in the growing envelope of the protoplanet and of the rate of solid accretion on formation timescales. Ida and Lin (2004) investigated analytically the migration of planets.

## 10.5 Recent results

There are two major studies based on the CAGC model. The ARC/UCSC group addressed the issue examining the conditions for which Jupiter could have formed with a low-mass core and a short formation timescale (HBL05). The Bern group considered the effects on giant-planet formation with the inclusion of migration and protoplanetary evolution in the core accretion formation model (Alibert *et al.*, 2005). Highlights of these two studies are presented below.

Simulations by the ARC/UCSC group of the growth of Jupiter were computed for three parameters shown in Paper 1 to affect planet formation. The opacity produced by grains in the protoplanet's atmosphere was varied and two different values for the initial planetesimal surface density (10 g cm$^{-2}$ and 6 g cm$^{-2}$) in the Solar Nebula were used. Additionally, halting the solid accretion at selected core-mass values during the protoplanet's growth was studied. Decreasing the atmospheric opacity due to grains emulates the settling and coagulation of grains within the protoplanetary atmosphere, and halting the solid accretion simulates the presence of a competing embryo. The effects of these parameters were examined in order to determine whether gas runaway can still occur for small-mass cores on a reasonable timescale.

Four series of simulations were computed in this latest study. Each series consisted of a run computed through the cooling and contracting of the protoplanet (i.e., Fig. 10.1), plus up to three runs with a cutoff of planetesimal accretion at a particular core mass. The mass as a function of time for these four basic cases is plotted in Fig. 10.2a.

The nomenclature for the simulations that denotes the parameters used in the computations is in the following form: $\sigma$-*opacity*-*cut*, where $\sigma$ is the initial surface density of planetesimals in the Solar Nebula with values 10 or 6 g cm$^{-2}$; *opacity* is denoted by either "L" for grain opacity at 2% of the interstellar value, "H" for the full interstellar value, or "V" for a variable (temperature dependent: $T < 350$ K ramping up to the full interstellar value for $T > 500$ K) grain opacity; and *cut* specifies the core mass (in units of $M_{\oplus}$) at which the planetesimal accretion rate is turned off. For cases with no solid-accretion cutoff, *cut* is set to $\infty$. As an example, the model labelled 10L$\infty$ signifies that the simulation was computed with $\sigma_{init,Z} = 10$ g cm$^{-2}$, the grain opacity is 2% of the interstellar value and there was no solid-accretion cutoff.

The results of this study demonstrate that the reduced grain opacities produce formation times that are less than half of that for models computed with full interstellar grain opacity values (see curves labelled 10H$\infty$ and 10L$\infty$ in Fig. 10.2). The reduction of opacity due to grains in the upper portion of the envelope with $T \leq 500$ K has the largest effect on the lowering of the formation time (see curves labelled 10H$\infty$, 10L$\infty$, and 10V$\infty$ in Fig. 10.2). Decreasing the surface density of planetesimals lowers the final core mass of the protoplanet, but increases the formation timescale considerably (see curves labelled 10L$\infty$ and 6L$\infty$ in Fig. 10.2a).

The effect of halting solid accretion is illustrated in Fig. 10.2b. The plot is of the mass as a function of time for the baseline case 10L$\infty$ and the three associated runs for which the solid planetesimal accretion is turned off at core masses 10, 5, and 3 $M_{\oplus}$. It is clearly demonstrated that the time needed for a protoplanet to evolve to the stage of runaway gas accretion is reduced, provided the cutoff mass is sufficiently large. The overall results indicate that, with reasonable parameters and with the assumptions in the ARC/UCSC CAGC model code, it is possible that Jupiter formed via the core accretion process in 3 Myr or less.

As mentioned in Section 10.4, migration is considered by planet scientists to be an integral factor in gas-giant formation, especially in the case of the hot Jupiter-type extrasolar planets (Jupiter-sized planets found in orbits very close to their central star). Recent work by Alibert *et al.* (2005) examines the effect of migration on the growth of gas giants.

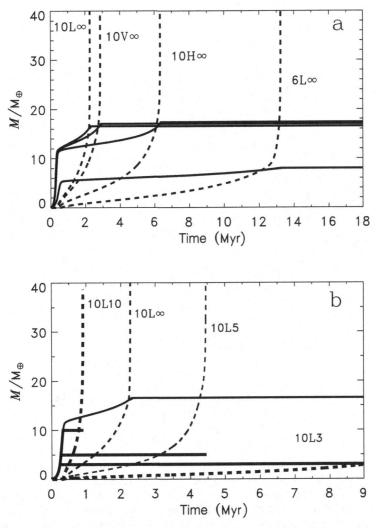

Fig. 10.2. a) The masses (units of $M_\oplus$) of the four basic cases are plotted as a function of time (units of Myr). The solid line denotes the mass of solids and the dashed line denotes the mass of gas. b) The masses of the baseline case 10L∞ and the associated cutoff cases are plotted as a function of time. Units and line designations are the same as in Fig. 10.2a.

Fig. 10.3 is a plot of the mass as a function of time for three simulations computed by Alibert *et al.* (2005). The models are for growing embryos that start their migration at 15 AU (dotted lines), 8 AU (solid lines) and one *in situ* model (dashed lines). The top three lines denote the evolution of the mass of accreted planetesimals (curves beginning at 0.6 $M_\oplus$), and the bottom three lines denote the mass of accreted gas, until the crossover mass is reached. Their results show that the

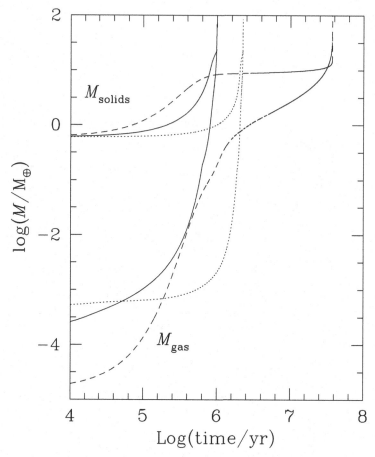

Fig. 10.3. The log plot of the masses (units of $M_\oplus$) as a function of time (units of yr) for three cases of Alibert *et al.* (2005). The top three curves denote the mass of the accreted planetesimals and the bottom three curves show the mass of the accreted gas. The dashed lines denote the *in situ* case, the dotted lines show the case when migration starts at 15 AU, and the solid lines denote the case when migration starts at 8 AU.

formation timescale of gas giants is much shorter, by a factor of ten, when migration is included compared to *in situ* formation. A migrating embryo starting at 8 AU will migrate to 5.5 AU and reach crossover mass in ∼ 1 Myr, whereas the same embryo at 5.5 AU without migration and without disk evolution (i.e. *in situ* formation) reaches crossover mass in ∼ 30 Myr, a factor of 10 greater. (Note: the *in situ* model computed by Alibert *et al.* (2005) should be compared to case J2 in Paper 1, not to newer models in HBL05.) The reason for this speed-up due to migration is quite simple. In CAGC formation models, the long formation timescale depends on the presence of phase 2 occurring after the core is isolated at the end of phase 1.

A migrating embryo will never suffer isolation and will go directly from phase 1 to phase 3, reducing the formation time.

## 10.6 Summary

Since Safronov's introduction of his planetesimal-accretion model for terrestrial planets and Mizuno's extension to the formation of Jupiter and Saturn, theoreticians have made substantial progress in understanding planet formation in the last few decades. As of the Ringberg Workshop 2004 the CAGC simulations have provided conclusions that can be summarized as follows:

(1) The opacity due to grains in the protoplanetary envelope has a major effect on the formation timescale but no effect on the core mass. It is *not* possible for a gas giant to form with a small solid core on a short timescale for models computed with the grain opacity equal to that for typical interstellar material. For models computed with the grain opacity below interstellar values, formation times are short but the final core mass is unaffected. The baseline case computed in HBL05 ($10L\infty$) shows that Jupiter can be formed at 5 AU in just over 2 Myr, but the core mass is 16 $M_\oplus$.

(2) Halting the planetesimal accretion provides for formation times to be in the range of 1–4.5 Myr and for a core mass consistent with that of Jupiter, if the initial solid surface density in the disk is three times that of the minimum-mass Solar Nebula.

(3) By reducing the initial solid surface density in the disk to two times that of the minimum-mass Solar Nebula, it is still possible to form Jupiter in less than 5 Myr if the core accretion is cut off at 5 $M_\oplus$.

(4) All models satisfy the constraint that the total heavy element abundance is less than or comparable to the value deduced from observations of Jupiter. A few models have low heavy-element abundance (3–5 $M_\oplus$), but it is quite reasonable to expect the planet to accrete more solids during or after rapid gas accretion, which is not taken into account in these models.

(5) The results of Paper 1 show that phase 2, the early gas accretion phase before crossover mass is reached, essentially determines the timescale for the formation of a giant planet. However, the results presented in HBL05 indicate that for low atmospheric opacity and/or a cutoff in accretion of solids, phase 2 can be relatively short. Thus, the time for phase 1, the solid core-accretion phase, may be the determining factor.

(6) Migration, which prevents the depletion of the feeding zone that occurs in *in situ* formation, appears to have a very important effect on the formation timescale by decreasing it by about a factor of ten (Alibert *et al.*, 2005) without having to consider massive disks (Lissauer, 1987).

Although some old problems relating to planet formation have been resolved there are others that still need investigating. Recent simulations of the core-accretion process, taking into account multiple embryos and a number of other physical processes (Inaba *et al.*, 2003; Thommes *et al.*, 2003; Kokubo and Ida, 2002), indicate that the core formation times are longer than those computed by the ARC/UCSC studies. Thus, the question remains as to how large an enhancement of solid surface density, as compared with that in the minimum-mass Solar Nebula, is needed to form a giant planet in a few Myr. Another problem is related to migration of the gas-giant planet. It seems difficult to form a planet and to prevent it from spiraling into its central star, according to type I migration calculations by Tanaka *et al.* (2002). Further problems that will be investigated by detailed models in the future include that of the formation of Uranus and Neptune, whose *in situ* core formation times seem to be too long for any reasonable disk model.

## Acknowledgments

The author would like to thank the organizing committee for the privilege of attending this Workshop and being able to present these results. The author would also like to thank P. Bodenheimer and J. Lissauer for their collaboration on this CAGC model project. This work is supported by NASA grants NAG5–9661 and NAG 5–13285 from the Origins of the Solar Systems Program.

# 11

# Properties of exoplanets: a Doppler study of 1330 stars

Geoffrey Marcy

*University of California, Berkeley, CA, USA*

Debra A. Fischer

*San Francisco State University, San Francisco, CA, USA*

R. Paul Butler

*Carnegie Institution of Washington, DTM, Washington DC, USA*

Steven S. Vogt

*UCO/Lick Observatory, University of California, Santa Cruz, CA, USA*

## 11.1 Overview of exoplanet properties and theory

In the past ten years, 170 exoplanets have been discovered orbiting 130 normal stars by using the Doppler technique to monitor the gravitational wobble induced by a planet, as summarized by Marcy *et al.* (2004) and Mayor *et al.* (2004). Multiple-planet systems have been detected around 17 of the 130 planet-bearing stars, found by superimposed multiple Doppler periodicities (Mayor *et al.*, 2004; Vogt *et al.*, 2005). Another five exoplanets have been found photometrically by the dimming of the star as the planet transits across the visible hemisphere of the star (Bouchy *et al.*, 2005; Torres *et al.*, 2005).

The Doppler surveys for planets have revealed several properties:

- Planet mass distribution: $dN/dM \propto M^{-1.5}$ (Fig. 11.1).
- Planet occurrence increases with semimajor axis (Fig. 11.2).
- Hot Jupiters ($a < 0.1$ AU) exist around 0.8% of FGK stars.
- Eccentric orbits are common (Fig. 11.3).
- Planet occurrence rises rapidly with stellar metallicity (Fig. 11.4).
- Multiple planets are common, often in resonant orbits.

The transiting planets permit measurement of planet radius and mass, demonstrating that they are "gas giants", as expected (Henry *et al.*, 2000; Charbonneau *et al.*, 2000). The properties of the 170 known exoplanets motivate theories of their formation and their dynamical interactions with both the protoplanetary disk and

*Planet Formation: Theory, Observation, and Experiments*, ed. Hubert Klahr and Wolfgang Brandner.
Published by Cambridge University Press. © Cambridge University Press 2006.

other planets (Bryden *et al.*, 2000; Laughlin and Chambers, 2001; Rivera and Lissauer, 2001; Chiang and Murray, 2002; Lee and Peale, 2002; Ford *et al.*, 2003; Ida and Lin, 2004). Interactions between planets and disks lead to inward migration of the planets, shaping the distribution of final orbital sizes and eccentricities (Trilling *et al.*, 2002; Armitage *et al.*, 2003; D'Angelo *et al.*, 2003a; Thommes and Lissauer, 2003; Ida and Lin, 2004; Alibert *et al.*, 2005). The migration can lead to interactions between pairs of planets, resulting in their capture into orbital resonances and in pumping of orbital eccentricities (Marzari and Weidenschilling, 2002; Nelson and Papaloizou, 2002; Goździewski, 2003; Kley *et al.*, 2005).

The theory of the growth of giant planets that is most widely accepted involves dust growth in a protoplanetary disk leading to rock–ice cores which subsequently accrete gas in their vicinity of the disk (Lissauer, 1995; Levison *et al.*, 1998; Bodenheimer *et al.*, 2003). Young gas giants will accrete most of the gas near them, and they will be surrounded by heated, extended envelopes. Both of these effects slow gas accretion, leading to predicted growth times of 5–10 Myr, uncomfortably longer than the observed ∼3 Myr lifetime of the disks themselves. Therefore the original, classic core-accretion model suffers from planet-growth times longer than the lifetimes of the disk material.

However, inward migration may bring giant planets to fresh, gas-rich regions (Alibert *et al.*, 2005; Armitage *et al.*, 2003), and new, lower opacities would allow the heat in the planets' envelopes to be radiated efficiently, shrinking them more quickly to enable accretion. The resulting planet-growth time is shortened to ∼1 Myr, well within the lifetime of protoplanetary disks (Chiang and Murray, 2002; D'Angelo *et al.*, 2003a; Trilling *et al.*, 2002; Armitage *et al.*, 2003; Thommes and Lissauer, 2003; Hubickyj *et al.*, 2004; Ida and Lin, 2004; Alibert *et al.*, 2005). A mystery remains regarding the origin of the orbital eccentricities which are presumably damped by the disk gas that is required for planet growth. If so, *the orbital eccentricities must arise after the major stage of gas accretion.* Gravitational interactions among planets, and between planets and protoplanetary disks may cause the observed orbital eccentricities (Bryden *et al.*, 2000; Laughlin and Chambers, 2001; Rivera and Lissauer, 2001; Chiang and Murray, 2002; Lee and Peale, 2002; Ford *et al.*, 2003; Ida and Lin, 2004; Ford, 2005).

## 11.2 The Lick, Keck, and AAT planet searches

The determination of the statistical properties of giant planets depends on a survey of planets that has well-understood detection thresholds in both mass and orbital period. We have carried out precise radial velocity measurements of 1330 FGKM dwarfs at the Lick, Keck 1, and Anglo-Australian telescopes. The majority of stars had their first high quality measurement between 1995 and 1998, giving a time coverage of ~7–10 years thus far. The target stars and their properties are listed in Wright *et al.* (2004) and Valenti and Fischer (2005), and were drawn from the Hipparcos catalog (ESA, 1997) with criteria that they have $B - V > 0.55$, reside no more than 3 mag above the main sequence (to avoid photospheric jitter seen in giants), and have no stellar companion within 2 arcsec (to avoid confusion at the entrance slit).

The target list also includes 120 M dwarfs, located mostly within 10 pc with declination north of $-30°$, listed in Wright *et al.* (2004). For the late-type K and M dwarfs, we restricted our selection to stars brighter than $V = 11$. All slowly rotating stars are surveyed with a Doppler precision of 3 m s$^{-1}$ to provide a uniform sensitivity to planets. Thus far, our Lick, Keck, and AAT surveys have revealed 101 planets and the orbital elements and masses of these exoplanets are regularly updated at: `http://exoplanets.org`.

## 11.3 Observed properties of exoplanets

We derive the statistical properties of planets from the 1330 FGKM target stars for which we have uniform precision of 3 m s$^{-1}$ and at least 6 yrs duration of observations. Detected exoplanets have minimum masses, $M \sin i$, between 6 M$_\oplus$ and ~15 M$_J$, with an upper limit corresponding to the diminishing tail of the mass distribution. The planet mass distribution is shown in Fig. 11.1 and follows a power law, $dN/dM \propto M^{-1.5}$ (Marcy and Butler, 2000; Marcy *et al.*, 2003), affected very little by the unknown $\sin i$ (Jorissen *et al.*, 2001). The paucity of companions with $M \sin i$ greater than 12 M$_J$ confirms the presence of a "brown dwarf desert" (Marcy and Butler, 2000) for companions with orbital periods up to a decade. The planets of lowest $M \sin i$ in our sample are Gl 876d, 55 Cnc d and Gliese 436 (Butler *et al.*, 2004; McArthur *et al.*, 2004) with $M \sin i = 6$, 15 and 21 M$_\oplus$, respectively. Lovis *et al.* (Chapter 13) describe the properties of hot Jupiters and Neptunes.

The observed semimajor axes span the range 0.03–6.0 AU. Close-in, "hot" Jupiters orbiting within 0.1 AU, exist around 0.8±0.2% of FGK main sequence stars (Marcy *et al.*, 2004; Jones *et al.*, 2004) on our survey. We find that 6.5% of all nearby main sequence stars harbor Saturn-mass and Jupiter-mass planets within 3 AU. The number of planets increases with distance from the star from 0.3 to 3 AU, shown in Fig. 11.2 (in logarithmic bins). A mild (flat) extrapolation suggests that a comparable population of yet-undetected Jupiters exists between 5–20 AU, bringing the occurrence of giant planets to roughly 12% within 20 AU (Fig. 11.2).

The eccentricities are plotted versus semimajor axis in Fig. 11.3 for the 101 detected exoplanets. Eccentricities span the range 0.0–0.8, and those orbiting within 0.1 AU are all in nearly circular orbits, presumably due to tidal circularization. For

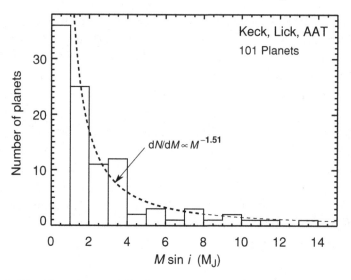

Fig. 11.1. The histogram of 101 planet masses ($M \sin i$) found in the uniform 3 m s$^{-1}$ Doppler survey of 1330 stars at the Lick, Keck, and AAT telescopes. The distribution of planet masses rises steeply with decreasing planet mass as $M^{-1.5}$ down to Saturn masses.

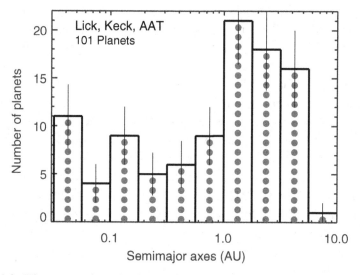

Fig. 11.2. Histogram of semimajor axes, $a$, found in the 101 exoplanets found from the Doppler survey at the Lick, Keck, and AAT telescopes. Note the equal logarithmic bins, $\Delta \log a$. There are increasing numbers of planets toward larger orbits beyond 0.5 AU. The occurrence of planets within 3 AU is 6.5%. Flat extrapolation from 3–20 AU suggests that ~12% of all nearby FGK stars have a giant planet with mass greater than Saturn.

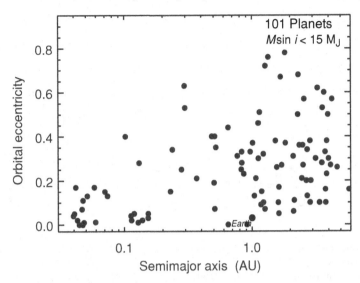

Fig. 11.3. Eccentricity versus semimajor axis, for the 101 planets discovered in the Lick, Keck, and AAT Doppler survey. Eccentricity ranges from 0 to 0.8 and no decline in eccentricity is observed beyond 3 AU. It remains likely that Jupiter-mass planets at 5.2 AU will also reside in eccentric orbits, unlike Jupiter in our Solar System.

the planets orbiting farthest from the host star, $a = 2$–$4$ AU, there is no tendency for them to have small eccentricities, as observed for the giant planets in our Solar System. In the upcoming years, a population of exoplanets at $a \approx 5.2$ AU will allow direct comparison of cosmic eccentricities with that of Jupiter, $e = 0.048$.

### 11.3.1 The planet-metallicity relationship

Planet occurrence correlates strongly with the abundance of heavy elements in the host star, as shown in Fig. 11.4. In our survey of FGK stars, ~25% of the most metal-rich stars, [Fe/H] > +0.3, harbor planets while fewer than 3% of the metal-poor stars, [Fe/H] < −0.5, have detected planets (Gonzalez, 1997; Reid, 2002; Santos *et al.*, 2004b; Fischer and Valenti, 2005). The physical mechanism for the observed planet-metallicity correlation is often cast as "nature or nurture." In the former case, high metallicity enhances planet formation because of increased availability of small particle condensates, the building blocks of planetesimals (Kornet *et al.*, 2005). In the latter case, enhanced stellar metallicity is due to the late-stage accretion of gas-depleted, dust-rich material, causing "pollution" of the star's convective zone (CZ). These two mechanisms leave different and distinguishable marks on the host stars. In the former case, the star is metal-rich throughout its interior. In the latter case, additional metals are mixed from the photosphere

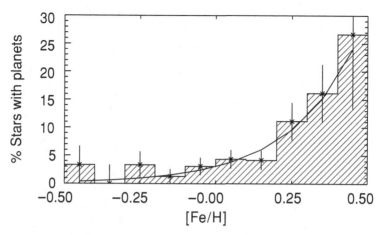

Fig. 11.4. The occurrence of exoplanets versus iron abundance [Fe/H] of the host star measured spectroscopically (Fischer and Valenti, 2005). The occurrence of observed giant planets increases strongly with stellar metallicity.

only throughout the convective zone, leaving the interior of the star with lower metallicity.

There is strong support for the former "nature" hypothesis. Of particular importance, the metallicities of stars are independent of both CZ depth and of evolution across the sub-giant branch (where dilution is expected due to a deepening CZ). While accretion of metals must occur for all pre-main-sequence stars, the stars with extrasolar planets appear to have enhanced metals extending below the CZ. Thus it is unlikely that the high metallicity of planet-bearing stars is caused by accretion. Furthermore, planet-bearing stars with super-Solar metallicity are more than twice as likely to have multiple planet systems than planet-bearing stars with sub-Solar metallicity. Taken together, these findings suggest that initial high metallicity enhances planet formation, providing support for the core accretion model of giant-planet formation.

## 11.4 The lowest-mass planets and multi-planet systems

Among all of the stars found to harbor planets by all Doppler efforts, 108 stars harbor a single planet, 14 stars have 2 known planets, 2 have 3 known planets (Upsilon And and HD 37124), and 1 has 4 detected planets (55 Cancri). Thus, multi-planet systems comprise 17 of 125 (14%) of known planet-bearing stars. This fraction is certainly a lower limit, as multiple systems demand more Doppler observations to extract the multiple signals. Additional planets continue to emerge in the set of stars with known planets as more observations are obtained and as Doppler precision improves.

Three Neptune-mass planets have been discovered to date having minimum masses between 14 and 21 $M_\oplus$, all with short periods of 2.5–10 days (Butler *et al.*, 2004; McArthur *et al.*, 2004; Santos *et al.*, 2004a). The Doppler method with state-of-the-art precision of 1 m s$^{-1}$ can reveal planets having masses as low as 10 $M_\oplus$ for periods less than 5 days. But astrophysical noise ("jitter") caused by stellar surface turbulence, spots, and stellar acoustic p-modes make the detection of planets below 10 $M_\oplus$ difficult, notably due to the unpredictable, stochastic interference of the acoustic p-modes in Solar-type stars (Lovis *et al.*, Chapter 13). The Doppler detection of Earth-mass planets orbiting a Solar-mass star at ∼1 AU will require a 6 m class dedicated telescope to detect the Doppler amplitude of only ∼0.1 m s$^{-1}$.

The Kepler and CoRoT missions are designed to photometrically detect Earth-mass planets during transits of the host star, providing the first measure of the occurrence of rocky planets and ice giants. However, the host stars will reside at typical distances beyond 250 pc, making imaging and spectroscopic follow-up of the planets difficult. A method is needed to detect Earth-mass planets around nearby stars, amenable to follow-up.

## 11.5 The Space Interferometry Mission

The Space Interferometry Mission, SIM, will have a 9 m baseline used at optical wavelengths to measure the positions of stars and point sources with a precision as high as 1 µas. With launch planned for 2011, SIM will carry out a variety of galactic and extragalactic projects that require high astrometric precision during its nominal 5 yr mission. One primary goal is to survey ∼200 stars at 1 µas, located within 20 pc and along the main sequence, for wobbles caused by planets of nearly Earth-mass that orbit within 2 AU.

The technical details of SIM are given by Shao (2003). Briefly, the SIM interferometer operates at wavelengths from 400–900 nm, obtaining fringes and measuring the optical path delays. SIM will be carried into an Earth-trailing Solar orbit via an expendable launch vehicle and will slowly drift away from the Earth at 0.1 AU yr$^{-1}$, reaching a maximum communication distance of 0.6 AU after 5.5 yr. During operation, the 9 m baseline vector between the two mirrors is fixed by a separate guide interferometer that monitors bright stars.

For the planet search, the position of each target star is measured relative to a minimum of three reference stars located within ∼1°. The triangle formed by the three reference stars should contain the target star to constrain the differential angular separations in both of two axes. The reference stars are ideally bright ($V < 10$), distant (∼1 kpc) giants so that the astrometric wobble due to planets around them is minimized. Candidate reference stars having brown dwarf or stellar companions within 10 AU are rejected by using repeated ground-based radial velocity measurements at a precision of 20 m s$^{-1}$ (Frink *et al.*, 2001), a reconnaissance effort that is currently ongoing. With the aid of wide-angle constraints on the baseline orientation, SIM will be able to generate a local reference frame over a 1° field of regard with 1 µas astrometric accuracy per one-dimensional orientation axis. At $V = 10$ mag, a ten-chop sequence between target and each reference star, with 30 s integrations per chop, will achieve the 1 µas precision, including anticipated systematic and photon-limited errors. Spots on the youngest, most active, stars will move their photocenters by $\sim 1$ µas, but stars older than 2 Gyr have spot-covering factors less than 0.2%, keeping this noise insignificant.

### 11.5.1 Finding Earth-mass planets with SIM

The SIM planet search is being carried out by three teams with principal investigators, Mike Shao, Chas Beichman, and Geoff Marcy. The Shao and Marcy teams have selected $\sim$200 nearby AFGKM-type target stars to search for Earth-mass planets. Most are located within 15 pc, and some were chosen for the large angular separation of their habitable zones, for later direct imaging. The preliminary target list is given at `http://www.physics.sfsu.edu/SIM/`

With astrometric precision of 1 µas, SIM is the only near-term mission that can detect planets with masses less than 10 M$_\oplus$ orbiting between 0.1 and 2 AU around nearby stars. The Doppler method is incapable of detecting such low-mass planets orbiting near 1 AU. SIM will determine the masses and orbits of such planets during its 5 yr mission, providing key reconnaissance for later imaging missions such as TPF and Darwin.

### 11.5.2 Low-mass detection threshold of SIM

Careful analyses of the detectability of planets with SIM have been carried out by Sozzetti *et al.* (2002, 2003) and by Ford and Tremaine (2003). Here we critically examine the low-mass detection threshold of SIM and compute the false-alarm probability (FAP).

We simulate a benchmark case of a planet in a circular orbit at 1 AU orbiting a Solar-mass star at 5 pc. We simulate 30 observations (in each of two dimensions) by SIM taken over 5 yr with errors of 1 µas. The simulations presented here were based on semi-random temporal spacing with a minimum separation in observations of 30 days since SIM is not sensitive to very short-period planets. We also included gaps in the 5 yr observing sequence corresponding to 4 month intervals when stars will not be observable because of the sun avoidance angle. In any case, Ford (2004) showed that the detection efficiency is not sensitive to the cadence of observations.

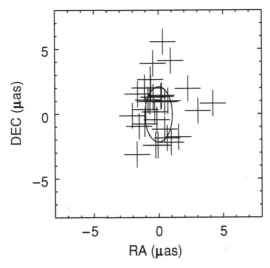

Fig. 11.5. Simulation of the astrometric signal caused by a planet of mass 3 $M_\oplus$ orbiting 1 AU from a Solar-mass star at 5 pc. The orbit inclination is 60°. The solid line shows the actual wobble. The crosses show simulated SIM measurements with errors of 1 μas over 5 yr (SIM lifetime). The scatter seems to overwhelm the signal, but the temporal coherence makes this planet detectable with high confidence.

While the current key projects have only been allocated sufficient time for 24 two-dimensional observations of 137 stars at 1 μas precision (relative to three reference stars), the use of adaptive scheduling algorithms such as those developed in Ford (2005) can allow observations to be preferentially allocated to the most promising target stars.

First, we consider the case of a 3 $M_\oplus$ planet. Figure 11.5 shows one such realization of the resulting SIM observations on the sky, for an assumed inclination of $i = 60°$. Figure 11.5 shows the actual wobble as the solid ellipse and the scattered astrometric measurements as crosses (assumed to be made in both baseline orientations contemporaneously). Detection of the 3 $M_\oplus$ planet seems dubious, as the scatter is large relative to the size of the orbit. However, the points carry a temporal coherence with orbital phase, making a detection possible. One approach is to fit the data with a Keplerian model to obtain $\chi^2$ and determine the associated FAP with Keplerian fits to mock velocity sets that contain noise but no planet, to determine the distribution of $\chi^2$.

Here, we simplify the estimation of FAP by computing a periodogram of the astrometric measurements along both axes (labelled "RA" and "DEC") and determining the FAP by Monte Carlo data sets that have no planet. Figure 11.6 shows (at the left) the simulated astrometric measurements (due to the 3 $M_\oplus$ planet and 1 μas noise) in both RA and DEC, the latter showing a clear periodicity to the eye

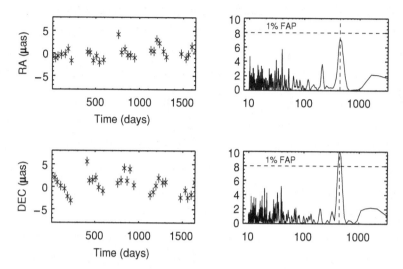

Fig. 11.6. Simulated SIM measurements in orthogonal directions (labeled RA and DEC) for the 3 $M_\oplus$ planet shown in Fig. 11.5. The periodicity in the DEC measurements (long axis) is apparent. The periodogram of position for the RA and DEC measurements is shown at right. For the DEC measurements, the signal has a FAP less than 1%, implying that planets of 3 $M_\oplus$ at 1 AU are detectable (star at 5 pc).

(the arbitrary inclination suppresses the signal of the former). At the right in Fig. 11.6, the periodogram of the astrometric measurements shows a clear peak residing above the 1% FAP threshold. *Thus, SIM can detect planets of 3 $M_\oplus$ orbiting at 1 AU around Solar-type stars at 5 pc.* These results agree with those of Sozzetti *et al.* (2002) who assumed somewhat fewer observations per star.

We now consider a 1.5 $M_\oplus$ planet orbiting, as before, at 1 AU around a Solar-mass star at 5 pc. Figure 11.7 shows one realization of the SIM observations on the sky (for $i = 60°$). The large scatter relative to the actual stellar wobble makes detection again seem dubious. However, Fig. 11.8 shows the simulated measurements along RA and DEC revealing a marginal periodicity to the eye. Indeed, at the right in Fig. 11.8, the periodogram of the astrometric measurements shows a marginal, but obvious, peak residing somewhat below the 1% FAP threshold. *Thus, SIM can marginally detect planets of 1.5 $M_\oplus$ orbiting at 1 AU around Solar-type stars at 5 pc.*

We have run 1000 realizations of the two cases, planets of 3 and 1.5 $M_\oplus$, orbiting at 1 AU around Solar-mass stars at 5 pc. The histogram of resulting FAP values is shown in Fig. 11.9. The histogram shows that planets of 3 $M_\oplus$ achieve an FAP of 0.05 or better in over 99% of the trials. Planets of 1.5 $M_\oplus$ achieve an FAP of 0.05 or better in over 63% of the trials. Thus SIM is capable of detecting planets of 1.5 $M_\oplus$ with 63% efficiency, but will incur false alarms in 5% of the stars. SIM

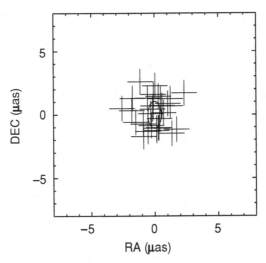

Fig. 11.7. Simulated astrometric signal caused by a planet of mass $1.5\,M_{\oplus}$ orbiting 1 AU from a Solar-mass star at 5 pc. The orbit inclination is 60°. The solid line shows the actual wobble. The crosses show simulated SIM measurements with errors of 1 μas over 5 yr (SIM lifetime). Despite large scatter, temporal coherence makes this planet marginally detectable.

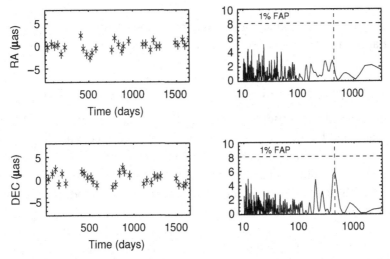

Fig. 11.8. Simulated SIM measurements in orthogonal directions (labeled RA and DEC) for the $1.5\,M_{\oplus}$ planet shown in Fig. 11.7. The periodicity in the DEC measurements (long axis) is barely apparent. The periodogram for the DEC measurements has a clear peak at P = 365 days, but with an FAP not quite 1%. Thus, planets of $1.5\,M_{\oplus}$ at 1 AU are detectable, but not securely (star at 5 pc).

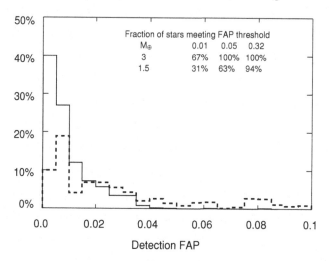

Fig. 11.9. Histogram of FAP, for two assumed planet masses, 1.5 (dashed) and 3.0 $M_\oplus$ (solid). Here, 30 observations with precision of 1 μas in each orthogonal direction were assumed. As in Figs. 11.5–11.8, a Solar-mass star is taken to be at 5 pc and the planet orbits at 1 AU. Integrating under the curve shows that planets of 3 $M_\oplus$ create a SIM signal that meets the 0.01 FAP threshold in 67% of trials. Planets of 1.5 $M_\oplus$ create a SIM signal that meets the 0.05 FAP threshold in 63% of trials. Thus, two-thirds of planets having masses as low as 1.5 $M_\oplus$ are detectable albeit with a false alarm rate of 0.05.

cannot securely detect planets of 1.5 $M_\oplus$, but it can identify stars highly likely to harbor Earths at 1 AU with only 5% false detections.

## 11.6 The synergy of SIM and Terrestrial Planet Finder (TPF)/Darwin

The simulations of SIM observations of Earth-mass planets show that 3 $M_\oplus$ planets are detectable and 1.5 $M_\oplus$ planets are marginally found at 5 pc. Thus, the SIM survey of 200 nearby stars will identify a subset that has planets of 3–10 $M_\oplus$ (should they be common) and another subset that is likely to have even lower mass planets, 1–3 $M_\oplus$, albeit with some false alarm interlopers. SIM can thus produce a list of nearby stars that is enriched in 1–3 $M_\oplus$ planets. Assuming, for example, that the fraction of stars with Earths in the habitable zone, $\eta_\oplus$, is 0.1, it is easy to show that SIM will produce an output list of stars that is enriched by a factor of three in habitable Earths over the original input sample of stars. Thus, SIM will provide TPF and Darwin with target stars having either strong or plausible evidence of rocky planets. SIM will also identify those stars that TPF and Darwin should avoid, notably those with a large planet near the habitable zone that renders any Earths dynamically unstable. Of course, any such Saturn or Neptune-mass planets within 2 AU will be valuable themselves for planetary astrophysics.

If $\eta_\oplus$ is indeed $\sim 10\%$, TPF/Darwin will be hard pressed to detect these few Earths because of their rarity and their faintness, $V \approx 30$ mag. Moreover, for modestly inclined orbital planes, TPF/Darwin will miss planets located angularly within the diffraction-limited "inner working angle" (IWA $= \sim 4\lambda/D = 0.065$ arcsec for TPF-C). A planet orbiting 1 AU from a star located 5 pc away will spend roughly one-third of its orbit inside the IWA, leaving it undetected. *Thus if the occurrence of earths in habitable zones is $\sim 10\%$, SIM will triple the efficiency of TPF and Darwin both by identifying the likely host stars and by predicting the orbital phase during which the Earth is farthest from the glare of the host star.*

The Space Interferometry Mission alone provides a wealth of planetary astrophysics, including the masses, orbital radii, and orbital eccentricities of rocky planets around the nearest stars. It will also find correlations between rocky planets and stellar properties such as metallicity and rotation. With a lifetime extended beyond 5 yr, SIM can detect planets of even lower mass, down to 1 $M_\oplus$.

Together SIM and TPF/Darwin, along with Kepler, provide a valuable combination of information about rocky planets: Kepler offers the occurrence rate of small planets; SIM provides the masses and orbits of planets around nearby stars, identifying the candidate earths; TPF/Darwin measure radii, chemical composition, and atmospheres. In some cases, images from TPF/Darwin may feed back on the analysis of old SIM data, helping orbit determination especially for multiple planetary systems.

## Acknowledgments

We thank the SIM Project team at JPL, especially Michael Shao, Chas Beichman, Steven Unwin, Shri Kulkarni, Chris Gelino, Joe Catanzerite, and Jo Pitesky. We thank C. Tinney, H. Jones, J. T. Wright, John Johnson, and Chris McCarthy their work on the Doppler planet search. We appreciate support by NASA grant NAG5-75005 and NSF grant AST-0307493.

# 12

# Giant-planet formation: theories meet observations

Alan Boss

*Carnegie Institution, Washington DC, USA*

## 12.1 Introduction

The two mechanisms that have been advanced for explaining the formation of giant planets are core accretion ("bottom up") and disk instability ("top down"). Core accretion, the conventional mechanism, relies on the collisional accumulation of planetesimals to assemble $\sim 10$ $M_\oplus$ solid cores, which then accrete massive gaseous envelopes from the disk gas (Mizuno, 1980; Lissauer, 1987; Pollack *et al.*, 1996; Kornet *et al.*, 2002; Inaba *et al.*, 2003). In this scenario, the ice-giant planets probably did not form *in situ* (Levison and Stewart, 2001), but rather formed between Jupiter and Saturn and then were scattered outward to their present orbits (Thommes *et al.*, 1999, 2002).

The alternative to core accretion is disk instability, where gas-giant protoplanets form rapidly through a gravitational instability of the gaseous portion of the disk (Cameron, 1978; Boss, 1997, 1998b, 2000, 2002a,b, 2003, 2004; Gammie, 2001; Mayer *et al.*, 2002, 2004; Nelson, 2000; Nelson *et al.*, 1998, 2000; Pickett *et al.*, 1998, 2000a,b, 2003; Rice and Armitage, 2003, Rice *et al.*, 2003a) and then more slowly contract to planetary densities. Solid cores form simultaneously with protoplanet formation by the coagulation and sedimentation of dust grains in the clumps of disk gas and dust. Ice-giant planet formation in this scenario (Boss *et al.*, 2002; Boss, 2003) requires the presence of a strong source of ultraviolet radiation (i.e. location in a region of high-mass star formation, such as Orion or Carina), sufficiently strong to photoevaporate the outer disk gas as well as the gaseous envelopes of the protoplanets outside a critical radius, roughly equal to Saturn's orbit for a Solar-mass parent star.

Trying to decide between these two different formation mechanisms purely on theoretical grounds is an entertaining but ultimately unconvincing exercise.

*Planet Formation: Theory, Observation, and Experiments*, ed. Hubert Klahr and Wolfgang Brandner.
Published by Cambridge University Press. © Cambridge University Press 2006.

However, the two mechanisms do yield different predictions for the frequency and structure of extrasolar planetary systems, and both are also intended to explain the formation of our own Solar System. The primary goal of this review is to begin to examine such testable predictions, in the light of current and future planet detection programs and knowledge of our own Solar System. While both processes may be able to produce giant planets, it could be that one process dominates over the other, and one can hope that future observations will reveal the identity of such a dominant mechanism.

## 12.2 Gas-giant planet census

We begin with a consideration of what sort of predictions these two mechanisms might have for ongoing and future extrasolar-planet searches. Ground-based radial velocity surveys (Mayor and Queloz, 1995; Marcy and Butler, 1998) have discovered nearly every extrasolar planet found to date, and we can expect their dominant role to continue for the next decade. With errors approaching a few m s$^{-1}$ (Butler *et al.*, 1996, 2001; Vogt *et al.*, 2000) or even 1 m s$^{-1}$ (Santos *et al.*, 2004a), high precision radial velocity surveys can detect the presence of sub-Saturn-mass planets orbiting at Earth's distance around Solar-mass stars. In fact, Neptune-mass objects orbiting around the distance (0.05 AU) of 51 Pegasi's planet (Mayor and Queloz, 1995) have now been discovered around three stars (Butler *et al.*, 2004; McArthur *et al.*, 2004; Santos *et al.*, 2004a). In addition, the longevity of these surveys means that planets with increasingly longer orbital periods are beginning to be discovered, including Jupiter analogs (e.g. Carter *et al.*, 2003). These surveys are thus on the verge of determining the frequency of Solar System-like planetary systems, where Jupiter-mass planets orbit on long-period, roughly circular orbits, in the absence of shorter-period, massive planets that would forestall the existence of Earth-mass planets orbiting close enough for liquid water to exist (i.e. in the habitable zone).

The relatively high frequency ($\sim$ 15%) of stars with giant planets with orbital periods less than a few years suggests that even longer-period planets may be quite frequent, with a frequency of perhaps $\sim$ 25% for Solar-System analogs with only long-period gas-giant planets (A. Hatzes, 2004, private communication). Hence perhaps $\sim$ 40% of nearby G-type stars appear to have gas-giant planets orbiting within about 10 AU. These results seem to require that the dominant gas-giant planet formation mechanism be relatively robust and efficient.

Core accretion seems to require a relatively long-lived disk in order to have sufficient time for the accretion of a massive gaseous envelope, a disk with a lifetime of a few Myr or more (Pollack *et al.*, 1996; Kornet *et al.*, 2002; Inaba *et al.*, 2003). Studies of the lifetimes of circumstellar disks in a variety of star-forming regions suggest that half the disks disappear within 3 Myr (Haisch *et al.*, 2001; Eisner and

Carpenter, 2003), or in even less time in regions of high-mass star formation (Bally *et al.*, 1998; Briceño *et al.*, 2001). Core accretion thus may have a problem with producing a high frequency of gas-giant planets, should that turn out to be the case. Disk instability, on the other hand, with its timescale of the order of 1000 yr for the formation of self-gravitating clumps, has no problem in forming gas-giant planets in even the shortest-lived protoplanetary disk.

In order to explain the formation of the ice-giant planets, however, disk instability requires the presence of massive (OB) stars to generate a high flux of extreme-ultraviolet (EUV) or far-ultraviolet (FUV) radiation. Given that most stars are believed to have formed in regions of high-mass star formation, such as Orion or Carina (Lada and Lada, 2003), disk instability would be consistent with a high frequency of gas-giant planets. Core accretion, however, seems to require a more quiescent setting with long-lived disks and a low EUV/FUV background, such as occurs in regions of low-mass star formation (Taurus, Auriga, Rho Ophiuchus). These regions are thought to contribute only a minor fraction of stars (Lada and Lada, 2003). Core accretion would then be more consistent with a low overall frequency of gas-giant planets.

## 12.3 Metallicity correlation

Disk instability appears to be relatively insensitive to the opacity of the disk, which is dominated by dust grains, and thus depends on the metallicity of the host star and its disk (Boss, 2002a). Hence even low-metallicity stars should host gas-giant planets if disk instability is operative. On the other hand, it is clear that core accretion is helped by higher metallicity, as raising the surface density of solids dramatically speeds the growth of cores (Pollack *et al.*, 1996), while the increased opacity in their envelopes, which makes it harder for the atmospheres to collapse onto the protoplanets, is a much weaker effect (Podolak, 2003).

Spectroscopic planet searches have concentrated on metal-rich target stars as a result of the expectation that such stars would be more likely to host planets. In fact, a strong correlation between metal-rich host stars and the presence of short-period (<3 yr) planets has been found (Laws *et al.*, 2003; Fischer *et al.*, 2004; Santos *et al.*, 2004b) and has been widely interpreted as a strong argument in favor of core accretion (e.g. Livio and Pringle, 2003). However, it is important to note that the fact that the lowest-metallicity bin (less than one-third of Solar metallicity) of the Fischer *et al.* (2004) analysis contains no stars with planets is largely a result of the small sample size of that bin (29 stars) compared to those of the two higher metallicity bins (347 and 378 stars, respectively). If the lowest-metallicity bin stars had planets at the same rate (5%) as the middle-metallicity bin (one-third of Solar to Solar metallicity), then 1.5 planets would have been found. The fact that none were found in the Fischer *et al.* (2004) sample could then be due to small number statistics. In fact, in the Santos *et al.* (2004b) analysis, even the stars in the lowest-metallicity bin have planets at a rate comparable to or higher than that of the stars with intermediate metallicities. Evidently even low metallicity stars can have short-period planets.

Nevertheless, there remains a strong correlation with finding short-period gas giants around the highest-metallicity stars, those with up to three times the Solar metallicity. Some of this enhancement appears to be due to the fact that spectro-scopic searches are easier when host stars have strong metallic absorption lines: the residual velocity jitter typically increases from a few m s$^{-1}$ to 5 to 16 m s$^{-1}$ for stars with one-quarter the Solar metallicity or less (D. Fischer, 2004, private communication). On the basis of signal-to-noise alone, then, fewer planets should be detected around lower-metallicity stars. It is also interesting that a spectroscopic search for hot Jupiters in the Hyades cluster, with a metallicity 35% greater than Solar, found no planets orbiting 98 stars, whereas about 10 hot Jupiters would have been predicted to be found, based on the frequency in the Solar neighborhood (Paulson *et al.*, 2004). Still, the correlation with high stellar metallicities in the Solar neighborhood deserves an explanation: e.g. is it due to formation by core accretion, or to inward orbital migration? Forming hot Jupiters *in situ* at 0.05 AU is nearly impossible with either formation mechanism – formation at distances of several AU or more followed by inward orbital migration is the preferred scenario for their origin.

Jones (2004, 2005) found that the average metallicity ([Fe/H]) of stars with planets increased from $\sim 0.07$ to $\sim 0.24$ for those stars hosting planets with semimajor axes of $\sim 0.03$ AU, compared to those with planets with semimajor axes of $\sim 2$ AU. This data suggests that there is a trend toward the shortest-period planets orbiting the most metal-rich stars. Similarly, Sozzetti (2004) showed that both metal-poor and metal-rich stars have more and more giant planets as the orbital period increases, but that only the metal-rich stars also had an excess of the shortest-period planets. These observations thus imply that the metallicity correlation has more to do with inward migration than with formation mechanisms: lower metallicity stars have giant planets, but only rarely do they end up migrating inward to become hot or warm Jupiters.

Inward planetary-orbital migration is thought to be caused primarily by gravita-tional interactions with the gaseous disk. Once the planet has grown large enough to clear a gap in the disk (roughly Jupiter-mass), the planet is locked into the disk, and evolves along with the disk in what is called type II migration (e.g. Nelson and Benz, 2003a, b). An increased effectiveness of type II migration is consistent with increased metallicity, because disk torques depend on the disk viscosity $v$, and in standard viscous accretion disk theory (e.g. Ruden and Pollack, 1991) $v = \alpha c_s h$, where $\alpha$ is a free parameter, $c_s$ is the sound speed, and $h$ is the disk thickness. As the disk metallicity decreases, the disk opacity decreases, leading to lower disk temperatures, lower sound speeds, and a thinner disk. Accordingly, $\nu$ decreases with lowered metallicity, and the time scales for type II migration increase. Ruden and Pollack (1991) showed that viscous disk evolution times lengthen by a factor of about 20 when $\nu$ decreases by a factor of 10. This trend is in the right direction, but it remains to be quantified by detailed models of the thermal structure of disks with greater variations in the opacity than have been explored to date (e.g. Boss, 1996, 1998b). Livio and Pringle (2003) suggested that a change in the metallicity by a factor of ten may only change the migration time by a factor of two.

The metallicity correlation may thus be more an indicator of the speed of orbital migration than of the formation mechanism. This suggests a second possible explanation for the correlation of metallicity with shorter period orbits found by Jones (2004, 2005) and Sozzetti (2004): higher metallicity means faster type II migration in either formation scenario, and hence more hot Jupiters. This migration hypothesis has readily testable consequences: if orbital migration is a major source of the metallicity correlation, then the "missing planets" around metal-poor stars should be found on long-period orbits, if disk instability is operative. A failure to find long-period gas-giant planets in orbit around low-metallicity stars would be an argument in favor of core accretion as the dominant mechanism.

The evidence for the existence of a long-period gas-giant planet in the M4 globular cluster (Sigurdsson *et al.*, 2003), where the metallicity is one-twentieth to one-thirtieth that of the Sun, depending on the element, while only a single, debatable example, suggests that disk instability is able to operate in low-metallicity disks, as it is doubtful if core accretion could produce a gas-giant planet in M4 with so few solids available. Combined with the absence of short-period Jupiters in the 47 Tucanae globular cluster (Gilliland *et al.*, 2000), a relatively metal-rich member of the population of metal-poor globular clusters (47 Tucanae has [Fe/H] $= -0.7$, or one-fifth the Solar metallicity) the M4 result instead suggests that long-period Jupiters may be common in globular clusters (and elsewhere), but that inward orbital migration is less important in such low-metallicity systems, leading to a lower frequency of short-period Jupiters.

## 12.4 Low-metallicity stars

The two formation mechanisms thus make clear predictions regarding the frequency of long-period planets in low-metallicity systems. Deciding this issue observationally will require maintaining the radial velocity observations of low-metallicity stars for decades or more, as well as initiating new searches which are sensitive to long-period planets, such as astrometric searches with the Keck Interferometer (KI) or the Very Large Telescope Interferometer (VLTI), which unfortunately will also require decades of observations, or direct-imaging searches sensitive to long-period Jupiter-mass planets with the Large Binocular Telescope or specialized space telescopes. Astrometric and direct-imaging searches are able to search metal-poor stars as readily as metal-rich stars. If the low-metallicity stars do not show evidence for giant planets on long-period orbits, this would be a strong argument in favor of core accretion.

It could also be that the metallicity correlation is partially a consequence of disk-lifetimes, at least in the disk-instability scenario. If 47 Tucanae initially contained high-mass stars, the outer regions of its protoplanetary disks may have been photo-evaporated before significant inward orbital migration of giant planets could have occurred, explaining the absence of short-period Jupiters. The hot Jupiters in the Solar Neighborhood may have formed in relatively long-lived disks in Taurus-like star-forming regions, where there was plenty of time for inward orbital type II migration, while the majority of nearby stars without short-period gas-giants may have formed in Orion- or Carina-like regions with short-lived disks. One observational test for this hypothesis would be the detection of a high frequency of long-period gas-giant planets around nearby main-sequence stars, most of which had to have been formed in regions where protoplanetary disks were relatively short-lived (Lada and Lada, 2003).

## 12.5 Gas-giant planets orbiting M dwarfs

While most of the target stars for the spectroscopic surveys have been G dwarfs, reflecting our anthropocentric bias, early M dwarfs have been added to the search lists as well. The M dwarf GJ 876 has already been shown to be orbited by a pair of gas-giant planets, and there are indications that other M dwarfs have short-period gas giants as well, though perhaps not as frequently as the G dwarfs.

Laughlin *et al.* (2004) pointed out that core accretion may not be able to account for the formation of gas-giant planets around M dwarfs (and presumably L and T dwarfs). The problem is that the collisional accumulation of the cores proceeds slower around lower-mass stars because of the longer orbital periods at a given semimajor axis, so that 10 Myr or more may be needed to form a gas giant by core accretion around a 0.4 $M_\odot$ star. Disk instability, on the other hand, can still proceed on a timescale much less than the mean disk lifetime of about 3 Myr, even if the instability proceeds ten times slower because of longer orbital periods. Disk instability, then, is likely to predict that M, L, and T dwarf stars should harbor gas-giant planets, though detailed calculations are needed to verify this hunch. Spectroscopic searches of late M, L, and T dwarfs are hampered by the faintness of the targets, so astrometric searches may be necessary to determine if the lower

*Alan Boss*

end of the main sequence and brown dwarfs support gas-giant planet formation or not.

## 12.6 Core masses of Jupiter and Saturn

The masses of Jupiter's and Saturn's cores are important clues to their origin. In the disk-instability mechanism, one Jupiter mass of disk gas has at most $\sim 6~M_{\oplus}$ of elements heavier than hydrogen and helium that could settle to the core prior to contraction of the protoplanet to planetary densities, after which heavy elements must remain trapped in the planet's envelope and cannot settle to the core. Hence if Jupiter's core is much more massive than $\sim 6~M_{\oplus}$, then it probably could not have formed by disk instability, unless it had lost part of its gaseous envelope by ultraviolet photoevaporation and then migrated inward to 5.2 AU. A Jupiter core much larger than $\sim 6~M_{\oplus}$ would then seem to support formation by core accretion.

The current state-of-the-art in modeling the interior structures of Jupiter is given in Guillot *et al.* (2004) and Saumon and Guillot (2004), which show the results of detailed models of hydrostatic planets subject to various constraints, such as the exterior gravitational field of the planet, as determined by robotic spacecraft. The equation of state (EOS) of hydrogen at high temperatures and pressures is a major uncertainty. The preferred EOS implies that Jupiter's core mass is less than $\sim 3~M_{\oplus}$ (Saumon and Guillot, 2004). Jupiter may even have no core at all. Even a core mass of $\sim 3~M_{\oplus}$ is insufficient to trigger the dynamic collapse of disk gas that is needed to convert an ice/rock core into a gas-giant planet. Inaba *et al.* (2003) found that Jupiter formation required a core mass of $\sim 21~M_{\oplus}$ (and also found that Saturn could not grow large enough to dynamically accrete a gaseous envelope).

Apparently core accretion could not have formed Jupiter, unless its initial core mass was $\sim 20~M_{\oplus}$ or more, and this initial core had been eroded away until its mass dropped to its present value. The possibility of core erosion is now under serious study (Guillot *et al.*, 2004; Saumon and Guillot, 2004). However, if core erosion appears to be a possibility, the core masses of the giant planets may lose much of their usefulness as constraints on their formation. A core inside a planet formed by disk instability presumably could also be subject to subsequent erosion.

Saturn's core mass is also of great interest. Current models imply that Saturn's core mass is considerably larger than that of Jupiter (Guillot, 1999; Saumon and Guillot, 2004), perhaps $\sim 15~M_{\oplus}$, which may be hard to reconcile with the smaller total mass of Saturn in the core-accretion scenario: presumably the more massive core would be the first one to undergo dynamic collapse of its atmosphere. Erosion of Jupiter's initial core might solve this problem for the core-accretion mechanism, provided that Saturn's core is not similarly subject to erosion. In the disk-instability scenario (Boss *et al.*, 2002), a more massive Saturnian core simply means that protoSaturn started with a total mass substantially larger than that of protoJupiter, with much of Saturn's excess gas being lost by EUV/FUV photoevaporation.

With either core accretion or disk instability, ongoing accretion of planetesimals/cometesimals associated with forming the Oort Cloud and clearing out the remaining small bodies would lead to the high metal (Z) contents of the gaseous envelopes of Jupiter and Saturn.

The moments of Jupiter's gravitational field (e.g. J4) are important to the Jovian interior models, and a Jupiter polar-orbiter mission intended to probe the planet's gravitational field might be needed to better constrain the Jovian interior. Even more important would be better laboratory data and improved theoretical understanding of the EOS of hydrogen at high temperatures and pressures. Understanding whether or not core erosion can occur is another key unknown.

## 12.7 Super-Earths and failed cores

If core accretion is the dominant formation mechanism, but seldom occurs in a disk long-lived enough for a complete gas-giant planet to form, then the typical outcome of core accretion may be a system of "failed cores," i.e. a system of ice-giant planets, unaccompanied by gas giants. If disk instability dominates, inner gas giants should be the rule, accompanied by outer ice-giant planets in systems which formed in Orion-like regions and experienced EUV/FUV photoevaporation of their gaseous envelopes. In Taurus-like regions, disk instability should produce only gas giants, unaccompanied by outer ice giants.

Ground-based radial velocity surveys for "hot Neptunes" have begun to shed light on the frequency of short-period Neptune-mass planets (Butler *et al.*, 2004; McArthur *et al.*, 2004; Santos *et al.*, 2004a), but it remains to be seen if they are ice giants (with mean densities of $\sim 2$ g cm$^{-3}$), or "super-Earths", i.e. Neptune-mass planets composed of rock, with a mean density of 5 g cm$^{-3}$ or more. The presence of at least two or more Jupiter-mass planets orbiting at even greater distances around $\mu$ Arae and $\rho^1$ Cancri implies that their hot Neptunes must have formed inside a distance of no more than a few AU, supporting the idea that they must be composed primarily of rock and should hence be super-Earths. Gravitational close encounters probably did not interchange the orbital distances of these planetary systems, because such interactions lead to the more massive planet becoming more tightly bound at the expense of the less massive planet, which is not the case here. Disk migration mechanisms are not thought to lead to passing of one planet by another, so the fact that the hot Neptunes orbit well inside gas giants implies that they formed in much the same way as the terrestrial planets in our Solar System. Given that roughly one in ten short-period planets is expected to transit its star, by the time that another seven or so hot Neptunes are discovered, the density of one of the objects might be determined, and the question of the true identity of at least one of the hot Neptunes will be settled. However, a space-based transit mission might be necessary in order to detect the transit of such a small planet.

The CoRoT and Kepler space missions will use high precision photometry to search for the presence of hot and warm Neptunes (and Earths) by the transit method. "Cold Neptunes" could be detected astrometrically by the KI or the VLTI, though their orbital periods would be much longer. Finding systems composed exclusively of Neptune-mass objects composed primarily of ice and rock, i.e. failed cores, would point to core accretion. Systems composed of ice giants orbiting outside of gas giants might be consistent with core accretion or with disk instability coupled with UV photoevaporation, though in the latter case the dividing line between the gas giants and the ice giants should vary with the mass of the host star: for lower-mass stars, the critical radius for photoevaporation is proportionally smaller, and proportionally larger for higher-mass stars. For lower-mass stars, gas giants would not be expected to be formed by core accretion (Laughlin *et al.*, 2004), just failed cores.

Note that both transits and spectroscopic or astrometric wobbles are needed in order to determine the planet's mean density and determine if a Neptune-mass planet is a super-Earth or an ice giant. Super-Earths would form in basically the same manner in either the core-accretion or disk-instability scenarios, though the epoch at which Jupiter forms can have an effect (J. Chambers, 2004, private communication).

## 12.8 Gas-giant planet formation epochs

Because core accretion is a much slower process than disk instability, if core accretion dominates, young stars should not show evidence of gas-giant companions until they reach ages of several million years or more. If disk instability dominates, however, even the youngest stellar objects may show evidence of gas-giant planets. Dating the epoch of gas-giant planet formation is thus another means to differentiate between these two contending theories (Boss, 1998b; Rice *et al.*, 2003b). There is indirect evidence for the formation of a gas-giant planet orbiting at $\sim$ 10 AU around the 1 Myr-old star CoKu Tau/4 (Forrest *et al.*, 2004), based on a spectral energy distribution from the Spitzer Space Telescope showing the absence of disk dust inside this radius. A number of other 1 Myr-old stars show similar evidence for rapid formation of gas-giant planets (D. M. Watson, 2004, private communication). Thus Spitzer may already be telling us that some gas-giant planets form in 1 Myr or less.

The radial velocity technique is prevented from dating the epoch of gas-giant planet formation because young stars typically have rapid rotation rates (and hence broad spectral lines unsuitable for high precision spectroscopy), chromospheric activity, and variability. However, several other techniques could be employed. The Space Interferometry Mission (SIM) will have sufficient astrometric precision to search for the wobbles of young stars caused by gas-giant companions on Jupiter-like orbits, though star spots may be a troublesome source of photocenter noise, and the limited life (5 to 10 yr) of SIM may prevent the detection of long-period orbits. Nearby low-mass star-forming regions such as Taurus and Ophiuchus would be the primary hunting grounds. Searching the very youngest stars, still embedded in their infalling clouds of gas and dust, may be difficult with an optical telescope such as SIM, but if SIM finds that the youngest stars are already wobbling, this would be additional good evidence for a rapid formation process.

Searches intended to determine the epoch of gas-giant planet formation could also be launched at longer wavelengths, where the youngest stellar objects can often be seen more clearly. The Atacama Large Millimeter Array (ALMA) will be able to form mm-wave images of protoplanetary disks with $\sim 1$ AU resolution and to search for disk gaps created by massive protoplanets. While the protoplanets themselves may not be resolvable by ALMA, the gaps they eventually create may well be. Disk instability appears to be capable of forming protoplanets at large distances from the host star, at least out to 30 AU or so (Boss, 2003), where core accretion does not appear to be capable of forming gas-giant planets *in situ* (Levison and Stewart, 2001). While core accretion might lead to orbital interactions that could populate the outer regions (Thommes *et al.*, 1999, 2002), the planets thereby kicked outward would be the least massive planets, not the gas giants. Gaps caused by gas-giant planets at large orbital radii would then seem to be an indicator of planet formation by disk instability.

The Spitzer Space Telescope does not have sufficient spatial resolution to search directly for young protoplanets in orbit around their parent stars, and instead must rely on the indirect evidence contained in disk spectra. The James Webb Space Telescope (JWST), however, with its much larger aperture and possible corona-graph, might be able to image hot young Jupiters around nearby stars, and to image gaps in more distant protoplanetary disks.

## 12.9 Planetary-system architectures

Further understanding about the architectures of planetary systems may only come about with the success of the Terrestrial Planet Finder (the TPF-C optical corona-graph and the TPF-I mid-infrared interferometer) and Darwin mid-infrared inter-ferometer missions, which will have the ability to image Earth-like planets around roughly 100 nearby stars, and to characterize their atmospheres. Given the diffi-culty of imaging Earth-like planets, TPF/Darwin is likely to have little trouble in imaging many of the other components of nearby planetary systems, including their gas- and ice-giant planets, assuming that their outer working diameters permit detecting planets at much larger orbital radii than a few AU. If extrasolar planetary systems are as well-ordered as the Solar System, with inner terrestrial planets, intermediate gas giants, and outer ice giants, this would imply their formation *in situ*, or at least an orderly, disk-driven inward orbital migration from somewhat more distant regions. However, if extrasolar planetary systems are more often highly disordered, this would imply that gravitational interactions between the protoplanets led to a phase of chaotic evolution where information about the primordial planetary orbits had been lost. In the latter case, it may be harder to place constraints on the formation mechanisms involved.

Both mechanisms of giant-planet formation appear to be tolerant of terrestrial-planet formation, assuming that the gas giant-planets are on sufficiently long-period orbits (Wetherill, 1996). Disk instability might even help to speed the growth of the Earth and other terrestrial planets (Kortenkamp *et al.*, 2001; J. Chambers, 2004, private communication). Hence in either case, we expect that Earth-like planets should be frequent components of extrasolar planetary systems. However, given that disk instability may be capable of forming planetary systems similar to the Solar System in the most common type of star-forming region (i.e. a region of high-mass star formation with relatively short-lived disks), it would appear that disk instability might produce a higher frequency of Solar System-analogs than core accretion. Kepler, SIM, and TPF/Darwin will test this and other predictions about the frequency of habitable planets.

## 12.10  Conclusions

This review has laid out some of the differences between the core-accretion and disk-instability mechanisms that lead to predictions that can be observationally tested. Undoubtedly there will be other observational predictions that can be made as theoretical work on these two mechanisms continues to increase our understanding of them and their consequences. In a decade or two we will have a much firmer understanding of how the largest members of planetary systems are formed than we do now, as well as a good estimate of the frequency of Earth-like planets in the Solar Neighborhood.

Supported in part by NASA Planetary Geology and Geophysics grant NAG5-10201 and by NASA Astrobiology Institute grant NCC2-1056.

# 13

# From hot Jupiters to hot Neptunes ... and below

Christophe Lovis, Michel Mayor, Stéphane Udry

*Observatoire de Genève, Switzerland*

## 13.1 Recent improvements in radial velocity precision

Since the first discovery of an extrasolar planet around a Solar-type star ten years ago (Mayor and Queloz, 1995), research in this field has been very productive and has led to the detection of more than 140 exoplanets. The vast majority of these discoveries has been made with the radial-velocity (RV) technique, i.e. the precise measurement of the RV wobble that a planet induces in its parent star due to its orbital movement. A major effort to improve the accuracy of the RV measurements has been undertaken by several groups, since this is absolutely necessary to detect the RV signatures of giant planets, in the range $1-100 \text{ m s}^{-1}$. Two main techniques were developed: one using a ThAr calibration simultaneously with each observation (Baranne *et al.*, 1996) to track instrumental drifts, and one using an iodine absorption cell, superimposing a reference spectrum on the stellar spectrum (Butler *et al.*, 1996). Both techniques have been able to deliver RV precision at the level of $\sim 3 \text{ m s}^{-1}$, opening the way to the discovery of many planetary systems.

Over the past decade, the exoplanet group at Geneva Observatory has been operating two high-resolution spectrographs able to achieve high RV precision, namely the ELODIE instrument mounted on the 1.93 m telescope at Observatoire de Haute-Provence (France), and the CORALIE instrument installed on the Swiss 1.2 m telescope at La Silla Observatory (Chile). Both ELODIE and CORALIE are high-resolution ($R = 50\,000$), fiber-fed echelle spectrographs. They are fed by two fibers, the first one carrying the stellar beam and the second one for the simultaneous recording of a ThAr reference spectrum. The very stable illumination of the spectrograph, together with the tracking of instrumental drifts, makes it possible to reach an RV accuracy of a few $\text{m s}^{-1}$.

*Planet Formation: Theory, Observation, and Experiments*, ed. Hubert Klahr and Wolfgang Brandner.
Published by Cambridge University Press. © Cambridge University Press 2006.

The ELODIE survey, started in 1993, has been monitoring a magnitude-limited sample of 350 F–K close dwarfs. This has led to the discovery of 19 planets with $m_2 \sin i < 10$ M$_J$. The CORALIE survey, started in 1998, has been targeting about 1650 FGKM main-sequence stars in a volume-limited sample, leading to the detection of $\sim$40 planets.

In 2001, a consortium led by Geneva Observatory started to build for the European Southern Observatory (ESO) an instrument capable of achieving the 1 m s$^{-1}$ precision level. For that purpose, a thorough understanding of the instrumental factors limiting the RV precision was mandatory. It was recognized that atmospheric pressure and temperature variations are the main causes of instrumental drifts during an observing night. In numbers, a RV drift of 1 m s$^{-1}$ corresponds to a pressure change of only 0.01 mbar, or a temperature change of 0.01 K. Although these drifts can be accurately corrected with the simultaneous ThAr reference, minimizing these influences is necessary to further improve the RV accuracy. Therefore, the new instrument HARPS was put under vacuum in a strictly temperature-controlled environment. Another critical point was the stability of the spectrograph illumination. A fiber link to the spectrograph with high light-scrambling properties, coupled to dedicated guiding software, was found to be the best solution to avoid spurious wavelength shifts due to changing illumination or varying instrumental profile. Further requirements to reach the desired precision included a large spectral coverage, high spectral resolution, high pixel sampling (3–3.5 pixels full width at half maximum) and the possibility to obtain high S/N data for a large number of stars to maximize the Doppler information content of the spectra. All these constraints led to the construction of an echelle spectrograph spanning the whole visible range (3800–6900 Å) at a resolution $R = 115\,000$, mounted on the ESO 3.6 m telescope at La Silla Observatory (Chile), operational since October 2003.

First results from HARPS included some asteroseismological time series demonstrating the unprecedented accuracy of this instrument (see Mayor *et al.*, 2003). Fig. 13.1 shows the oscillations in radial velocity due to acoustic p-modes in four different stars, as observed with HARPS during the commissioning period. The amplitude of these modes depends on spectral type and evolution stage, early-G and evolved stars having the largest amplitudes (up to 10 m s$^{-1}$ peak-to-peak) whereas late-K dwarfs show only weak modulations (1–2 m s$^{-1}$). Periods also vary along the main sequence, ranging from $\sim$4 to $\sim$15 minutes between K and early-G stars. These RV measurements have an accuracy better than 40 cm s$^{-1}$ (including photon noise and guiding errors), demonstrating the extraordinary capabilities of HARPS. Long-term stability is found to be at the level of 1 m s$^{-1}$ or better (see Lovis *et al.*, 2005, for a detailed analysis).

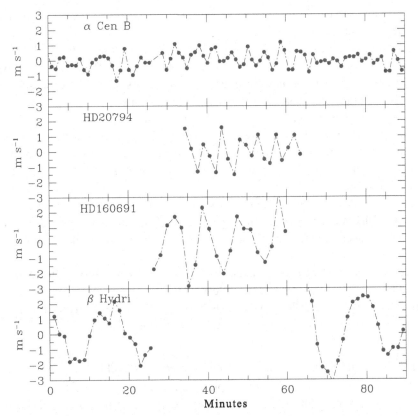

Fig. 13.1. Radial-velocity time series for four stars with spectral types K1V, G8V, G3IV-V and G2IV (from top to bottom). Amplitude and period of the oscillations vary along the main sequence, earlier spectral types showing larger and longer RV modulations.

## 13.2 Detecting planets down to a few Earth masses

The scientific objectives of the HARPS consortium can be summarized as follows:

- *Search for very low-mass planets*: we have selected a sample of $\sim$400 stars from the larger CORALIE sample which will be closely monitored at the 1 m s$^{-1}$ precision level. These stars have been chosen for being chromospherically quiet and slowly rotating so that activity-induced RV jitter does not hide possible planetary signals down to the 1–2 m s$^{-1}$ level.
- *Better knowledge of statistical properties of exoplanets*: we have built an extension to the CORALIE volume-limited sample, including most late-F to M dwarfs between 50 and 60 pc. These stars will be monitored at the 3 m s$^{-1}$ precision level (photon-noise limited), with the objective of increasing the total number of stars monitored by RV

planet-search surveys to improve our knowledge of exoplanet properties, such as mass, period or eccentricity distributions.

- *Search for planets around M-dwarfs*: we have set up a catalog of 120 close, single M-dwarfs that we are following at the 3 m s$^{-1}$ precision level. The aim of this survey is to investigate the planet occurrence around these low-mass stars to better understand the dependence of planet formation on the mass of the parent star.

- *Search for planets around metal-deficient stars*: we are following 100 stars from the galactic halo with metallicities between −2.0 and −0.5 dex. This survey should be able to refine the now well-established relation between planet occurrence and metallicity, providing stronger constraints at low metallicity (see Santos *et al.*, 2001, 2004b; Fischer and Valenti, 2005).

- *Planet occurrence in "twin" binary stars*: We have built a sample of visual, physical binaries with both stars having similar spectral type. This survey will provide new information on planet formation in binaries and test whether similar initial conditions lead to similar planetary systems.

- *Follow-up of CoRoT planetary candidates*: The CoRoT space mission, to be launched in 2006, will detect transiting planetary candidates using very precise photometry. We will try to confirm and follow up these candidates with radial-velocity measurements to establish the true nature of these objects and derive their masses.

The search for very low-mass planets has already yielded a very interesting result: the discovery of a 14 M$_\oplus$ extrasolar planet around $\mu$ Ara (HD 160691) (see Santos *et al.*, 2004a). The faint signal of this planet was first detected during an asteroseismological run on this star. Apart from the expected, short-period ($\sim$8 min (see Bouchy *et al.*, 2005)) p-mode oscillations, another modulation was clearly present, with a period at least as long as the span of the asteroseismological data (see Fig. 13.2). Follow-up observations confirmed the presence of a periodic signal with period $P = 9.5$ days and RV amplitude $k = 4.1$ m s$^{-1}$, which is best explained by the Keplerian movement of an exoplanet with a minimum mass of 14 M$_\oplus$ (see Fig. 13.3). $\mu$ Ara was already known to host one giant planet and another substellar companion (McCarthy *et al.*, 2004), making this star extremely interesting for planet formation theories and dynamical studies. Further high-precision observations are needed to better constrain the orbital parameters of this system.

The detection of $\mu$ Ara c was made simultaneously with two other important discoveries, the very low-mass planets 55 Cnc e (McArthur *et al.*, 2004) and Gl 436 b (Butler *et al.*, 2004). All three objects have probable masses between 14 and $\sim$25 M$_\oplus$, making them the first members of a new class of exoplanets, the hot Neptunes. The question about their composition and internal structure is still open, although theoretical arguments seem to favor the rocky-planet hypothesis. In the core-accretion model of planet formation, these objects might actually have started their formation within the ice boundary, accreting mainly solid materials and

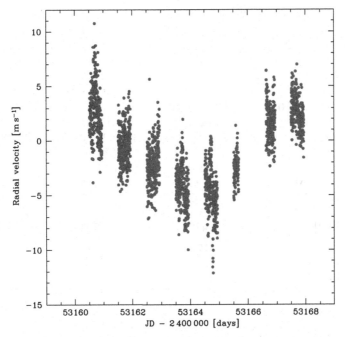

Fig. 13.2. RV time series of $\mu$ Ara obtained during an asteroseismological run (8 nights). The p-mode oscillations are responsible for the high-frequency RV variations, but cannot explain the night-to-night RV changes.

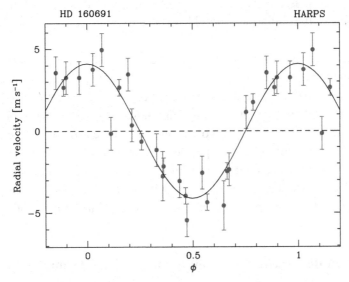

Fig. 13.3. Phased radial velocities for $\mu$ Ara after removing the long-term trends due to the giant planets in the system. The remaining signal has a period $P = 9.5$ days and a semi-amplitude $k = 4.1$ m s$^{-1}$, corresponding to a minimum mass of only 14 M$_\oplus$ for the hot Neptune.

Table 13.1. *Orbital and physical parameters for three planets recently detected with HARPS (Lovis et al., 2005)*

| Parameter | | HD 93083 b | HD 101930 b | HD 102117 b |
|---|---|---|---|---|
| $P$ | [days] | 143.58±0.60 | 70.46±0.18 | 20.67±0.04 |
| $T$ | [JD-2400000] | 53181.7±3.0 | 53145.0±2.0 | 53100.1±0.1 |
| $e$ | | 0.14±0.03 | 0.11±0.02 | 0.00 (+0.07) |
| $V$ | [km s$^{-1}$] | 43.6418±0.0004 | 18.3629±0.0003 | 49.5834±0.0003 |
| $\omega$ | [deg] | 333.5±7.9 | 251±11 | 162.8±3.0 |
| $K$ | [m s$^{-1}$] | 18.3±0.5 | 18.1±0.4 | 10.2±0.4 |
| $a_1 \sin i$ | [10$^{-3}$ AU] | 0.239 | 0.116 | 0.019 |
| $f(m)$ | [10$^{-9}$ M$_\odot$] | 0.088 | 0.042 | 0.0023 |
| $m_2 \sin i$ | [M$_J$] | 0.37 | 0.30 | 0.14 |
| $a$ | [AU] | 0.477 | 0.302 | 0.149 |
| $N_{\mathrm{meas}}$ | | 16 | 16 | 13 |
| $Span$ | [days] | 383 | 362 | 383 |
| $\sigma$ (O–C) | [m s$^{-1}$] | 2.0 | 1.8 | 0.9 |

becoming "super-Earths." Alternatively, these planets might have formed further away in the disk, starting with a solid core and then going through the gas-accretion phase. Migration would have simultaneously brought them closer to the star before they became too massive. The gaseous envelope could then have been evaporated by the strong stellar radiations, provided the semimajor axis became sufficiently small.

From the observational point of view, the detection of such objects turns out to be challenging, but not impossible. It requires a large number of measurements and some care in dealing with the stellar oscillations. Asteroseismological "noise" can be greatly reduced by integrating long enough to cover one or more typical oscillation periods. Furthermore, evolved and early-type stars should be avoided as they exhibit large-amplitude, long-period oscillations which are more difficult to average out. With the accumulation of high-precision observations in the coming months and years, we should be able to tell whether hot Neptunes are very common or rather rare, bringing important new constraints on planet-formation theories.

### 13.3 New discoveries and implications for planet-formation theories

In this section we want to present and discuss the properties of three planets recently detected by HARPS (see Lovis *et al.*, 2005). These planets orbit around the stars HD 93083 (K3V, $V = 8.30$), HD 101930 (K1V, $V = 8.21$) and HD 102117 (G6V, $V = 7.47$). Their orbital and physical parameters are summarized in Table 13.1.

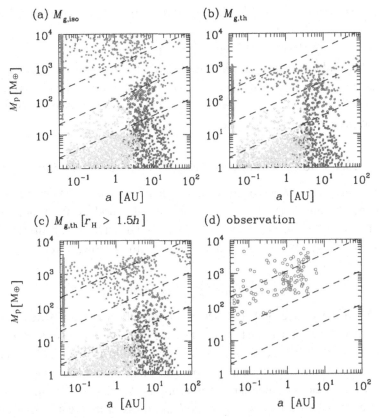

Fig. 13.4. Mass-period diagrams taken from Ida and Lin (2004), showing the presence of a "desert" at intermediate distances for masses between 0.05 and 0.5 M$_J$. Three planets recently detected with HARPS do fall in this area of the diagram, raising questions about their formation history.

All three planets orbit at intermediate distances from their parent star (0.15–0.5 AU) and have minimum masses in the Saturn-mass regime and below. These characteristics can be compared to theoretical models of planet formation that try to predict the planet distribution in the mass-period diagram. We consider here in particular the models recently proposed in Ida and Lin (2004). Fig. 13.4, taken from their paper, shows the expected distribution of planets as a function of orbital period and mass. Interestingly, the newly discovered planets fall in a region of these diagrams that seems difficult to populate by models. The reason is to be found in the fast and efficient core-accretion and migration processes. Indeed, after the formation of a solid core of about 10 M$_\oplus$, the runaway gas-accretion phase should lead very rapidly to the formation of a gas giant with a mass above ~0.5 M$_J$. On the other hand, interactions with the disk will make the planet migrate inwards to distances below 0.05–0.1 AU, and thereby become a hot Jupiter. As a result, there will be a planet-desert in the mass-period diagram, as explained in Ida and Lin (2004).

The discovery of planets in this desert and the obtention of statistically meaningful numbers on planet occurrence in this region will therefore represent a very important test for planet-formation theories, especially for the standard core-accretion and migration scenarios. The increasing sensitivity of radial-velocity surveys will bring strong constraints on that issue in the near future.

## 13.4 Update on some statistical properties of exoplanets

Statistical properties of known extrasolar planets allow us to better understand the mechanisms of planet formation and to constrain the models. With more than 140 exoplanets known to date, some very interesting trends have already drawn attention in the past few years. We review some of these in the following sections.

### 13.4.1 Giant-planet occurrence

A few RV surveys have now reached a sufficiently advanced stage to provide precise numbers on giant-planet occurrence around Solar-type stars. Results from the ELODIE planet search (see Section 13.1), started in 1993, have been analyzed in that respect in Naef *et al.* (2004). The data analysis is based on Monte Carlo simulations on the actual calendar of observations, which lead to realistic detection probabilities as a function of orbital period and mass of the planet. Knowledge of these probabilities then makes it possible to correct the actual number of detections to obtain the total number of giant planets orbiting the stars in the sample. Fig. 13.5 shows the detection probabilities for different masses as a function of orbital period. The ELODIE survey should basically have detected all brown dwarfs with periods below $\sim 3000$ days. For 1 $M_J$-planets, the detection probability is 50% at $\sim 500$ days. Given the average measurement precision of $10\,\mathrm{m\,s^{-1}}$ (including photon noise and stellar jitter), the sensitivity decreases rapidly for planets lighter than Saturn with periods longer than 10 days. Interesting features of these probability curves include the influence of stellar activity, which has a significant impact on planet detection, and also some sampling effects, well visible at 1–2 days and 1–2 years, which locally lower the detection probability.

A total of 19 planets were found around the 350 stars in the ELODIE planet search sample. When correcting these numbers by taking into account the effective detection probabilities, two important results come out:

- The fraction of stars harboring a planet with a mass higher than 0.5 $M_J$ is 7.3±1.5% for orbital periods up to 3900 days.
- The fraction of stars harboring a hot Jupiter ($m > 0.5$ $M_J$, $P < 5$ days) is $0.7 \pm 0.5\%$.

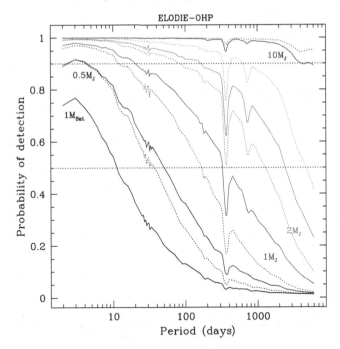

Fig. 13.5. Detection probabilities as a function of orbital period and planet mass for the ELODIE survey. Dashed lines indicate probabilities in the case of negligible activity-induced stellar jitter. Note the sampling effects at 1–2 days and 1–2 years, which locally lower the detection probabilities.

These numbers are in good agreement with other studies (e.g. Marcy *et al.*, 2004). Note that the scarcity of hot Jupiters and the absence of very hot Jupiters ($P = 1$–2.5 days) in present RV surveys is probably not in contradiction with the recent discoveries of such close-in planets by the OGLE photometric transit survey (Udalski *et al.*, 2004; Bouchy *et al.*, 2004; Pont *et al.*, 2004; Konacki *et al.*, 2003). Indeed, the number of stars observed by transit searches is larger by 2–3 orders of magnitude compared to the number of stars observed in RV searches. Moreover, transit searches are particularly sensitive to very short orbital periods, whereas their efficiency decreases rapidly for periods longer than a few days (Gaudi *et al.*, 2005).

### 13.4.2 *Mass and period distributions*

When examining the mass distribution of all known stellar and sub-stellar companions on a logarithmic scale (see Fig. 13.6), the most striking feature is probably the presence of the so-called brown-dwarf desert, situated approximately between 0.01 and 0.1 $M_\odot$. The distribution is rising on both the stellar and the planetary sides

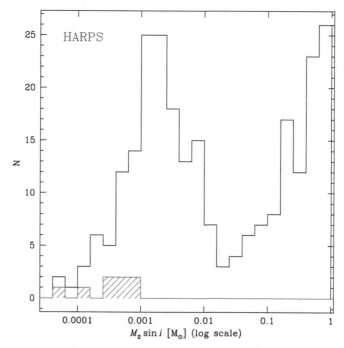

Fig. 13.6. Secondary mass distribution for stellar and substellar companions. The famous brown-dwarf desert marks the separation between planetary and stellar populations. The lack of low-mass planets is an observational bias due to the limited precision of radial-velocity searches. The dashed histogram indicates low-mass planets recently discovered with HARPS, showing the potential of this instrument to explore this mass regime.

of the desert, indicating that we are facing two different populations and probably two different mechanisms for star and planet formation.

Focusing now on the planetary-mass regime alone, the most striking feature is the rapid rise in planet frequency towards low masses, despite the increasing difficulty of detecting planets with lower masses. This mass distribution can be fitted approximately by a power law $f(m) \sim 1/m^p$ with $p = 1$–2. This trend certainly reflects an important aspect of the statistical properties of exoplanets. An extrapolation of this law at the low-mass end of the distribution leads to the conclusion that there are probably many more Neptune- and Earth-like bodies in planetary systems than Jupiter-size objects.

The period distribution of known extrasolar planets exhibits a minimum between 10 and 100 days, the "period valley" (Udry *et al.*, 2003) (see Fig. 13.7). This feature is probably related to the different migration processes that were acting on young giant planets at different distances from the star. On the short-period side of the

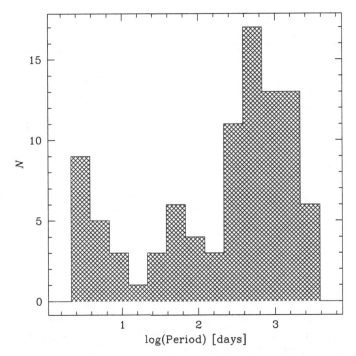

Fig. 13.7. Distribution of orbital periods for the presently known extrasolar planets. A period "valley" is visible between 10 and 100 days.

distribution, the "period wall" at $\sim$ 3 days is now well-established and can be viewed as a kind of migration stop, whose origin is still not understood.

Other interesting statistical properties include the absence of massive planets ($m > 2$ M$_J$) on periods shorter than 100 days, except in the case of binaries (see Udry *et al.*, 2003). A possible explanation could be that migration does not stop for massive planets, which are eventually engulfed by the star. The influence of a stellar companion might change this behavior and allow massive planets to survive on short-period orbits. There seems to be no such mass segregation at larger distances from the star ($P > 100$ days), where massive planets are commonly found.

### 13.4.3 Eccentricity distribution

The observed eccentricities of known extrasolar planets are surprisingly high. The average eccentricity is around 0.3, higher than for any Solar-System planet. This is not due to an observational bias: on the one hand, large eccentricities make RV amplitudes larger and therefore detections easier, but on the other hand, planets on eccentric orbits move quicker at periastron, making this orbital phase more difficult

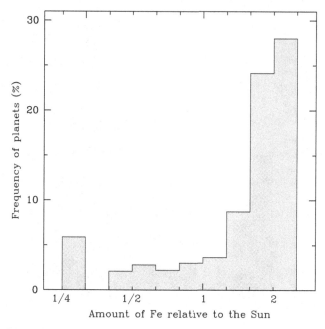

Fig. 13.8. Planet frequency as a function of the Fe content of the parent star. Note the sharp increase above Solar metallicity.

to catch. Both effects approximately compensate each other, making the detection bias very small.

An interesting feature of exoplanet eccentricities is the observed circularization period at ∼ 6 days. The circularization process seems to be at work only during the formation of the system (Halbwachs *et al.*, 2005), as opposed to binaries where it continues over the whole lifetime of the system. Beyond the circularization limit, exoplanet eccentricities tend to be smaller than for binaries. These discrepancies probably indicate two different processes for binary and planet formation.

### 13.4.4 Metallicity of planet-host stars

There is a now well-established correlation between planet occurrence and metallicity of the host star. Comparisons between planet-host stars and stars without (giant) planets have been performed based on the volume-limited CORALIE sample by Santos *et al.* (2001, 2004b). Using a homogeneous method to determine abundances on high-resolution spectra, they find an important increase in planet frequency at high metallicities, with a threshold at Solar metallicity (see Fig. 13.8). At [Fe/H] = 0.3 dex, almost 25% of the stars in the Solar Neighbourhood have giant planets. These results have been confirmed by Fischer and Valenti (2005) in their analysis of the Keck, Lick and AAT planet search samples.

The origin of this excess in heavy elements has been discussed for several years. However, the primordial scenario, in which the metallicity excess is of primordial origin, seems to be favored compared to the pollution scenario, where the absorption of planetesimals by the star increases its metallicity. If the latter theory was correct, there should be a correlation between the depth of the stellar convection zone and metallicity in planet-host stars. This effect is not observed at the moment.

# 14

# Disk–planet interaction and migration

Frederic Masset

*SAp, CE-Saclay, France and IA-UNAM, Mexico*

Wilhelm Kley

*Universität Tübingen, Institut für Astronomie und Astrophysik, Germany*

## 14.1 Introduction

Directly after the discovery of the very first extrasolar planets around main-sequence stars it has become obvious that the new planetary systems differ substantially from our own Solar System. Amongst other properties one distinguishing feature is the close proximity of several planets to their host stars (hot Jupiters). As it is difficult to imagine scenarios to form planets so close to their parent star it is generally assumed that massive, Jupiter-like planets form further away, and then migrate inwards towards the star due to disk–planet tidal interactions. Hence, the mere existence of hot Jupiters can be taken as clear evidence of the occurence of migration. Interestingly, theoretically the possibility of migrating planets has long been predicted from the early 1980s.

Another observational indication that some migration of planets must have occured is the existence of planets in mean motion resonances. Due to converging differential migration of two planets both embedded in a protoplanetary disk they can be captured in a low-order mean motion resonance. The most prominent example is the system GJ 876 where the planets have orbital periods of roughly 30 and 60 days.

In this review we focus on the theoretical aspects of the disk–planet interaction which leads to a change in the orbital elements of the planet most notably its semimajor axis. We only treat systems with a single planet and do not consider planetery systems containing multiple planets.

## 14.2 Type I migration

We consider in this section the tidal interaction between a low-mass protoplanet orbiting a central star within a laminar gaseous Keplerian disk. This planet exerts

*Planet Formation: Theory, Observation, and Experiments*, ed. Hubert Klahr and Wolfgang Brandner.
Published by Cambridge University Press. © Cambridge University Press 2006.

a force on any disk fluid element, which produces a wake in the disk. This wake in turn exerts a force on the planet, which leads to a change in the planet's orbital elements. The purpose of this section is to specify the wake properties and to evaluate its impact on the orbit of the planet.

### 14.2.1 Evaluation of the tidal torque

The problem of determining the impact of disk interaction on the evolution of the planet orbit amounts to an evaluation of tidal torques. For a sufficiently small planet mass (an upper limit of which will be specified later) one can perform a linear analysis which consists of considering the Fourier decomposition in azimuth of the planet tidal potential $\phi$,

$$\phi(r, \varphi, t) = \sum_{m=0}^{\infty} \phi_m(r) \cos\{m[\varphi - \varphi_p(t)]\}, \tag{14.1}$$

where $\varphi_p = \Omega_p t$ is the azimuth angle of the planet. One is then left to evaluate the torques $\Gamma_m$ exerted on the disk by the $m$-folded potential components $\phi_m(r) \cos[m(\varphi - \Omega_p t)]$. In the linear regime, the total torque is then recovered by summing over all $m$.

We do not reproduce here the expression and the derivation of these torques, which can be found in Goldreich and Tremaine (1979) or Meyer-Vernet and Sicardy (1987). We rather give a few comments on their physical meanings.

An $m$-folded external forcing potential $\phi_m(r, \varphi)$ which rotates with a pattern frequency $\Omega_p$ in a disk with angular velocity profile $\Omega(r)$ triggers a response wherever the potential frequency as seen in the matter frame $\omega = m(\Omega - \Omega_p)$ matches either 0 or $\pm\kappa$ ($\kappa$ being the epicyclic frequency). The first case corresponds to a corotation frequency (since it implies $\Omega = \Omega_p$, hence fluid elements corotate with the forcing potential) while the second case corresponds to a Lindblad resonance (outer Lindblad resonance for $\omega = \kappa$ and inner Lindblad resonance for $\omega = -\kappa$). The denomination of Lindblad resonance has historical reasons and comes from galactic dynamics (see Binney and Tremaine, 1987).

### 14.2.1.1 Torque at Lindblad resonances

In a Keplerian disk, the torque exerted on the disk by the external planetary potential is positive at an outer Lindblad resonance, and negative at an inner Lindblad resonance.

The torque value is independent of the physical processes at work in the disk (Meyer-Vernet and Sicardy, 1987). In the simplest case of a non-self-gravitating, pressureless and inviscid disk, the corresponding angular momentum exchange occurs at the exact location of the Lindblad resonances, and angular momentum accumulates there. To achieve a steady state, some additional physics is required to get rid of the angular momentum deposited at the resonance by the external perturber. Meyer-Vernet and Sicardy (1987) showed that dissipation (provided for example by a simple drag law: $-Qv$, or by a shear or bulk viscosity), pressure effects or self-gravity can help the disk to transfer the angular momentum away from the resonance, thus allowing a steady state and hence a torque value constant in time. Remarkably, the torque value is not altered by the underlying physics, which only modifies the shape and width of the resonant region.

These authors also worked out the case of a satellite "switched on" at $t = 0$ in a disk and found that, after a transient stage as short as $2/(3\Omega_p)$, i.e. one-tenth of an orbit, the torque adopts a constant value as long as the perturbation remains linear. This result may be useful to bear in mind by numericists who perform simulations of embedded planets. Any torque variation that lasts longer than the first orbit cannot be accounted for by the Lindblad torque transients and has to be attributed either to the corotation torque oscillations (due to the libration of material in the co-orbital region, which occurs on a much longer timescale) or to a gradual change in the disk profiles.

### 14.2.1.2 Differential Lindblad torque

In order to sum up the Lindblad torques at inner and outer resonances, one must know the effective location of these resonances, which slightly differs from their nominal position owing to pressure effects. The nominal location of the Lindblad resonances can be identified with the WKB turning point in the dispersion relation of pressure-supported density waves in a differentially rotating disk, which reads

$$m^2(\Omega - \Omega_p)^2 = \kappa^2 + c_s^2 \left( \frac{m^2}{r^2} + k_r^2 \right), \tag{14.2}$$

where $c_s$ is the sound speed and $k_r$ the radial wavevector.

The effective position of these resonances, which can be found by setting $k_r = 0$ in Eq. (14.2), is shifted with respect to the nominal one. In particular, when $m \to \infty$, Lindblad resonances pile up at locations given by

$$r = r_c \pm \frac{\Omega}{2A} H, \tag{14.3}$$

where $A = 1/2 \, rd\Omega/dr$ is the first Oort constant, instead of at the orbit. These points of accumulation correspond to the minimum distance from corotation at which the flow is supersonic (Goodman and Rafikov, 2001). In the case of a Keplerian disk, they lie at $\pm(2/3)H$ from corotation. This has the important consequence that the high-$m$ torque components ($m \gg r/H$) undergo a sharp cut-off (Artymowicz,

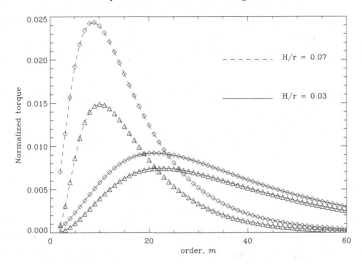

Fig. 14.1. Individual inner and outer torques (in absolute value) in an $h = 0.07$ and $h = 0.03$ disk, as a function of $m$. For each disk thickness, the upper curve shows the outer torque and the lower one the inner torque. These torques are normalized to $\Gamma_0 = \pi q^2 \Sigma a^4 \Omega_{\mathrm{p}}^2 h^{-3}$. Since one-sided Lindblad torques scale with $h^{-3}$, the total area under each of the four curves has same order of magnitude.

1993), since high-$m$ potential components are localized in narrow annuli around the perturber orbit, away from the accumulation points, while their width tends to 0 as $m \to \infty$.

We define the outer (resp. inner) Lindblad torque as the sum of individual Lindblad torques on the outer (resp. inner) Lindblad resonances,

$$\Gamma_{\mathrm{OLR(ILR)}} = \sum_{m=1(2)}^{+\infty} \Gamma_m^{\mathrm{OLR(ILR)}}, \qquad (14.4)$$

where the sum begins at $m = 2$ for the inner torque, since an $m = 1$ rotating potential has no ILR in a Keplerian disk. We shall also refer to these torques as one-sided Lindblad torques. They scale with $h^{-3}$, where $h = H/r$ is the disk aspect ratio (Ward, 1997).

Fig. 14.1 illustrates a number of properties of the one-sided Lindblad torques. In particular, one can see that the torque cut-off occurs at larger $m$ values in the thinner disk (the outer torque value peaks around $m \sim 8$–$9$ for $h = 0.07$, while it peaks around $m \sim 21$–$22$ for $h = 0.03$). Also, there is for both disk aspect ratios a very apparent mismatch between the inner and the outer torques, the former being systematically smaller than the latter. If we consider the torque of the disk acting on the planet, then the outer torques are negative and the inner ones positive, and the total torque is therefore negative. As a consequence migration is directed inwards and leads to a decay of the orbit onto the central object (Ward, 1986).

One can note in Fig. 14.1 that the relative mismatch is larger for the thicker disk. Indeed, it can be shown that this relative mismatch scales with the disk thickness (Ward, 1997). Since one-sided torques scale as $h^{-3}$, the migration rate scales with $h^{-2}$.

There are several reasons for the torque asymmetry which conspire to make the differential Lindblad torque a sizable fraction of the one-sided torque in an $h = O(10^{-1})$ disk (Ward, 1997). In particular, for a given $m$ value, the inner Lindblad resonance lies further from the orbit than the corresponding outer Lindblad resonance.

### 14.2.1.3 The pressure buffer

At first glance the negative sign of the differential Lindblad torque might appear as a fragile result. One may think that a steep surface density profile should reverse this torque, as it should strengthen the inner Lindblad torque and weaken the outer one. However, increasing the surface-density gradient (in absolute value) also increases the pressure gradient. The disk rotational equilibrium then implies that the angular-velocity profile drops so that the sum of the centrifugal force and pressure force cancels out the gravitational force from the central object. As a consequence of the smaller disk angular velocity, all resonances are shifted inwards with respect to a shallower surface-density profile case. This effect plays against the aforementioned surface-density weighting of the inner and outer torques. Quantitatively, both effects appear to nearly cancel each other, and migration is always directed inwards for realistic power-law surface-density and temperature profiles. This effect is known as the pressure buffer (Ward, 1997).

### 14.2.1.4 Torque at a corotation resonance

The angular momentum exchange at a corotation resonance corresponds to different physical processes than at a Lindblad resonance. At the latter the perturbing potential tends to excite epicyclic motion, and, in a protoplanetary disk, the angular momentum deposited is evacuated through pressure-supported waves. Conversely, these waves are evanescent in the corotation region, and are unable to remove the angular momentum brought by the perturber (Goldreich and Tremaine, 1979).

Again, it is instructive to bear in mind a number of properties of the corotation torque exerted on a disk by an $m$-folded external perturbing potential. This torque scales with the gradient of $\Sigma/B$ at the corotation radius. Since $B$ is half the flow vorticity, the corotation torque scales with the gradient of (the inverse of) the specific vorticity, sometimes also called the vortensity. The corotation torque

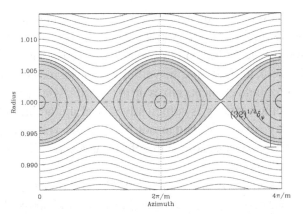

Fig. 14.2. Streamlines in the $(\varphi, r)$ plane at an $m = 2$ corotation resonance. The gray shaded regions show the libration islands. One can notice that the outer and inner disk streamlines (in the white regions) are circulating, and exhibit radial oscillations with an amplitude that decreases with the distance to the corotation radius ($r = 1$). At the same time they do not show any winding, i.e. all these streamlines reach their maximum distance to the corotation radius at a constant azimuth $\varphi = 0$, $2\pi/m$, ... This behavior corresponds to the evanescent pressure-supported waves in the corotation region, which have a purely imaginary radial wavevector (no winding and an exponential decay on the disk pressure scalelength).

therefore cancels out in a $\Sigma \propto r^{-3/2}$ disk, such as the minimum-mass Solar Nebula (MMSN).

The physical picture of the flow at a corotation resonance with azimuthal wavenumber $m$ is a set of $m$ eye-shaped libration islands, in which fluid elements librate on closed streamlines. Such islands are depicted in Fig. 14.2. The libration timescale is much larger than the orbital timescale. As a consequence, the motion of librating fluid elements in a non-rotating frame can be considered, on the orbital timescale, as a circular motion. Therefore, these fluid elements carry an amount of specific angular momentum that only depends on their radial position. As they librate, their radial position oscillates about the corotation radius over the libration period, which implies that they periodically give and take back angular momentum from the perturber. As librating fluid elements remain in a radially bounded interval, the angular momentum they exchange with the perturber averages out to zero over a timescale large compared to their libration timescale.

The libration period furthermore depends on the streamline. This implies phase mixing, which makes the corotation torque tend to zero after a few libration timescales, not only on average, but in instantaneous value as well. This is known as the saturation of the corotation torque. It can be avoided if fluid elements have the possibility to exchange angular momentum not only with the perturber, but also with the rest of the disk. Viscous stress can act to extract angular momentum from the libration islands and prevent saturation.

The corotation-torque saturation can also be described as follows: when the disk viscosity tends to zero, the flow specific vorticity is conserved along a fluid-element path. The libration of fluid elements mixes up the specific vorticity over the libration islands. Once specific vorticity is sufficiently stirred up, an infinitesimally small amount of viscosity suffices to flatten out the specific vorticity over the whole libration island. The corotation torque therefore cancels out, i.e. saturates, since it scales with the vorticity gradient.

In order to avoid saturation, the viscosity must be sufficient to prevent the vortensity profile flattening out across the libration islands. This is possible if the viscous timescale across these islands is smaller than the libration timescale, as shown by Ogilvie and Lubow (2003) and Goldreich and Sari (2003).

Finally, it should be noted that saturation properties are not captured by a linear analysis, since saturation requires a finite libration time, hence a finite resonance width. In the linear limit, the corotation torque appears as a discontinuity at corotation of the advected angular-momentum flux, which corresponds to infinitely narrow, fully unsaturated libration islands.

### *14.2.2 Corotation torque*

All corotation resonances are located at the same corotation radius for a planet in a circular orbit. The flow topology in the co-orbital region of a planet offers striking similarities with the reduced three-body problem of gravitational dynamics. In particular, one can identify a horseshoe region, which is the set of fluid elements librating about the opposite of the planet direction. A sketch of the horseshoe region is given in Fig. 14.3.

Ward (1991, 1992) has identified the full corotation torque with the horseshoe region drag. The physical ingredient that can prevent corotation-torque saturation is viscous stress. In particular, at low $\nu$, the corotation torque is proportional to $\nu$ (Balmforth and Korycansky, 2001), while at high $\nu$ one finds the horseshoe drag of an unperturbed disk profile (Masset, 2001).

Under most standard circumstances, the corotation torque can be safely neglected when evaluating an order of magnitude of the migration timescale in the linear regime. Indeed, even the fully unsaturated corotation torque amounts at most to a few tens of % of the differential Lindblad torque (Ward, 1997; Tanaka *et al.*, 2002), while Korycansky and Pollack (1993) found through numerical integrations that the corotation torque is an even smaller fraction of the differential Lindblad torque than given by analytical estimates.

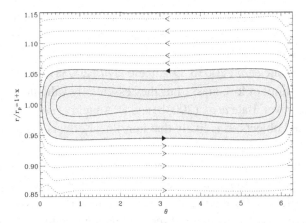

Fig. 14.3. Horseshoe region around the corotation of a planet in circular orbit (shaded area). The planet is located at $r = 1$ and $\theta = 0$ or $2\pi$.

There are, however, noticeable exceptions for which neglecting the corotation torque can be misleading:

- A subgiant planet (typically Saturn sized) in a massive disk may result in runaway migration (also referred to as type III migration), based upon the action of the corotation torque, see Section 14.4.
- The corotation torque can also dominate over the differential Lindblad torque when a planet is located at a sharp surface-density gradient in a disk, such as the edge of a cavity.
- Finally, it seems that for Neptune-sized objects, the total torque estimate obtained from numerical simulations is much lower than the differential Lindblad torque (D'Angelo *et al.*, 2003a; Masset, 2002) and can even reverse migration if the disk is sufficiently thin (Masset, 2002), see Section 14.3.3. Additional work is required on the role of the corotation torque for this mass range, which could correspond to the onset of non-linear effects.

### 14.2.3 Type I migration-drift rate estimates

There is a number of estimates of the type I migration rate in the literature (see Ward, 1997 and refs. therein). The most recent linear calculations by Tanaka *et al.* (2002) take into account three-dimensional effects, and are based upon the value of the total tidal torque, including the corotation torque (fully unsaturated since it is a linear estimate). It reads (Tanaka *et al.*, 2002):

$$\tau \equiv a/\dot{a} = (2.7 + 1.1\alpha)^{-1} q^{-1} \frac{M_*}{\Sigma a^2} h^2 \Omega_{\rm p}^{-1}, \qquad (14.5)$$

for a surface-density profile $\Sigma \propto r^{-\alpha}$.

For an Earth-mass planet around a Solar-mass star at $r = 1$ AU, in a disk with $\Sigma = 1700$ g cm$^{-2}$ and $h = 0.05$, this translates into $\tau = 1.6 \times 10^5$ years.

This analytical estimate has been verified by means of three-dimensional numerical simulations (Bate *et al.*, 2003b; D'Angelo *et al.*, 2003a). Both find an excellent agreement in the limit of low-mass, thus they essentially validate the linear analytical estimate. However, Bate *et al.* (2003b) and D'Angelo *et al.* (2003a) results differ for Neptune-sized objects. It is likely that non-linear effects begin to be important at this mass.

The type I migration timescale is very short, much shorter than the build up time of the $M_{\rm p} \sim 5$–15 $M_\oplus$ solid core of a giant planet. Hence, type I migration constitutes a bottleneck for the accretion scenario for these massive cores. To date this remains an unsolved problem, but see the recent work by Alibert *et al.* (2004). Some recent attempts to include more detailed physics of the protoplanetary disk, such as opacity transitions and their impact on the disk profile (Menou and Goodman, 2004), or radiative transfer and the importance of shadowing in the planet vicinity (Jang-Condell and Sasselov, 2005), have lead to lower estimates of the type I migration rates which might help resolve the accretion/migration timescale discrepancy.

## 14.3 Type II migration

When the planet grows in mass it cannot be treated as a small perturbing object anymore. The gravitational interaction with the disk becomes non-linear and the wake of the planet turns into a shock. Eventually, for planetary masses around 1 $M_{\rm J}$ the density at the planetary orbit is lowered and an annular gap forms. In this regime numerical methods are used primarily to analyze the dynamical planet–disk interaction, the density structure of the disk, and the resulting gravitational torques acting on the planet. This non-linear regime is called type II migration.

### 14.3.1 Numerical modeling

The first modern high-resolution hydrodynamical calculations of planet–disk interaction were performed by Kley (1999), Bryden *et al.* (1999), and Lubow *et al.* (1999). Since protoplanetary accretion disks are assumed to be vertically thin, these first simulations used a two-dimensional ($r - \varphi$) model of the accretion disk. The vertical thickness $H$ of the disk was incorporated by assuming a given radial temperature profile $T(r) \propto r^{-1}$ which makes the ratio $H/r$ constant. Typically the simulations assumed $H/r = 0.05$ which refers to a disk where at each radius the Keplerian speed is 20 times faster that the local sound speed. Initial density profiles typically had power laws for the surface density $\Sigma \propto r^{-s}$ with $s$ between 0.5 and

1.5. Later also fully three-dimensional models were calculated which still use a simple isothermal equation of state today.

For the anomalous viscosity of accretion disks a Reynolds stress tensor formulation (Kley, 1999) is used typically where the kinematic viscosity $\nu$ is either constant or given by an $\alpha$-prescription $\nu = \alpha c_s H$, where $\alpha$ is constant and $c_s$ is the local sound speed. From observations, values lying between $10^{-4}$ and $10^{-2}$ are inferred for the $\alpha$-parameter of protoplanetary disks. Full magnetohydrodynamic (MHD) calculations have shown that the viscous stress-tensor ansatz may give (for sufficiently long time averages) a reasonable approximation to the *mean* flow in a turbulent disk (Papaloizou and Nelson, 2003). The embedded planets are assumed to be point masses (using a smoothed potential), and together with the star they are treated as a classical $N$-body system. The disk also influences the orbits through gravitational torques. This is the desired effect to study as it will cause the orbital evolution of the planets. The planets may also accrete mass from the surrounding disk (Kley, 1999). To enhance resolution in the vicinity of the planet, the computations are typically performed in a rotating frame. Numerically, a special treatment of the Coriolis force has to be incorporated to ensure angular momentum conservation (Kley, 1998).

### *14.3.2 Viscous laminar disks*

The type of modeling outlined in the previous section yields, in general, smooth density and velocity profiles, and we refer to those models as *viscous laminar disk* models, in contrast to models which do not assume an a priori given viscosity and rather model the turbulent flow directly.

A typical result of such a viscous computation obtained with a $128 \times 280$ grid (in $r - \varphi$) is displayed in Fig. 14.4. Here, the planet with mass $M_p = 1\,M_J$ and semi-major axis $a_p = 5.2$ AU is *not* allowed to move and remains on a fixed circular orbit, an approximation which is typical in many simulations. Clearly seen are the major effects an embedded planet has on the structure of the protoplanetary accretion disk. The gravitational force of the planet leads to a spiral wave pattern in the disk. In this calculation (Fig. 14.4) there are two spirals, in the outer disk and inner disks. The tightness of the spiral arms depends on the temperature (i.e. $H/r$) of the disk. The lower the temperature the tighter the spirals. The second prominent feature is the density gap at the location of the planet. It is caused by the deposition of positive (at larger radii) and negative (at smaller radii) angular momentum in the disk. The spiral waves are corotating in the frame of the planet, and hence their pattern speed is faster (outside) and slower (inside) than the disk material. Dissipation by shocks or viscosity leads to the deposition of angular momentum, and pushes material away from the planet. The equilibrium width of the gap is determined by

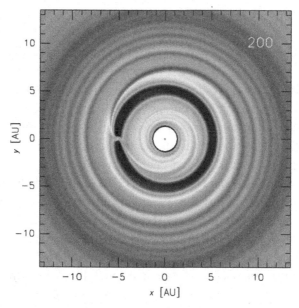

Fig. 14.4. Surface-density profile for an initially axisymmetric planet–disk model after 200 orbits of the planet.

the balance of gap-closing viscous and pressure forces and gap-opening gravitational torques. For typical parameters of a protoplanetary disk, a Saturn-mass planet will begin to open a visible gap. More details on gap-opening criteria are given in the review by Lin and Papaloizou (1993), and see also Bryden *et al.* (1999).

To obtain more insight into the flow near the planet and to calculate accurately the torques of the disk acting on the planet, a much higher spatial resolution is required. As this is necessary only in the immediate surroundings of the planet, a number of nested-grid and also variable grid-size simulations have been performed (D'Angelo *et al.*, 2002, 2003a; Bate *et al.*, 2003b). Such a grid system is not adaptive, as it is defined at the beginning of the computation and does not change with time. The planet is placed in the center of the finest grid.

The result for a two-dimensional computation using six grids is displayed in Fig. 14.5; for more details see also D'Angelo *et al.* (2002). The top left base grid has a resolution of $128 \times 440$ and each sub-grid has a size of $64 \times 64$ with a refinement factor of two from level to level. It is noticeable that the spiral arms inside the Roche-lobe of the planet are detached from the global outer spirals. The two-armed spiral around the planet extends deep inside the Roche-lobe and allows for the accretion of material onto the planet. The nested-grid calculations have recently been extended to three dimensions and a whole range of planetary masses have been investigated, starting from $1 \, M_\oplus$ to a few $M_J$ (Kley *et al.*, 2001; D'Angelo *et al.*, 2003a). In the three-dimensional case the strength of the spiral

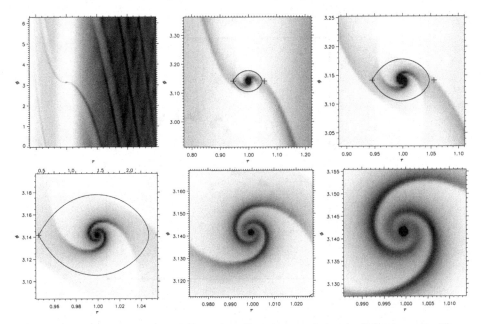

Fig. 14.5. Density structure of a 1 $M_J$ planet on each level of the nested grid system, consisting of six grid levels in total. The top left panel displays the total computational domain. The line indicates the size of the Roche lobe.

arms is weaker and accretion occurs primarily from regions above and below the midplane of the disk.

### *14.3.3 The migration rate*

Such high-resolution numerical computations allow for a detailed computation of the torque exerted by the disk material onto the planet, and its mass-accretion rates. Figure 14.6 gives the inverse $1/\tau_M$ of the migration for three-dimensional nested-grid calculations as a function of the planet mass, given in units of the mass of the central star, $q = M_p/M_\odot$. The straight line by Tanaka *et al.* (2002) assumes a three-dimensional flow and takes the corotation torques into account. The symbols refer to different approximations of the potential of the planet. It can be seen that for low masses $q \approx 10^{-5}$ and intermediate masses $q \approx 8 \times 10^{-5}$ the numerical results fit well to the linear theory. In the intermediate range of about $q \approx 3 \times 10^{-5}$ the migration rates are about an order of magnitude smaller (D'Angelo *et al.*, 2003a). This effect may be caused by the onset of gap formation in the mass range of about 10–15 $M_\oplus$. Here non-linear effects begin to set in and modify the physics. These results should be compared to those obtained by Bate *et al.* (2003b) and Masset (2002).

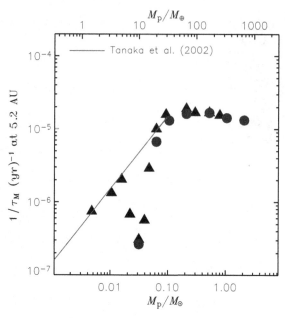

Fig. 14.6. The inverse migration rate for different planet masses. The symbols denote different approximations (smoothing) for the potential of the planet. The solid line refers to linear results for type I migration from Tanaka *et al.* (2002).

The consequences of accretion and migration have been studied by numerical computations which do not hold the planet fixed at some radius but rather follow the orbital evolution of the planet (Nelson *et al.*, 2000), allowing planetary growth. The typical migration and accretion timescales are of the order of $10^5$ yr, while the accretion timescale may be slightly smaller. This is in very good agreement with the estimates obtained from the models using a fixed planet. These simulations show that during their inward migration they grow up to about 4 $M_J$.

The consequence of the inclusion of thermodynamic effects (viscous heating and radiative cooling) has been studied by D'Angelo *et al.* (2003b) in two-dimensional calculations, where an increased temperature in the circumplanetary disk has been found. This has interesting consequences for the possible detection of an embedded protoplanet (Wolf *et al.*, 2002). The effect that the self-gravity of the disk has on migration has been analyzed through numerical simulations (Nelson and Benz, 2003a,b). For typical disk masses the influence is rather small.

### 14.3.4 Inviscid disks

To investigate the influence of viscosity we also studied recently inviscid models with no physical viscosity added. Only a numerical bulk viscosity has to be added

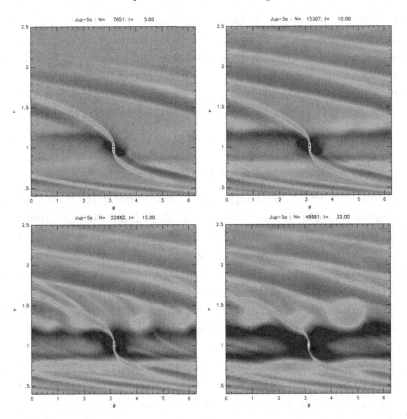

Fig. 14.7. Results of an inviscid disk computation with an embedded 1 $M_J$ planet. Results are displayed at four different times (5, 10, 15, and 32) in units of the orbital period of the planet. Shown are gray-scale plots of the surface density in an $(r - \varphi)$-coordinate system.

to ensure numerical stability (Kley, 1999). However, this has no influence on the physical effects of the simulations. In Fig. 14.7 we display the density structure of such an inviscid model for a 1 $M_J$ planet. The planet is not allowed to move and for stability purposes its mass is gradually switched on during the first five orbits. Results are displayed at four different times (5, 10, 15, 32). After only five orbits the spiral waves are already clearly visible, as they form on a dynamical timescale. The gap is just beginning to clear. At the edges of the gap high density blobs (*vortices*) are forming which are reduced in number through merging. Eventually only one big blob remains. Also inside the gap, some detailed structure is visible. In comparison, fully viscous simulations show these vortices mostly as transient features at the beginning of the simulations. They vanish later on as a result of the viscosity. Vortices in a inviscid disk with an embedded planet have also been studied by Koller *et al.* (2003).

If accretion and also migration of the planet are considered, these moving vortices may create some more complex time dependence. However, the details of such inviscid models will have to be studied in future work. Vortices in accretion disks have sometimes been considered to enhance planet formation either through triggering a direct gravitational instability or by trapping particles in it (Godon and Livio, 2000; de la Fuente Marcos and Barge, 2001; Klahr and Bodenheimer, 2003).

## 14.4 Type III migration

Thus far, the torque acting on a migrating planet was considered to be independent of its migration rate. This is true for the differential Lindblad torque. However, the corotation torque implies material that crosses the planet orbit on the U-turn of the horseshoe streamlines. In a non-migrating case, only the trapped material of the horseshoe region participates in these U-turns, but in the case of an inwards (resp. outwards) migrating planet, material of the inner disk (resp. outer disk) has to flow across the co-orbital region and executes one horseshoe U-turn to do so. By doing this, it exerts a corotation torque on the planet that scales with the drift rate. We give below a simplified derivation of the corotation torque dependency upon the drift rate $\dot{a}$. A more accurate derivation can be found in Masset and Papaloizou (2003).

We will call $x_s$ the half radial width of the horseshoe region. The amount of specific angular momentum that a fluid element near the separatrix takes from the planet when it switches from an orbit with radius $a - x_s$ to $a + x_s$ is $4Bax_s$.

The torque exerted on the planet in steady migration with drift rate $\dot{a}$ by the inner or outer disk elements as they cross the planet orbit on a horseshoe U-turn is therefore, to the lowest order in $x_s/a$,

$$\Gamma_2 = (2\pi a \Sigma_s \dot{a}) \cdot (4Bax_s), \tag{14.6}$$

where we keep the same notation as in Masset and Papaloizou (2003), and where $\Sigma_s$ is the surface density at the upstream separatrix. The first bracket in the above equation represents the mass flow rate from the inner disk to the outer one (or vice versa, depending on the sign of $\dot{a}$). As the system of interest for which we evaluate the sum of external torques, we take the system composed of the planet and all fluid elements trapped in libration in its co-orbital region, namely the whole horseshoe region (with mass $M_{HS}$) and the Roche lobe content (with mass $M_R$), because all of these parts perform a simultaneous migration.

The drift rate of this system is then given by

$$(M_p + M_{HS} + M_R) \cdot (2Ba\dot{a}) = (4\pi a x_s \Sigma_s) \cdot (2Ba\dot{a}) + \Gamma_{LR}, \tag{14.7}$$

which can be rewritten as

$$m_p \cdot (2Ba\dot{a}) = (4\pi a \Sigma_s x_s - M_{HS}) \cdot (2Ba\dot{a}) + \Gamma_{LR}, \qquad (14.8)$$

where $m_p = M_p + M_R$ is the mass content of the Roche lobe, including the planet, which for short we also refer to as the planet mass, assuming that the material orbiting the circumplanetary disk "belongs" to the planet. The first term of the first bracket of the r.h.s. corresponds to the horseshoe region surface multiplied by the upstream separatrix surface density, hence it is the mass that the horseshoe region would have if it had a uniform surface density equal to the upstream surface density. The second term is the actual horseshoe region mass. The difference between these two terms is called, in Masset and Papaloizou (2003), the coorbital mass deficit and denoted $\delta m$. Note that we could also have included the Roche lobe mass in the coorbital mass deficit and kept the planet mass as the mass of the point-like object at the center of the Roche lobe alone. This choice would lead to an equivalent formulation, and to the same runaway criterion. Equation (14.8) yields a drift rate

$$\dot{a} = \frac{\Gamma_{LR}}{2Ba(m_p - \delta m)}. \qquad (14.9)$$

This drift rate is faster than the standard estimate in which one neglects $\delta m$. This comes from the fact that the co-orbital dynamics alleviates the differential Lindblad torque task by displacing fluid elements from the upstream to the downstream separatrix. The angular momentum they extract from the planet by doing so favors its migration.

As $\delta m$ tends to $m_p$, most of the angular momentum lost by the planet and its co-orbital region is gained by the orbit-crossing circulating material, making migration increasingly cost effective.

When $\delta m \geq m_p$, the above analysis, assuming a steady migration ($\dot{a}$ constant), is no longer valid. Migration undergoes a runaway, and has a strongly time-varying migration rate, that increases exponentially over the first libration timescales. An analysis similar to the above calculation may be performed, in which the corotation torque depends on the migration rate, except that one now has to introduce a delay $\tau$ between the mass inflow at the upstream separatrix and its consequence on the corotation torque. This delay represents the feedback loop latency,

$$\Gamma_{CR}(t) = 2Ba\delta m\dot{a}(t - \tau). \qquad (14.10)$$

A Taylor expansion in time of $\dot{a}(t - \tau)$ yields a first order differential equation for $\dot{a}$ (Masset and Papaloizou, 2003). The linear dependence of the corotation on the drift rate remains valid as long as the semi-major-axis variation over a horseshoe

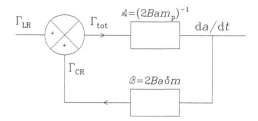

Fig. 14.8. Schematic representation of the positive feedback loop. The loop latency is $\sim \tau_{lib}$.

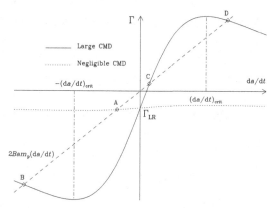

Fig. 14.9. The solid curve shows the total torque on the planet in a massive disk (hence with a large co-orbital mass deficit) as a function of the drift rate. For $|\dot{a}| \ll \dot{a}_{crit}$ the torque exhibits a linear dependence in $\dot{a}$. The dotted line shows the torque in a low mass disk (i.e. with a negligible co-orbital mass deficit), in which case the torque is almost independent of the migration rate and is always close to the differential Lindblad torque $\Gamma_{LR}$. The dashed line represents the planet angular-momentum-gain rate as a function of $\dot{a}$, assuming a circular orbit. For a given situation, the migration rate achieved by the planet is given by the intersection of the dashed line with the torque curve. In the low-mass disk case, the intersection point, A, is unique, and stable. It yields a negative drift rate controlled by the differential Lindblad torque. In the high-mass disk case (runaway case), there are three points of intersection (B, C and D). The central point (C) is unstable, while the extreme ones (B and D) are stable and correspond to the maximum runaway drift attained by the planet, either inwards (point B) or outwards (point D).

libration time is smaller than the horseshoe zone width, i.e.

$$|\dot{a}| < \dot{a}_{crit} = \frac{Ax_s^2}{2\pi a}. \tag{14.11}$$

The corotation torque then reaches a maximum and slowly decays for larger values of $\dot{a}$ (see Fig. 14.9).

We should comment that the mechanism upon which type III (or runaway) migration is based can be described by the standard formalism of positive feedback loops, as shown in Fig. 14.8.

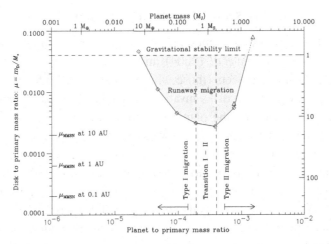

Fig. 14.10. Runaway-limit domain for a $H/r = 0.04$ and $\nu = 10^{-5}$ disk, with a surface density profile $\Sigma \propto r^{-3/2}$. The variable $m_D = \pi \Sigma r^2$ features on the y-axis. It is meant to represent the local disk mass, and it therefore depends on the radius. Runaway is most likely for Saturn-mass planets. These would undergo runaway in disks no more massive than a few times the minimum-mass Solar Nebula. Runaway is impossible for massive planets ($M > 1\ M_J$) as the surface density is very low at the upstream separatrix. Low mass objects ($M < 10\ M_\oplus$) do not deplete their co-orbital region and therefore cannot undergo a runaway.

The open loop gain is $G' = \mathcal{AB} = \delta m / m_p$, so the system stability condition, which is $G' < 1$ reads here $\delta m < m_p$. Above this threshold, one gets a runaway. Below, the closed loop gain, which is given by

$$G = \frac{\mathcal{A}}{1 - \mathcal{AB}}, \tag{14.12}$$

yields Eq. (14.9).

A standard bifurcation analysis can be performed on this feedback loop, as illustrated in Fig. 14.9. The transition from one case to the other (one intersection point to three intersection points) occurs when the angular-momentum-change rate line (which has slope $2Bam_p$) and the torque curve near the origin are parallel. Since this latter has a slope $2Ba\delta m$ near the origin, the transition occurs for $m_p = \delta m$.

The disk critical mass above which a planet of given mass undergoes a runaway depends on the disk parameters (aspect ratio and effective viscosity). The limit has been worked out by Masset and Papaloizou (2003) for different disk aspect ratios and a kinematic viscosity $\nu = 10^{-5}$. We reproduce in Fig. 14.10 the type III-migration domain for a disk with $H/r = 0.04$.

## 14.5 Other modes of migration

We shall briefly mention another mode of migration recently studied by Nelson and Papaloizou (2004), Laughlin *et al.* (2004), and Nelson (2005), which deals with the evolution of small-mass objects (those which would undergo type I migration in a laminar disk) embedded in disks invaded by magnetorotational turbulence. The torque of the disk acting on the planet is then a strongly time-varying quantity, which endows the planet with a random walk motion in the semimajor axis, rather than yielding the monotonous decrease observed in type I migration. This random walk motion, sometimes referred to as "diffusive" or "stochastic" migration, involves even very low-mass objects, which would undergo a negligible migration in laminar disks. Also, as superimposed onto the global drift under the wake action, it exacerbates the problem of type I migration in the sense that it lowers the timescale expectancy for a given protoplanet to reach the central object, but, being probabilistic in nature, it also seems to leave room for a small fraction of objects which never reach the central object over the turbulent-disk lifetime. A more conclusive outcome would require computations that are currently not possible. The interested reader should refer to the above references for further details.

## 14.6 Eccentricity driving

The population of extrasolar giant planets with orbital periods larger than ten days displays an important scatter in eccentricity, which can reach, for some systems, values as large as 0.9. Thus far the origin of these eccentricities has not been elucidated. Planet–planet interactions (Ford *et al.*, 2001) and disk–planet interactions (Goldreich and Sari, 2003) have been proposed to be responsible for the excitation of giant-planet eccentricities. Here we sum up the basics of the driving of eccentricity through disk–planet resonant interactions, and we comment on recent developments on this subject.

In Section 14.2 we outlined the angular-momentum exchange between a planet on a circular orbit and the disk at the principal Lindblad resonances and at the co-orbital corotation resonances. A planet on an eccentric orbit interacts with the disk at additional resonances. Indeed, in addition to the potential components with pattern frequency $\Omega_p$ (see Eq. 14.1), other potential components appear with pattern frequencies $\Omega_p + \frac{k}{m}\kappa_p$, and with amplitudes scaling as $e^{|k|}$. Assuming that the eccentricity is low, one can restrict the analysis to the terms that scale with e. The resonances associated with these terms are called respectively the first-order Lindblad and corotation resonances. While the principal Lindblad resonances negligibly contribute to the eccentricity driving or damping, the first-order Lindblad resonances can lead to substantial eccentricity variation over a timescale much smaller than the migration timescale. Specifically, the first-order Lindblad resonances can be divided into two groups: the coorbital Lindblad resonances, which damp the eccentricity, and the external Lindblad resonances, which excite the eccentricity. In the case of a giant protoplanet that opens a clean gap, the co-orbital Lindblad resonances fall in the middle of the gap and are "switched off," hence as a net result the first-order Lindblad resonances excite the eccentricity. In addition, it can be shown that the first-order corotation resonances damp the eccentricity. The damping timescale can be directly compared to the excitation timescale by the first-order Lindblad resonances (Goldreich and Tremaine, 1980). One finds that eccentricity damping dominates by a small margin of $\sim 5\%$. This would be the end of the story if corotation resonances did not saturate. However, if for a given planet eccentricity, the disk viscosity is low enough, the corotation resonances can be sufficiently saturated to reverse the eccentricity balance. The analyses of Ogilvie and Lubow (2003) and Goldreich and Sari (2003), which assume a perfectly clean gap (i.e. no damping by the co-orbital Lindblad resonances), suggest that a Jupiter-sized protoplanet embedded in a disk with $H/r = 4$–5% and a viscosity $\alpha \sim 10^{-3}$ can undergo an eccentricity excitation if its initial eccentricity is as low as 0.01. These analytical works rely upon simplifying assumptions, as they consider the saturation of isolated corotation resonances, therefore neglecting the overlap between successive corotation resonances. They also neglect the fact that the first-order corotation resonances share their location with the principal Lindblad resonances. However, a numerical study of the two-resonance case (either corotation-corotation or corotation-Lindblad) show that the saturation properties of a non-isolated corotation resonance are very similar to the properties of an isolated one (Masset and Ogilvie, 2004). This seems to indicate that the above analytical studies essentially provide a correct evaluation of the saturation of the first-order corotation resonances. Nevertheless, no eccentricity excitation for planetary-mass objects has ever been observed in numerical simulations of disk–planet interactions (Papaloizou *et al.*, 2001). This could be a mesh-resolution effect (Masset and Ogilvie, 2004), but more recent simulations performed at higher resolution do not exhibit eccentricity excitation either. One explanation is that the gap opened by a Jupiter-sized planet in a disk such as the one mentioned above is far from being clean, so that co-orbital Lindblad resonances still sizably contribute to eccentricity damping. Another issue that needs to be properly dealt with by numerical simulations is the secular exchange of eccentricity between the planet and the disk (see Masset and Ogilvie, 2004, and refs. therein). This exchange occurs on extremely long timescales ($\gg 10^3$ orbits), which constitute a challenge for hydrodynamic numerical simulations of disk–planet interactions.

# 15

## The brown dwarf–planet relation

Matthew R. Bate

*School of Physics, University of Exeter, United Kingdom*

### 15.1 Introduction

Given that brown dwarfs are usually much more massive than planets (see Section 15.6), it is somewhat surprising that the first incontrovertible discovery of a brown dwarf (Nakajima *et al.*, 1995) and the discovery of the first extrasolar planet (Mayor and Queloz, 1995) were announced simultaneously in 1995. Over the past decade, the rapid progress made in both fields has been extraordinary. There are now more than 150 extrasolar planets known, including more than a dozen multiple-planet systems. The first brown dwarf, Gliese 229B, was found in orbit around an M-dwarf, but in the same year other candidates, later confirmed to be free-floating brown dwarfs, were announced (e.g. Teide 1 by Rebolo *et al.*, 1995), along with PP1 15 which was later discovered to be a binary brown dwarf (Basri and Martin, 1999). Observations now suggest that brown dwarfs are as common as stars, although stars dominate in terms of mass (e.g. Reid *et al.*, 1999).

Since the rest of this book is devoted to the topic of planets, in this chapter I will review the properties and potential formation mechanisms of brown dwarfs, comparing and contrasting them with planets, but referring the reader to the other chapters of the book for detailed information on planets.

### 15.2 Masses

The most fundamental parameter of a brown dwarf is its mass. In particular, a brown dwarf can be considered to be a failed star, in that it may form in a similar manner to a star (Section 15.5) but its mass is insufficient for it to join the hydrogen-burning main-sequence (Hayashi and Nakano, 1963; Kumar, 1963). Note that brown dwarfs with masses somewhat lower than this mass may burn hydrogen briefly, but not

*Planet Formation: Theory, Observation, and Experiments*, ed. Hubert Klahr and Wolfgang Brandner.
Published by Cambridge University Press. © Cambridge University Press 2006.

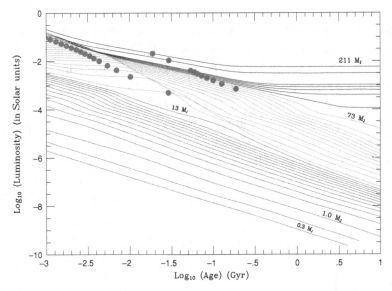

Fig. 15.1. The time evolution of luminosity for low-mass stars, brown dwarfs and giant planets (taken from Burrows *et al.*, 2001, Figure 1). The first set of dots mark when 50% of the deuterium has burned, while the second set denote when 50% of the lithium has burned. Note that deuterium burning delays the fading of brown dwarfs with masses greater than 13 $M_J$.

at a rate high enough to compensate for their radiant energy losses. The critical hydrogen-burning main-sequence mass depends weakly on the metallicity of the object (see Burrows *et al.*, 2001 for a review), ranging from approximately 0.07 $M_\odot$ for Solar metallicities to approximately 0.092 $M_\odot$ for zero metallicity gas. Thus, a working definition is that any object that forms like a star but has a mass lower than approximately 70 times the mass of Jupiter ($M_J$) is a brown dwarf.

When high-mass brown dwarfs are very young, their luminosities are almost indistinguishable from those of low-mass stars since both objects release energy gained from gravitational contraction on their Kelvin-Helmholtz timescales (Fig. 15.1). However, whereas low-mass stars eventually settle onto the hydrogen-burning main sequence which lasts for many Hubble times, brown dwarfs lack the required energy source and, thus, fade in brightness (Burrows *et al.*, 1997). Together, the faintness of old brown dwarfs and the similarity of young high-mass brown dwarfs to young stars, were the reasons that the first *incontrovertible* brown dwarf was not discovered until 1995; although many candidates had been found earlier in star-forming regions it was not possible to determine their masses accurately. The first brown dwarf, Gliese 229B, was an old brown dwarf that was clearly cool enough not to be a star but also too luminous and at too large a distance from its companion star to be a giant planet.

Over the past decade, the use of large telescopes in the infrared has allowed us to probe the mass function of brown dwarfs in young star-forming regions (i.e. the low-mass end of the "stellar" initial mass function, IMF). Spectroscopically determined mass functions down to $\approx 20$ $M_J$ have been obtained in Taurus (Briceño *et al.*, 2002; Luhman *et al.*, 2003a), IC 348 (Luhman *et al.*, 2003b), and Orion (Slesnick *et al.*, 2004). Other surveys, in particular 2MASS and DENIS, have allowed the local Solar Neighborhood to be probed for old brown dwarfs. In the local Solar Neighborhood, brown dwarfs seem to be roughly as numerous as stars (Reid *et al.*, 1999). Thus, while very common numerically, brown dwarfs do not play a significant role in the mass budget of our Galaxy. In particular, brown dwarfs are not a significant source of dark matter. The parameterisation (see Fig. 15.6) of the IMF given by Chabrier (2003) of

$$\frac{dN}{d\log M} \propto \begin{cases} 0.158 \exp\left(-\dfrac{(\log m - \log 0.079)^2}{2 \times (0.69)^2}\right), & m \leq 1\,M_\odot, \\[3mm] 0.044 m^{-1.3}, & m \geq 1\,M_\odot, \end{cases} \tag{15.1}$$

which is based on objects within 8 pc of the Sun also gives a reasonable fit (when appropriately normalized) to the mass functions down to 20 $M_J$ observed in most of the star-forming regions, with the possible exception of Taurus.

Currently, the lowest observed masses of brown dwarfs are $\approx 0.01$ $M_\odot$. Theoretically, the minimum mass of a brown dwarf is thought to be set by the opacity limit for fragmentation (Low and Lynden-Bell, 1976; Rees, 1976). The initial phases of the collapse of self-gravitating molecular gas (e.g. Larson, 1969) are thought to occur almost isothermally. This allows the possibility of fragmentation of a collapsing cloud into lower-mass objects since the Jeans mass

$$M_{\text{Jeans}} \propto \frac{c_s^3}{\sqrt{G^3 \rho}}, \tag{15.2}$$

which depends on the sound speed $c_s$ and density $\rho$ of the gas, decreases with increasing density. However, when the rate of energy released during the collapse exceeds the rate at which the gas can radiate that energy away, the collapsing gas begins to heat up and the Jeans mass increases. This results in the formation of a pressure-supported hydrostatic object with an initial mass of the order of a few $M_J$ and an initial radius of a few AU. This object is known as the first hydrostatic core. This object accretes gas from its surrounding envelope until its central temperature reaches $\approx 2000$ K, at which stage molecular hydrogen begins to dissociate, absorbing energy and resulting in a nearly isothermal collapse of the first core from the inside out to form a stellar core. Fragmentation during this second collapse is thought to be inhibited by the high degree of thermal pressure and, if non-axisymmetric structure does develop, by gravitational torques (Boss, 1989; Bate, 1998). The result is that if brown dwarfs form via the dynamic collapse of molecular gas, they are expected to have minimum masses of a few $M_J$ (Low and Lynden-Bell, 1976; Rees, 1976; Silk, 1977a,b; Boss, 1988; Masunaga and Inutsuka, 1999; Boss, 2001a). The density at which heating occurs during the first collapse and, thus, the minimum mass depends on the opacity of the gas, hence the term "opacity limit for fragmentation." However, at least for dusty gas (i.e. population I and II stars), the minimum mass is only thought to vary with metallicity $Z$, as $Z^{-1/7}$ (i.e. a very weak dependence; Low and Lynden-Bell, 1976).

## 15.3 Evolution

As described above, the defining difference between a star and a brown dwarf is that a brown dwarf does not join the hydrogen-burning main sequence. Thus, the main evolution a brown dwarf undergoes is simply to cool with time. In this respect, a brown dwarf and a giant planet are very similar with the main difference in temperature at a given age depending simply on the mass of the object.

However, although most brown dwarfs do not burn hydrogen, objects with masses $\gtrsim 13$ $M_J$ do burn their deuterium. This has little long-term effect on the brown dwarf, but it does mean that for the first 6 to 30 million years (masses $\approx 70$ to 13 $M_J$, respectively) they do not cool as fast as they would without a fusion energy source (see Fig. 15.1 and the review of Burrows *et al.*, 2001 from which this figure is taken). It also leads to a potentially useful distinction between a planet and a brown dwarf because it so happens that the brown dwarf desert observed around Solar-type stars applies to objects with masses in the range $\approx 15$–$80$ $M_J$.

## 15.4 The multiplicity of brown dwarfs

The first confirmed brown dwarf was the companion to a star. From the theoretical point of view, the multiplicity of brown dwarfs and the properties of those multiple systems give us many more constraints for models of brown-dwarf formation than the mass function, which only depends on a single parameter.

Multiplicity can be divided into two main categories. The first is the occurrence of binary (and multiple) brown dwarf systems and their comparison with multiple stellar systems. The second is the occurrence of brown dwarfs as stellar companions. The latter is of particular interest when it comes to comparing brown dwarfs and planets.

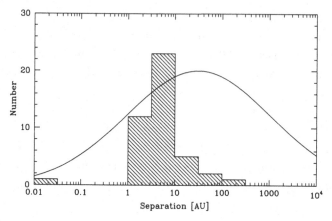

Fig. 15.2. The distribution of orbital separations of binary brown dwarfs (histogram) taken from Siegler *et al.* (2005) compared with a parameterisation of the separations of Solar-type binary stars (log-normal curve) taken from Duquennoy and Mayor (1991). Surveys of binary brown dwarfs are currently incomplete for separations less than a few AU, but the median separation is clearly smaller for brown-dwarf binaries than stellar binaries.

### *15.4.1 Binary brown dwarfs*

One of the earliest brown dwarfs to be discovered, PPl-15 in the Pleiades, turned out to be a 5.8 day spectroscopic binary brown dwarf with component masses of approximately 60 and 70 $M_J$ (Basri and Martin, 1999). However, to date, this object remains the only confirmed spectroscopic binary brown-dwarf system. Most known binary brown dwarfs have been discovered through adaptive optics surveys of nearby field brown dwarfs. They find that brown dwarfs with projected separations $\gtrsim 1$ AU have a frequency of $\approx 15\%$ (Reid *et al.*, 2001; Close *et al.*, 2002, 2003; Bouy *et al.*, 2003; Burgasser *et al.*, 2003; Gizis *et al.*, 2003; Martin *et al.*, 2003; Siegler *et al.*, 2005). The vast majority of these have separations less than 20 AU. In fact, for a long time, binary brown dwarfs with separations $\gtrsim 20$ AU appeared not to exist. However, recently Luhman (2004a) reported the discovery of a very wide binary brown-dwarf system (projected separation $\approx 240$ AU). There are also two other wide very low-mass or brown-dwarf binaries with projected separations of 35 and 55 AU (Siegler *et al.*, 2005 and references within).

Comparing binary brown dwarfs with stellar binaries (see Fig. 15.2), it is clear that the typical binary brown dwarf has a smaller separation. Duquennoy and Mayor (1991) found the median binary separation for Solar-type stars to be $\approx 30$ AU whereas for very low-mass and substellar binaries the median separation is less than 5 AU (Close *et al.*, 2003; Siegler *et al.*, 2005). As we will see in Section 15.5, the separation distribution of binary brown dwarfs may be crucial to determining how brown dwarfs form.

The mass ratios of binary brown dwarfs also appear to be very different from those of Solar-type stars. Duquennoy and Mayor (1991) found that the frequency of stellar binaries increases towards more unequal mass binaries with a peak at $M_2/M_1 \approx 1/4$. However, there is a clear preference for binary brown dwarfs to have mass ratios near unity (Close *et al.*, 2003).

### 15.4.2 Brown dwarfs as companions to stars

To date, most extrasolar planetary systems have been discovered using the radial velocity (Doppler) method to detect an object in orbit around a star. This method is more sensitive to higher-mass companions. Thus, any radial velocity searches for extrasolar planets should also be capable of detecting brown-dwarf companions.

However, although more than 150 planets have now been discovered (mainly orbiting Solar-type stars), very few brown-dwarf companions have been found. This has led to the term "brown-dwarf desert" when describing the frequencies of brown dwarfs with orbital periods $\lesssim 5$ years around Solar-type stars. Brown dwarfs have a frequency of $\lesssim 1\%$ (Marcy and Butler, 2000; Halbwachs *et al.*, 2000), much less than either giant planets (frequency $\gtrsim 6\%$; Marcy and Butler, 2000; Zucker and Mazeh, 2001) or stellar companions (frequency $\approx 11\%$; Duquennoy and Mayor, 1991; Zucker and Mazeh, 2001). In many ways, this is a useful "coincidence" since, at least at small separations around Solar-type stars, it makes the distinction between planets and brown dwarfs almost irrelevant (see Section 15.6).

The existence of the brown-dwarf desert naturally provokes two questions. Does the desert also exist for wider brown-dwarf companions? Does it also apply to lower-mass stars?

Recent searches for wide brown-dwarf companions to stars have begun to answer the former question. Gizis *et al.*, (2003) reported that brown-dwarf companions to stars at separations $\gtrsim 1000$ AU are just as frequent as stellar companions. However, at all other separations they appear to be much less frequent. From an infrared coronagraphic survey, McCarthy and Zuckerman (2004) estimate the frequency with separations between 75 and 300 AU to be $\sim 1\%$, similar to the brown-dwarf frequency at small separations. For lower-mass stars, the question has barely begun to be investigated. However, the frequency of binary brown dwarfs of $\approx 15\%$ shows that, at least at sufficiently low primary masses, the brown-dwarf companion desert disappears.

## 15.5 Formation mechanisms

In answering the question "What is the difference between a planet and a brown dwarf?," an obvious question is "Do they form via different mechanisms?"

There are two main mechanisms for the formation of giant planets. The first is the core-accretion model (see Chapters 8 and 10) where, in its simplest form, a solid core grows through planetesimal agglomeration until it is sufficiently massive to accrete a massive gaseous atmosphere from the protoplanetary disk in which it forms. The second possibility is that the protoplanetary disk undergoes a gravitational instability whereby collapse to form a giant planet occurs directly (Chapter 12).

By contrast, there are several models for the formation of a brown dwarf. Part of this is due to the fact that brown dwarfs may form as companions to stars or other brown dwarfs, or in isolation. Those that form in isolation are expected to form in a similar manner to stars, but those that form as companions may form from disks in a similar manner to giant planets. However, the rest of the problem is that we do not yet understand the origin of the IMF, and brown dwarfs (at least the isolated ones) are simply the low-mass end of the IMF.

### 15.5.1 *Brown dwarfs from the collapse of low-mass molecular cores*

The first possibility is that the IMF originates from the mass distribution of gravitationally unstable molecular-cloud cores in molecular clouds. In this picture, each gravitationally unstable core collapses to form a single object, or perhaps a small multiple system, but the final mass of the object is primarily determined by the mass of the pre-stellar core. Therefore, brown dwarfs would be formed from the collapse of very dense low-mass molecular cloud cores while stars would form from higher-mass cores (Elmegreen, 1999).

This idea has received observational support from surveys of molecular clouds from which a mass distribution of cores is extracted. Motte *et al.* (1998) decomposed molecular maps of the $\rho$ Ophiuchus molecular cloud into an ensemble of cores. Their core mass function displayed a striking similarity to the stellar IMF. Other surveys have found similar core mass functions in other molecular clouds (Testi and Sargent, 1998; Johnstone *et al.*, 2000, 2001; Onishi *et al.*, 2002). However, it is not clear that all of the cores identified in such maps are gravitationally bound, especially the lower-mass cores.

On the theoretical side, Padoan and Nordlund (2002) argued that the IMF was determined from the distribution of core masses in a turbulent molecular cloud. When applied to brown dwarfs, their model predicts that brown dwarfs should be more common in denser or more turbulent molecular clouds (Fig. 15.3). However, apart from predicting the overall abundance and general form of the IMF, these models have little predictive power for the properties of brown dwarfs (or stars). In particular, there are no predictions for how multiplicity or the properties of protoplanetary disks should vary with mass.

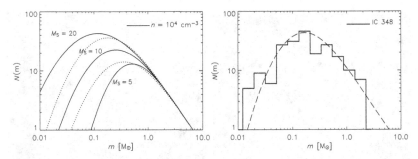

Fig. 15.3. Examples of the initial mass functions given by the model of Padoan and Nordlund (2002) in which stellar/brown-dwarf masses are determined by the distribution of core masses in turbulent molecular clouds. The IMF depends on the turbulent Mach number $M_s$, and the gas density $n$ (left panel). In the right panel, the model is compared with the IMF observed in the young cluster IC 348.

### 15.5.2 Brown dwarfs from the competition between accretion and ejection

Another possibility is that the IMF results from the competition between accretion and ejection of stars and brown dwarfs forming in small groups and clusters. In this picture, protostars are born primarily in groups and clusters and accrete competitively for the molecular gas (e.g. Bonnell *et al.*, 1997; Klessen *et al.*, 1998). The accretion onto them is terminated when they are ejected from the dense gas through dynamic interactions. Reipurth and Clarke (2001) proposed that brown dwarfs were those objects that happened to be ejected before they reached stellar masses.

Recent large-scale hydrodynamic simulations of star-cluster formation (Fig. 15.4) support this hypothesis (Bate *et al.*, 2002a, 2003a; Bate and Bonnell, 2005). These simulations follow the collapse and fragmentation of "turbulent" molecular clouds, resolving masses down to below the opacity limit for fragmentation, binaries with separations as small as 1 AU, and circumstellar disks with radii $\gtrsim$20 AU. Bate *et al.* (2002a) found that beginning with conditions typical of dense molecular clouds, these simulations naturally produced roughly equal numbers of stars and brown dwarfs. Roughly three-quarters of the brown dwarfs originated from the fragmentation of massive protostellar disks while the remainder formed in collapsing molecular filaments. However, the key to them ending up with substellar masses was that they were formed in or nearby small groups of protostars and were ejected from these unstable multiple systems quickly, before they were able to accrete to stellar masses.

In this picture, stars and brown dwarfs form in the same manner. Both begin as opacity-limited fragments with masses of a few times that of Jupiter and accrete to higher masses. However, those that end up as stars happen to accrete to stellar masses before their accretion is terminated, while those that become brown dwarfs are ejected from the dense gas before they reach stellar masses. In their most recent

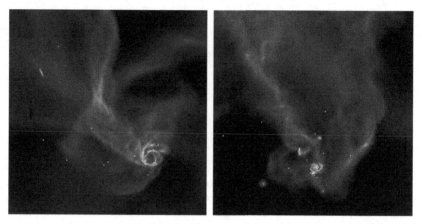

Fig. 15.4. Stars and brown dwarfs forming in a hydrodynamic simulation of a turbulent molecular cloud (Bate *et al.*, 2003a). The object in the lower left-hand corner of the second panel is an ejected brown dwarf with a 60-AU radius disk.

paper, Bate and Bonnell (2005) clearly demonstrated this mechanism by plotting the time an object spends accreting versus its final mass. These figures are reproduced in Fig. 15.5. Bate and Bonnell (2005) went on to propose an accretion/ejection model for the origin of the stellar and substellar IMF, in which objects begin as opacity-limited fragments and accrete at a characteristic rate determined by the sound speed in the molecular cloud until they are ejected, stochastically, through dynamic interactions. This simple model reproduces well the IMFs obtained from the hydrodynamic calculations. As with Padoan and Nordlund's (2002) turbulent-cores model of the IMF, the accretion/ejection model predicts that brown dwarfs should be more common (see Fig. 15.6) in molecular clouds with a lower value of the mean thermal Jeans mass (i.e. clouds that are more dense or cooler). There is some evidence for such a relation between the density of a molecular cloud and the abundance of brown dwarfs. Briceño *et al.* (2002) found that brown dwarfs were roughly a factor of two less abundant in the low-density Taurus star-forming region than in the Orion Trapezium Cluster, a high-density star-forming region. More recent results (Slesnick *et al.*, 2004; Luhman, 2004b) reduce this difference to a factor of $\approx 1.5$, but the difference does seem to be real and is of the same magnitude as that predicted by the hydrodynamic simulations (Bate and Bonnell, 2005).

Since the hydrodynamic calculations are able to resolve binaries and circumstellar disks, more information than simply the abundance of brown dwarfs is available. In particular, the calculations produce binary brown dwarfs. The first hydrodynamic calculation (Bate *et al.*, 2003a) produced one binary brown-dwarf system out of $\approx 20$ brown dwarfs, giving a frequency of binary brown dwarfs $\sim 5\%$. The

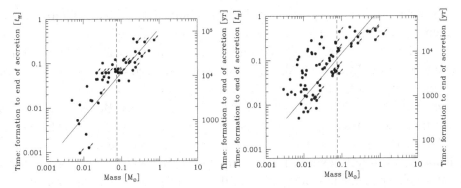

Fig. 15.5. The time between the formation of a protostar (a pressure-supported core) and the termination of its accretion (or the end of the calculation) versus the object's final mass (from the two hydrodynamic simulations presented by Bate and Bonnell, 2005). Stars and brown dwarfs *both* begin containing just a few $M_J$, but those that end up as brown dwarfs are ejected shortly after they begin to form so that their accretion is terminated before they reach a stellar mass.

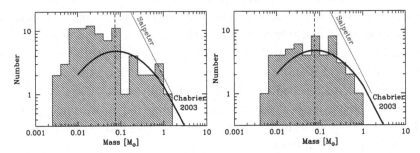

Fig. 15.6. The initial mass functions (IMF) from two hydrodynamic simulations (Bate and Bonnell, 2005) compared with the observed IMF as parameterized by Chabrier (2003) and the Salpeter (1955) slope. The star-forming cloud that produced the IMF in the right panel had a mean thermal Jeans mass three times lower than that on the left, resulting in a lower median mass and many more brown dwarfs.

second large-scale simulation produced many more brown dwarfs, including three binary brown dwarfs and two systems each consisting of a very low-mass star ($< 0.09\ M_\odot$) and a brown dwarf. Combining the first two hydrodynamic simulations gives a frequency of very low-mass binaries of $\approx 8\%$. This is roughly a factor of two lower than observed. Interestingly, four of the six systems have separations $< 20$ AU, in agreement with the observation that most binary brown-dwarf systems have separations $< 20$ AU.

Of the 42 stars produced in the two hydrodynamic calculations, only one star ($> 0.10\ M_\odot$) had a close ($< 40$ AU) brown-dwarf companion. This system was a member of an unstable septuple system and was still accreting when the calculation was stopped, so its long-term survival was in doubt. Thus, the calculations are in agreement with the results of radial velocity searches for planets that show that stars tend not to have brown-dwarf companions at small separations. By contrast, the frequency of close (< 10 AU) stellar binaries produced in the calculations is in excellent agreement with the value observed for Solar-type stars (Duquennoy and Mayor, 1991) of ≈ 20% (Bate *et al.*, 2002b, 2003a; Bate and Bonnell, 2005). The formation mechanisms of close binary systems are discussed in detail by Bate *et al.* (2002b). There are several reasons that stars have frequent close stellar companions but not close substellar companions. When a binary system is accreting gas from an infalling envelope the long term effect of this accretion is to drive the components towards equal masses, especially for closer systems (Bate, 2000). If the binary accretes from a circumstellar disk, mass equalization also occurs. Finally, if a close dynamic interaction occurs between a binary and a single object, the usual result is that the object with the lowest mass is ejected (van Albada, 1968). If the binary consists of a star and a brown dwarf, it is likely the single object will be a star which will exchange into the binary, ejecting the brown dwarf. All of these effects together result in the low frequency of brown dwarfs as close companions to stars.

On the other hand, hydrodynamic calculations produce many brown-dwarf companions as wide components of stellar multiple systems. Delgado-Donate *et al.* (2004) performed calculations of low-mass clouds which could be followed until the multiple systems reached their final dynamically stable states. They showed that although many wide brown dwarfs are lost through dynamic interactions there should, nevertheless, be a significant population of wide brown dwarfs as companions to stellar multiple systems.

### 15.5.3 *Brown dwarfs from evaporated cores*

Finally, it has been proposed that brown dwarfs may be formed when the accretion from a dense molecular-cloud core is terminated prematurely because radiation from a massive star evaporates the envelope (Hester *et al.*, 1996; Whitworth and Zinnecker, 2004). Whitworth and Zinnecker showed that such evaporation operates over a wide range of Lyman continuum luminosities, molecular-cloud core densities, and molecular-cloud temperatures.

One of the advantages of this formation mechanism is that binary brown dwarfs are formed as naturally as binary stars in that whether or not a protobinary embedded in the dense core in which it formed becomes a stellar or brown-dwarf binary simply depends on whether its envelope is destroyed quickly enough or not. On the other hand, the reason why most binary brown dwarfs have separations less than the median separation of (Solar-type) binary stars is unclear (Sterzik *et al.*, 2003).

Another problem is the efficiency of the mechanism. In low-mass star-forming regions where there are no OB stars, such a mechanism cannot operate. In high-mass star-forming regions, high-mass stars are usually accompanied by many low-mass stars (e.g. Clarke *et al.*, 2000). This implies that the radiation from the OB stars does not destroy the dense cores until many stellar-mass objects have formed. Once the ionizing radiation begins to disrupt the cloud, the timescale for destruction of the cloud may be very short and the number of objects that are caught in the protostellar phase but with masses less than the hydrogen-burning limit is likely to be small. Thus, while it seems almost certain that this mechanism can form some brown dwarfs, it is unlikely that this mechanism alone can produce similar numbers of stars and brown dwarfs as is observed (Reid *et al.*, 1999).

## 15.6 Planet or brown dwarf?

The minimum mass of brown dwarfs is thought to be only a few $M_J$, where as planets in tight orbits around Solar-type stars are observed to have masses as great as $\approx$ 10–20 $M_J$. Radial migration theory (see Chapter 14) also naturally leads to a maximum planet mass of $\approx 10\,M_J$ (Lubow *et al.*, 1999; Bate *et al.*, 2003b) because as a planet grows in mass it opens a broader gap in the disk so that its accretion rate decreases while its migration rate is approximately constant. Therefore, there is a region of overlap in the masses of brown dwarfs and planets.

As mentioned in Section 15.4.2, it is in many ways a fortunate coincidence that there exists a brown-dwarf desert at small separations around Solar-type stars. This clearly indicates that there is a fundamental difference in the formation mechanisms of planets and stars/brown dwarfs. However, the desert is not completely devoid of objects. Moreover, it is likely to disappear or perhaps move to lower masses when considering lower-mass stars. Similarly, when we have been able to map out the masses of planets and brown dwarfs at large separations (orbital periods longer than 10 yr) around Solar-type stars the distinction may not be so clear.

This leads to the problem of what is a planet and what is a brown dwarf? There are several possibilities for definitions. First, the distinction could be based on the way the object formed. Second, the distinction could be based on circumstance. Third, the distinction could be based on evolution. None of these, or even a combination of these, are entirely perfect as will become clear in the following discussion.

Distinguishing between planets and brown dwarfs based on formation mechanism may seem to be the ideal solution. However, the assumption here is that there are two different formation mechanisms. If brown dwarfs form from collapsing molecular-cloud cores and giant planets form via core accretion, this distinction is viable. However, as discussed in Chapter 12, a competing model for giant-planet formation is the direct formation via gravitational instability of a massive gaseous protoplanetary disk while, for brown dwarfs, the calculations of Bate *et al.* (2002a, 2003a) show that many may be formed via gravitational instabilities of massive gaseous protostellar disks while the protostars are undergoing rapid accretion from their surrounding envelopes. If these formation mechanisms are correct, the distinction becomes blurred.

However, even if there is a clear distinction based on formation mechanism, there is the problem of determining which formation mechanism applies to a given object. Current models of the interior structure of Jupiter allow for the possibility that Jupiter does not have a rocky core (Saumon and Guillot, 2004). If we cannot determine whether Jupiter formed from core accretion or direct gravitational instability by sending robotic probes to it to examine it in detail, what hope is there of determining the formation mechanism of an object around another star?

The second possibility is a distinction based on circumstance. This allows us to distinguish between two objects of planetary mass (i.e. in the range 1–10 $M_J$), one of which is isolated, the other is in orbit around a star. With this definition, a planet must be orbiting a star while anything not orbiting a star is a brown dwarf. However, there are several problems here. What is a 5 $M_J$ object orbiting a 25 $M_J$ object (see Chauvin *et al.*, 2004)? Is this a binary brown dwarf or a giant planet orbiting a brown dwarf? Furthermore, what do we call a 3 $M_J$ object that formed as a planet orbiting a star but was subsequently ejected from the planetary system because the system was unstable?

The third possibility is a distinction based on characteristics. By this we mean mass, shape, and composition, though for gas-giant planets in practice this simply means whether they undergo nuclear fusion. As mentioned in Section 15.3, brown dwarfs with masses greater than approximately 13 $M_J$ are thought to undergo deuterium fusion. Since 13 $M_J$ is also, coincidently, very close to the low-mass end of the brown-dwarf desert around Solar-type stars, this is a convenient distinguishing mass. Here an object that does not fuse hydrogen but does fuse deuterium is a brown dwarf, while a lower-mass object is a planet. However, this distinction also has problems. Perhaps the most obvious is that, to the general public a planet is something that orbits a star. An isolated object, regardless of its mass, is not really a planet.

A record of a debate held at IAU Symposium 211 on the subject of the nomenclature surrounding planets and brown dwarfs is provided in Boss *et al.* (2003). At the debate, Gibor Basri proposed what is, perhaps, the best method for designating an object as a star, a planet, or a brown dwarf based on the combination of circumstance and characteristics. A star is an object that joins the hydrogen-burning main sequence at some point in its lifetime. A brown dwarf is an object that is not a star but which undergoes fusion. A planet is an object that is massive enough that its gravity dictates that it be spherical in shape (as opposed to an asteroid), that orbits a star or a brown dwarf but is not massive enough to undergo fusion (i.e. less massive than the deuterium-burning mass). These definitions solve many of the above problems and allow us to ascribe a designation to every object *whose mass is currently well determined.*

However, one can foresee problems arising in the future. First, what is the designation of an isolated object with a mass less than 13 $M_J$, since such an object could be formed as a star (i.e. the tail end of the IMF) or it might have formed as a planet but then been ejected from an unstable young planetary system. Furthermore, if young planetary systems eject objects, it also seems likely that there will exist ejected Earth-type objects. Almost certainly such objects must have formed as planets, but can they be called planets once they have been ejected? Second, we know that around 15% of brown dwarfs are binary brown dwarfs. Assuming these objects form like stars and that the opacity limit for fragmentation allows objects with masses less than 13 $M_J$ to form, there is no reason to assume that there will not be binary "brown dwarf" systems in which both components have masses of only a few Jupiter masses. Since neither of these objects would undergo fusion, such a system has no designation though, from a formation point of view, it is clearly the low-mass end of the binary brown-dwarf sequence. Thus, it seems that new designations will need to be made for systems such as "binary planetary-mass objects" and "free-floating planetary-mass objects" (which is currently used to describe very low-mass "brown dwarfs") in the future.

## 15.7 Conclusions

I have examined the relationship between giant planets and brown dwarfs, comparing and contrasting their properties, circumstances, and proposed formation mechanisms. Observationally, both fields are ten years old and although much progress has been made, we are to a large degree still ignorant of their properties. Better characterization of the distributions of masses and orbital parameters of planets and brown dwarfs is crucial if we are to determine how they form and to define what we mean when we denote an object as a planet or a brown dwarf.

## Acknowledgments

MRB is grateful for the support of a Philip Leverhulme Prize and for the use of figures provided by Adam Burrows and Paolo Padoan.

# 16

## Exoplanet detection techniques – from astronomy to astrobiology

Wolfgang Brandner

*Max-Planck-Institut für Astronomie, Heidelberg, Germany*

### 16.1 Introduction: planet detection and studies in the historical context

Beyond Earth, only five planets have been known since historical times. These are the three "terrestrial planets," Mercury, Venus and Mars, and the two "gas giants," Jupiter and Saturn. The systematic studies of the skies starting in the seventeenth century marked the beginning of a new era. The ice giant Uranus, the next outer planet beyond Saturn, was first cataloged as *34 Tau* by John Flamsteed in 1690, but not recognized as a "wanderer" in the skies. In 1781 William Herschel was the first to spatially resolve the disk of Uranus, initially classifying it as a comet. Uranus also played an important role in the discovery of Neptune. Neptune, the eighth planet in the Solar System, and the outermost of the ice giants, was first observed in 1612/1613 by Galileo Galilei. At that time, Neptune was in close conjunction with Jupiter. Galileo recognized the moving source in the vicinity of Jupiter (see Fig. 16.1), but decided not to get distracted from his studies of the orbital motions of Jupiter's four largest moons, which we now refer to as "Galilean Moons." Another 230 years passed till precise measurements of Uranus' orbit indicated the presence of an additional outer planet. Based on predictions by Urbain Leverrier (which were in close agreement with independent calculations by John Couch Adams) in 1846, Johann Gottfried Galle finally identified and resolved Neptune. Thus Neptune deserves the honour of being the first planet detected by indirect methods, namely by its astrometric disturbance of the orbit of Uranus.

The discovery of the outer Solar System, in particular the Kuiper Belt, which constitutes a remnant of the planet formation period in the Solar System, by Tombaugh in 1930 and Jewitt and Luu (1993) marked another important milestone in our search for planets, namely the start of systematic wide-area and sensitive surveys.

*Planet Formation: Theory, Observation, and Experiments*, ed. Hubert Klahr and Wolfgang Brandner.
Published by Cambridge University Press. © Cambridge University Press 2006.

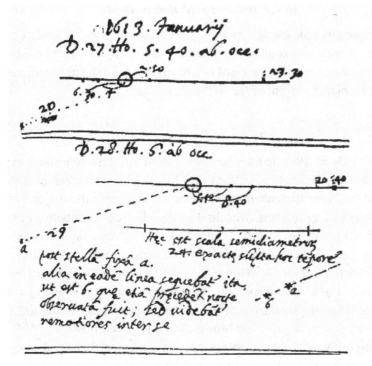

Fig. 16.1. Excerpt from Galileo Galilei's log book. On January 28, 1613 he remarks that the source seen in conjunction to Jupiter is moving, and hence not a fixed star (*stella fixa*). Galileo's work from December 1612 and January 1613 constitute the first recorded observations of Neptune.

A brief summary of the history of searches for planets around other star ("exoplanets") can be found in the contribution by Peter Bodenheimer (Chapter 1). In the following we focus on the detection methods and observables for exoplanets.

## 16.2 Observing methods and ground/space projects

Because of the enormous difference in luminosity between a star and a planet (Jupiter is about $10^9$ times fainter than the Sun), and the fact that distances of a few AU correspond to angular separations of less than 1 arcsec even for nearby stars, the direct detection of photons originating from an exoplanet proves to be highly challenging.

Indirect methods, which probe for a modulation in the signal from a star due to the effects of an exoplanet, also require high-precision measurements, yet can be mastered more easily than the direct methods.

### *16.2.1 Indirect detection methods*

Indirect methods include the detection of reflex motion of a star around its common center of mass with an exoplanet by means of the Doppler effect or astrometric "wobble," and the direct modulation of flux received from a star due to microlensing or transits of a planet in front of the stellar disk.

#### *16.2.1.1 Reflex motion*

Doppler methods in 1992 led to the detection of the first confirmed exoplanet system around the millisecond pulsar PSR1257+12 by Wolszczan and Frail (see contribution by Peter Bodenheimer, Chapter 1), and were the major workhorse in the discovery of exoplanets over the last decade (see contributions by Marcy *et al.*, Chapter 11, and Lovis *et al.*, Chapter 13). The systematic surveys reveal the frequency of (massive) exoplanets, their orbital periods and eccentricities, as well as the mass function ($M \sin i$). Stellar surface activity limits the Doppler methods to accuracies of $\geq 1$ m s$^{-1}$, corresponding to planetary masses of $\approx 10\,M_{\oplus}$ for Solar-type stars. Still lower planetary masses can be probed by extending the Doppler surveys to very-low mass stars and brown dwarfs (e.g. Martín *et al.*, 2004).

Astrometry aims at the same observables as the Doppler methods, yet the astrometric orbit also yields the inclination of the orbit, and hence the mass (rather than the mass function) of the exoplanet. Since the quality of the astrometric measurement is much less affected by stellar surface activity, astrometric methods in principle can probe lower-mass exoplanets (possibly down to a few $M_{\mathrm{Mars}}$) around nearby stars. The most promising technique is optical, long-baseline interferometry, either from the ground (like, for example, ESO VLTI (European Southern Observatory Very Large Telescope Interferometer) / PRIMA (Phase Referenced Imaging and Microarcsecond Astrometry), Schöller and Glindemann, 2003) or from space (Space Interferometry Mission, Shao, 2004). Finally, the GAIA (Global Astrometric Interferometer for Astrophysics) satellite should be able to detect 10 000 to 50 000 exoJupiters within 50 pc of the Sun (Lattanzi *et al.*, 2002).

#### *16.2.1.2 Flux modulation*

The detection of flux modulation of a star can be used as another tracer for the presence of a planet in orbit around a star.

Microlensing takes place when a (foreground) star passes the line of sight between Earth and a more distant star, thereby acting as a gravitational lens and thus magnifying/amplifying the signal. The presence of a planet around the lens modifies the signal, which provides information on the frequency of planets, as well as the mass and orbital radius of the planet (Bond *et al.*, 2004).

Transit refers to the dimming of a star due to the passage of a planet in front of the stellar disk star, or – at thermal infrared wavelengths – the dimming of the star–planet system when the planet is passing behind the star (Charbonneau *et al.*, 2005; Deming *et al.*, 2005). Since this puts quite narrow constraints on the orbital inclination, transits in combination with Doppler methods yield precise mass estimates for planets. In addition, the amount of dimming is directly proportional to the size of the occulting body, hence transits nicely constrain the radius of the planet. Ground-based optical surveys like OGLE (Optical Gravitational Lensing Experiment) and MOA (Microlensing Observations in Astrophysics), or upcoming space missions like CoRoT (Convection, Rotation and Planetary Transits) and Kepler hence open a new discovery space in exoplanet research down to the regime of terrestrial planets (see contribution by Lecavelier des Etangs and Vidal-Madjar, Chapter 9).

### 16.2.1.3 Circumstellar disks as tracers

Due to its high surface brightness,"zodi clouds" (dust disks) in exoplanetary systems can severely limit the ability to directly detect exoplanets. The gravitational interaction of an exoplanet with a circumstellar disk can, however, also create "wakes" and "gaps" in the disk, which potentially can be discovered much more easily than the planet itself, in particular at sub-mm wavelength using larger interferometers like ALMA (Atacama Large Millimeter Array) (Wolf and D'Angelo, 2005). In addition to orbital parameters like semimajor axis, eccentricity and period, the properties of the gaps and wakes might also provide an estimate of the mass of the planet.

## 16.2.2 Direct detection methods

The direct detection and identification of photons originating from planets around other stars opens a new realm in our study of exoplanets.

### 16.2.2.1 Differential imaging

Ground-based telescopes equipped with adaptive optics as well as the Hubble Space Telescope already now provide the optical quality and the contrast which is required to detect brown-dwarf companions to nearby and/or young stars. Residual wavefront errors inherent to optical systems, and the fact that these wavefront errors evolve with time, however, ultimately limit the contrast and sensitivity achievable. A simple way to improve the contrast even further is simultaneous differential imaging. Here one simultaneously observes two (or more) physical states of the electromagnetic radiation emanating from a star–planet system, like, for example, two (orthogonal) polarization states, or two neighboring wavelength regions. If the planet is significantly brighter in one state than in the other, while its host star exhibits the same brightness in both states, the planetary signal stands out in the difference measurement between both states.

Possible applications of this method are observations at two neighboring wave-length regions located in and out of a methane band (with the aim to detect Jupiter-type giant planets), or in two polarization states (with the aim to detect planets in the optical by means of radiation reflected off their surface or upper-cloud layers).

Prototype spectral differential imaging (SDI) instruments have already been implemented and are producing the first science results (Biller *et al.*, 2004; Close *et al.*, 2005; Marois *et al.*, 2005). The next generation of planet-finder instruments at ground-based 8 m class telescopes based on SDI and PDI (polarimetric differential imaging) techniques are currently in the study phase (Gratton *et al.*, 2004; Macintosh *et al.*, 2004).

### 16.2.2.2  Coronagraphs

Coronagraphs have the potential to suppress the light from a central source, thereby enhancing the contrast and the detectability of faint, nearby companions. Since exoplanets are in general located at small angular separation from their host star, the focal plane masks in simple Lyot-type coronagraphs should not cover more than the inner $\leq 5$ Airy rings. Simulations as well as experiments carried out at the Palomar 5 m telescope indicate that Strehl ratios $> 80\%$ are required for diffraction-limited coronagraphs (Sivaramakrishnan *et al.*, 2001), i.e. this requires the use of so-called extreme adaptive optics systems and/or observations at long wavelengths. The next generation of extremely large telescopes (ELTs) might be able to observe both Jovian and terrestrial planets in nearby extrasolar planetary systems in the thermal infrared.

### 16.2.2.3  Nulling

Similar to coronagraphs, nulling aims at suppressing the light from the star in order to enhance the detectability of exoplanets. While nulling experiments like GENIE (Ground-based European Nulling Interferometer Experiment) at the ESO-VLTI are limited (largely due to residual wavefront errors) to the detection and characterization of relatively bright dust disks around nearby stars (Absil *et al.*, 2003), space-based nulling interferometers like DARWIN/TPF-I clearly have the potential to detect extrasolar terrestrial planets (Fridlund, 2003).

## 16.3  Outlook: planet mapping and bio-signatures

The direct detection of photons from terrestrial planets around nearby stars means that standard remote-sensing techniques can be applied to further characterize these planets. Brightness/albedo variations can be used to trace the length of the day/night cycle and to probe for the presence of seasonal variations, like changes in cloud

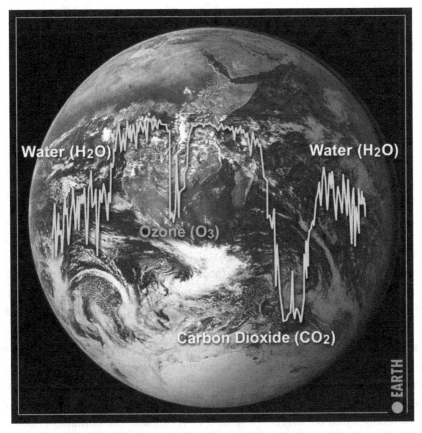

Fig. 16.2. Spectroscopic signatures for $H_2O$, $CO_2$ and the potential bio-marker $O_3$ as accessible with DARWIN overlaid on an image of planet Earth. Reprinted by permission from ESA.

coverage or extent of polar ice caps. Changes in the albedo and/or degree of polarization with varying phase angle yield information of the atmospheric composition and surface constitution. Ultimately, spectroscopic studies aim at characterizing the atmospheric composition. Instruments like DARWIN might be able to probe for the presence of water vapor or carbon dioxide, and potential biomarkers like ozone (see Fig. 16.2).

The next step beyond the planet-detection and spectroscopic missions would be an exoplanet mapper, which should be able to resolve the disks of exoplanets, and, for example, identify the distribution of oceans and continents.

# 17

# Overview and prospective in theory and observation of planet formation

Douglas N. C. Lin

*UCO/Lick Observatory, Santa Cruz, USA*

This conference truly reflects a microcosm of an explosive revolution in the quest to understand the origin of planet and star formation. The diverse nature of this wide-open field necessitates a multi-facet attack on all relevant issues. In this pursuit, it is particularly important to find the missing links between the many seemingly independent observations as circumstantial clues around a global picture. The development of a comprehensive coherent interpretation requires an integrated approach to identify the dominant physical processes which determine the physical characteristics of planets and the dynamic architecture of planetary systems.

On the basic concept of planetary origin, there is very little difference between modern theories and the original Laplacian hypothesis. The coplanar geometry of all the major planets' orbits hardly needs any extrapolation for theorists to postulate the scenario that the planets formed long ago in a rotational flattened disk which is commonly referred to as the Solar Nebula. Today, we have direct images and multi-wavelength spectra of protostellar disks within which planet formation is thought to be an ongoing process. Perhaps the biggest advancement in the past decade is the discovery of over 100 planets around nearby stars other than the Sun. For the first time in this scientific endeavor, the Solar System reduces its unique importance to a single entry in the rapidly growing database of planetary-system census. But the four decades of space-age exploration have also provided us with intricate details of our planetary environment and relic clues which serve as the Rosetta Stones of modern cosmogony.

The gross summary of this vast influx of data can be briefly recapitulated into the following main conclusions:

*Planet Formation: Theory, Observation, and Experiments*, ed. Hubert Klahr and Wolfgang Brandner.
Published by Cambridge University Press. © Cambridge University Press 2006.

(1) All the stellar material was processed through protostellar disks before it was accreted onto young stellar objects.

(2) The active phase of protostellar disks lasts 3–10 Myr during which the gas accretion rate reduces from $10^{-7}$ to $10^{-10}$ $M_\odot$ $yr^{-1}$. The aspect ratio in the direction normal to the disk is typically a few to 10%. The mass of the gas is not directly measurable at the present moment.

(3) Heavy elements condense into dust grains which are distributed over 10–100 AU in a disk. In active disks, the total dust mass in these disks is comparable to that inferred from the minimum-mass nebula, i.e. a few % of that accreted onto their central stars. The dust signatures decline below the detection limit on the same timescale that gas accretion decreases.

(4) In the Solar System, the most primitive grains, the calcium-aluminium-rich inclusions (CAIs) formed well within 1 Myr. But most common relic grains are in the form of chondrules within the size range of 0.1 to a few mm. Individual chondrules were formed within 1–2 Myr after the formation of the CAIs. They have experienced repeated partial melting episodes with minute-to-hours duration. They were assembled, no sooner than 1 Myr after their formation, into the parent bodies of meteorites which reside near the present-day asteroid belt.

(5) The satellites of gas-giant planets are composed of highly volatile ice. Although Mars contains little ice, its elemental composition is more volatile than that of the Earth and Venus which are primarily composed of silicates. With Mercury mostly composed of iron, the sublimation sequence of the Solar System is consistent with that inferred for protostellar disks.

(6) The heavily cratered surfaces of Mercury, the Moon, and asteroids leaves little doubt that they were assembled through cohesive collisions. The formation timescales inferred from the radioactive isotopes suggest that Mars and the Earth differentiated and formed their cores within 2–4 Myr and 10–30 Myr, respectively, after the CAIs.

(7) Jupiter has a very low-mass core ($< 5$ $M_\oplus$) and a strongly metal-enriched envelope whereas Saturn has a 15 $M_\oplus$ core. Uranus and Neptune also have similar mass cores with a 1 $M_\oplus$ gaseous envelope around them. The heavy-element abundances in all giant planets are at least 4–5 times that of the Sun.

(8) There is a large population of Kuiper-Belt objects which have been scattered beyond 40 AU by the gas- and ice-giant planets. Along the way, Neptune seems to have migrated and during its migration, Neptune captured several bodies, including Pluto, into its resonance.

(9) Jupiter-mass planets are found around at least 6% of all nearby mature, F-G-K stars on the main sequence. In cases where it is possible to measure, the density of these planets indicate that they are gas giants. They appear to have a logarithmic period distribution ranging from one day to the detection limit of several years. There are indications of additional planets with longer periods and lower masses.

(10) The fraction of stars with planets is a rapidly increasing function of their metallicity.

(11) The eccentricity of the known planets is distributed uniformly between 0 and 0.7.

(12) A large fraction of stars with known planets show signs of additional planets around them. In some multiple systems, the orbits of planets are in mean motion resonances whereas in other systems, very short-period planets coexist with long-period siblings.

(13) Dust grains produced from collisional fragmentation are found around a faction of stars with ages older than 10 Myr. The dispersion within these signatures is large, i.e. some disks have intense dust signatures whereas others show little evidence for them. The upper limit of the dust reprocessed radiation declines as a power law of the stellar age.

(14) The remarkable lack of abundance dispersion among mature stars in the Pleiades cluster, with age $\sim$ 100 Myr, places an upper limit on the mass of protostellar disks of less than two or three times that of the minimum-mass nebula.

These and many other related observational clues can be interpreted with a coherent model. For example, the observed properties (5)–(8) have been used to develop a sequential accretion scenario in which the Solar-System planets formed through the condensation of grains, coagulation of planetesimals, emergence of isolated embryos, onset of gas accretion, the termination of growth through gap formation, the propagation of gas giants' secular resonances during the epoch of disk depletion, the final assemblage of terrestrial planets, and the dynamic scattering of residual planetesimals beyond the confines of the gas-giant planets. The cosmochemical properties under (3) and (5) place a chronological formation sequence in the Solar System: (a) infall of gas and formation of the Sun occurred on the timescale of 0.1 Myr; (b) the first generation solid, less-than-Mars-mass embryos emerged in the inner Solar System in less than 1 Myr; (c) Jupiter attained most of its mass within $<$ 2 Myr and Saturn within $\sim$ 3 Myr; (d) the growth of Uranus and Neptune was limited by gas depletion in the Solar Nebula which proceeded on the timescale of 3–10 Myr; (e) the environmental impact of Jupiter's secular resonance cleared out the asteroid belt in 1 Myr, and promoted the formation of chondrules and their assemblage into the parent bodies of present-day asteroids in 2 Myr; (f) Jupiter's sweeping secular resonance also induced the final buildup of Mars in $\sim$4–10 Myr and the Earth in $\sim$10–30 Myr through giant impacts which also led to the formation of the Moon in 30–50 Myr; (g) tidal interaction between the emerging planets and the diminishing gas in their nascent disks circularized the orbit of the terrestrial planets despite the giant impacts; and (h) non-linear interaction led to a slight outward migration of Neptune in 10 Myr and the formation of the Kuiper Belt in 10–100 Myr.

There are of course many unresolved issues associated with this overall picture. Some of the most serious unresolved issues include: (a) the angular-momentum transport mechanisms which determined the structure and evolution of gas in the disk; (b) the dominant process through which grains evolved into planetesimals in a gaseous environment; (c) the retention efficiency of heavy elements in the Solar

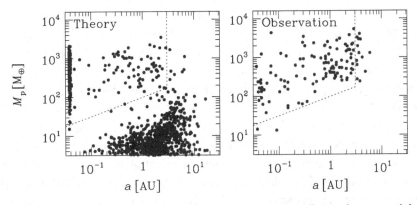

Fig. 17.1. Mass–semimajor axis distribution of planets according to the sequential-accretion scenario. The left panel represents the results of simulations carried out by Ida and Lin (2004). The right panel represents the known extrasolar planets. The dashed line indicates the current detection limit.

Nebula; (d) the survival of planetesimals during their collisions; (e) the effect of tidal interaction between embryos and their background gas; (f) the mechanisms and efficiency of heat transfer in the envelopes around cores of a few $M_\oplus$ which regulate the onset and timescale of efficient gas accretion; (g) the termination of gas accretion onto Jupiter and Saturn; (h) the processes which led to the clearing of the disk gas; and (i) the extent of post-formation dynamic evolution.

In order to address these unresolved issues, it is tempting to extrapolate the above sequential-formation scenario to the extrasolar planets with a grand unified theory. Figure 17.1 represents an attempt in this direction. Using the above data under items (1)–(3) and the sequential planet-formation scenario, it is possible to simulate the expected mass–period distribution of planets around other Solar-type stars. Although the observed properties (9)–(12) only apply to gas giants, they provide a set of strong constraints on our working hypothesis. For example, the discovery of short-period planets revives a notion that gas giants emerge in the active phase of relatively massive disks and undergo migration as they tidally interact with their ambient evolving nascent disks. Figure 17.1 highlights the implications of various processes on the statistical properties of the gas giants. For example: (a) an up-turn of their period distribution at 2–3 years is consistent with the expectation that gas giants are preferentially formed near the snow line; (b) the logarithmic period distribution (between 3 days and 3 yr) can be reproduced under the assumption that the migration and gas depletion occur on comparable timescales; (c) many gas giants migrated to and probably perished in the proximity of their host stars; (d) the size of planetary systems is determined by the limited available timescale for building sufficiently massive cores to accrete gas and by the location outside which the local Keplerian speed falls well below the surface escape speed of these critical-mass cores; (e) the emerged planets naturally fall into three categories, rocky cores, gas and ice giants; (f) interior to the snow line, the upper mass limit for the rocky cores is determined by the critical condition for the onset of efficient gas accretion whereas the lower mass limit of the gas giants is determined by the diminishing gas supply through either global or local depletion of the disk; and (g) outside the snow line, the smooth mass distribution indicates that intermediate-mass ice giants form through giant impacts after the depletion of the disk gas.

Despite these successes, perhaps the biggest challenge to the sequential-accretion scenario remains the ubiquity of extrasolar gas-giant planets which must form in a gaseous environment. The outstanding issue is that the signatures of disks appear to diminish over the timescale of 3–10 Myr. But according to the first models constructed for this scenario, the timescale for the onset of efficient gas accretion appears to be longer than this especially if: (a) there are insufficient heavy elements to build up cores with at least a few $M_{\oplus}$; or (b) the gas accretion is stalled by an efficient transfer of the heat which is released from the accreted gas. Several resolutions have been suggested and they include: (a) accelerated growth of cores through either an elevated surface density of planetesimals near the snow line or core migration to bypass the boundaries of the feeding zone; and (b) enhanced heat-transfer efficiency in the radiative zone through grain sedimentation, disk accretion, circulation, and hydrodynamic instabilities. The core of Jupiter may also have been eroded by its convective envelope, however, one must address the structural difference between Jupiter and Saturn in the core-erosion scenario. Of course, the observational signatures cited as evidence for disk depletion are generally associated with the dust. It remains a challenge to secure a reliable constraint on the depletion timescale for the gas, especially cold molecular hydrogen.

Another approach to resolve the existing timescale paradox is to assume that the gas-giant planets formed through gravitational instability. Since it does not provide a direct link between gas giant and terrestrial-planet formation, the frequency of gas giants cannot be used to extrapolate that of the terrestrial planets in this scenario. The occurrence of this instability requires: (a) the rapid buildup of initial stable disks; (b) efficient cooling in the disk; and (c) metal enhancement through a substantial evaporation of the original gaseous envelope. All of these processes are theoretically challenging. The metallicity–planet frequency correlation and the apparent homogeneity of stellar metallicity (properties (10) and (14)) are also difficult to reconcile with this scenario. Nevertheless, there are observational tests that can be made to differentiate these scenarios. In the sequential-accretion scenario, isolated cores with masses comparable to or less than that of Mars are expected interior to all emerging gas giants. Along the path of these migrating planets, they capture failed cores onto their evolving mean motion resonance, bring them to the stellar proximity and induce them to form ultra-short-period terrestrial planets.

According to the alternative gravitational-instability scenario, such planets should not exist. The search for these hot Earths can be conducted with highly accurate radial-velocity surveys and transit searches. The sequential-accretion scenario also naturally predicts a limited size for planetary systems (see Fig. 17.1). In contrast, gravitational instability is more likely to occur in disk regions far from their host stars. The frequency of widely separated massive planets is best obtained through the gravitational-lensing search program.

Another test for the two scenarios is the ubiquity of multiple-planet systems and their dynamic structure. In the sequential-accretion scenario, the emergence of one gas giant is expected to induce a gap in the disk gas distribution which provides a natural harbor to collect dust and planetesimals just beyond the outer edge of the gap. This enhancement promotes the formation of subsequent gas giants. The coexistence of resonant planetary systems such as that around GJ 876 and systems which contain both short- and long-period non-resonant planets suggests that the formation interval between successive gas giants is comparable to the migration timescale. These complex configurations can be more easily achieved with the sequential-accretion than the gravitational-instability scenario. Nevertheless, the nearly uniform eccentricity distribution of the extrasolar planets may be less difficult to establish in the gravitational-instability scenario.

Finally, the enhanced metallicity of Jupiter and the correlation between the planets' fraction and the metallicity of the host stars (observed properties (7) and (10)) is naturally implied from the sequential-accretion scenario because solid cores of heavy elements must form prior to gas accretion. A natural extrapolation of this correlation indicates that planets form earlier and migrate more extensively around disks with a large amount of planet-building material. Since migrating planets sweep through the disk and clean up the residual planetesimals, this process naturally self-regulates the retention efficiency of heavy elements in the planets and preserves the stellar-metallicity homogeneity. Nevertheless, the gravitational-instability scenario must be more effective for metal-deficient globular clusters or around metal-poor stars where planets have been found. There are also protoplanetary candidates found at 50–100 AU from very young stars. These are potential systems in which planets may have formed through gravitational instability.

On balance, the sequential-accretion scenario is more well developed and it is compatible with most observational data. Nevertheless, there are many unresolved theoretical and observational issues associated with the grand unification sequential-accretion scenario. These include: (a) what process sets the lower mass limit on the amount of planet-building block heavy elements in protostellar disks; (b) what is the direct observational evidence for the transformation of dust to planetesimals in protostellar disks; (c) how long does a significant amount of gas remain in the disk; (d) what process determines the surface-density magnitude and distribution of planetesimals in the disk; (e) what condition leads to the onset and quenching of efficient gas accretion; (f) what are the observational signatures of emerging protoplanets; (g) what process initiates and stalls protoplanetary migration; (h) what is the mass-period distribution in the low-mass and long-period range; (i) what is the persistent time and depletion mechanism of disk gas; (j) is the dynamical structure of the Solar System unique or common among extrasolar planets; and (k) how do multiple planets evolve dynamically during and after the removal of the disk gas?

This brief summary provides an interpretation of the present state of observational knowledge and theoretical concepts. There are many outstanding issues out of which only a few subsets are listed above. These issues need to be resolved in order for us to develop a robust and deterministic model for planet formation and planetary-system evolution. But with the anticipated progress on both observation and theoretical fronts, we can look into a bright future in this exciting discipline of modern astronomy and astrophysics.

# References

Absil, O., Kaltenegger, L., Eiroa, C. *et al.* (2003). Can GENIE characterize debris disks around nearby stars? *Towards Other Earths: DARWIN/TPF, and the Search for Extrasolar Terrestrial Planets*, ed. Fridlund, M. and Henning, Th. *ESA*, **SP-539**, 323–8.

Adachi, I., Hayashi, C., and Nakazawa, K. (1976). The gas drag effect on the elliptical motion of a solid body in the primordial Solar Nebula. *Prog. Theoret. Phys.*, **56**, 1756–71.

Adams, F. C. and Benz, W. (1992). Gravitational instabilities in circumstellar disks and the formation of binary companions. In *Complementary Approaches to Double and Multiple Star Research, IAU Coll. 135*, ed. H. A. McAlister and W. I. Hartkopf. San Francisco: Atro. Soc. of the Pacific.

Adams, F. C., Hollenbach, D., Laughlin, G., and Gorti, U. (2004). Photoevaporation of circumstellar disks due to external far-ultraviolet radiation in stellar aggregates. *Astrophys. J.*, **611**, 360–79.

Adams, F. C., Lada, C. J., and Shu, F. H. (1987). Spectral evolution of young stellar objects. *Astrophys. J.*, **312**, 788–806.

Aigrain, S. and Favata, F. (2002). Bayesian detection of planetary transits. A modified version of the Gregory–Loredo method for Bayesian periodic signal detection. *Astron. Astrophys.*, **395**, 625–36.

Aigrain, S. and Irwin, M. (2004). Practical planet prospecting. *Mon. Not. R. Astron. Soc.*, **350**, 331–45.

Alexander, R. D., Clarke, C. J., and Pringle, J. E. (2004a). On the origin of ionizing photons emitted by T Tauri stars. *Mon. Not. R. Astron. Soc.*, **348**, 879–84.

(2004b). The effects of X-ray photoionization and heating on the structure of circumstellar discs. *Mon. Not. R. Astron. Soc.*, **354**, 71–80.

(2005). Constraints on the ionizing flux emitted by T Tauri stars. *Mon. Not. R. Astron. Soc.*, **358**, 283–90.

Alexander, D. R. and Ferguson, J. W. (1994). Low-temperature Rosseland opacities. *Astrophys. J.*, **437**, 879–91.

Alfvén, H. (1954). *On the Origin of the Solar System*. Oxford: Clarendon Press.

Alibert, Y., Mordasini, C., and Benz, W. (2004). Migration and giant planet formation. *Astron. Astrophys.*, **417**, L25–8.

Alibert, Y., Mordasini, C., Benz, W., and Winisdoerffer, C. (2005). Models of giant planet formation with migration and disk evolution. *Astron. Astrophys.*, **430**, 1133–8.

Allègre, C. J., Manhès, G., and Göpel C. (1995). The age of the Earth. *Geochim. cosmochim. acta*, **59**, 1445–56.

Allègre, C. J., Manhès, G., and Lewin, E. (2001). Chemical composition of the Earth and the volatility control on planetary genetics. *Earth Planet. Sci. Lett.*, **185**, 49–69.

Alonso, R., Brown, T. M., Torres, G. *et al.* (2004). TrES-1: The transiting planet of a bright K0 V star. *Astrophys. J.*, **613**, L153–6.

Amelin, Y., Krot, A. N., Hutcheon, I. D., and Ulyanov, A. A. (2002). Lead isotopic ages of chondrules and calcium-aluminum-rich inclusions. *Science*, **297**, 1678–82.

Anders, E. and Grevesse, N. (1989). Abundances of the elements: meteoritic and solar. *Geochim. cosmochim. acta*, **53**, 197–214.

Anders, E. and Zinner, E. (1993). Interstellar grains in primitive meteorites: diamond, silicon carbide, and graphite. *Meteoritics*, **28**, 490–514.

André, P., Ward-Thompson, D., and Barsony, M. (1993). Submillimeter continuum observations of Rho Ophiuchi A – the candidate protostar VLA 1623 and prestellar clumps. *Astrophys. J.*, **406**, 122–41.

Arlt, R. and Urpin, V. A. (2004). Simulations of vertical shear instability in accretion discs. *Astron. Astrophys.*, **426**, 755–65.

Armitage, P. J., Clarke, C. J., and Palla, F. (2003). Dispersion in the lifetime and accretion rate of T Tauri discs. *Mon. Not. R. Astron. Soc.*, **342**, 1139–46.

Armitage, P. J., Livio, M., and Pringle, J. E. (2001). Episodic accretion in magnetically layered protoplanetary discs. *Mon. Not. R. Astron. Soc.*, **324**, 705–11.

Artymowicz, P. (1993). On the wave excitation and a generalized torque formula for Lindblad resonances excited by external potential. *Astrophys. J.*, **419**, 155.

Bacciotti, F., Ray, T. P., Eislöffel, J. *et al.* (2003). Observations of jet diameter, density and dynamics. *Astrophys. Sp. Sci.*, **287**, 3–13.

Backer, D. C. (1993). A pulsar timing tutorial and NRAO observations of PSR1257 + 12. In *Planets around Pulsars*, ASP Conference Series, Vol. 36, ed. J. A. Phillips, S. E. Thorsett, and S. R. Kulkarni. San Francisco: ASP.

Bailes, M., Lyne, A. G., and Shemar, S. L. (1991). A planet orbiting the neutron star PSR 1829 – 10. *Nature*, **352**, 311–13.

Balbus, S. A. and Hawley, J. F. (1991). A powerful local shear instability in weakly magnetized disks. I – Linear analysis. II – Nonlinear evolution. *Astrophys. J.*, **376**, 214–33.

(1998). Instability, turbulence, and enhanced transport in accretion disks. *Rev. Mod. Phys.*, **70**, 1–53.

(2000). Solar nebula magnetohydrodynamics. In *From Dust to Terrestrial Planets*, ed. W. Benz, R. Kallenbach, and G. W. Lugmair. Dordrecht: Kluwer.

Balbus, S. A., Hawley, J. F., and Stone, J. M. (1996). Nonlinear stability, hydrodynamical turbulence, and transport in disks. *Astrophys. J.*, **467**, 76–86.

Balbus, S. A. and Papaloizou, J. C. B. (1999). On the dynamical foundations of alpha disks. *Astrophys. J.*, **521**, 650–8.

Balbus, S. A. and Terquem, C. (2001). Linear analysis of the Hall effect in protostellar disks. *Astrophys. J.*, **552**, 235–47.

Ball, R. C. and Witten, T. A. (1984). Causality bond on the density of aggregates. *Phys. Rev. A*, **29**, 2966–7.

Balmforth, N. J. and Korycansky, D. G. (2001). Non-linear dynamics of the corotation torque. *Mon. Not. R. Astron. Soc.*, **326**, 833–51.

Bally, J., Testi, L., Sargent, A., and Carlstrom, J. (1998). Disk mass limits and lifetimes of externally irradiated young stellar objects embedded in the Orion Nebula. *Astron. J.*, **116**, 854–9.

Baranne, A., Queloz, D., Mayor, M. *et al.* (1996). ELODIE: A spectrograph for accurate radial velocity measurements. *A&AS*, **119**, 373.

Barge, P. and Sommeria, J. (1995). Did planet formation begin inside persistent gaseous vortices? *Astron. Astrophys.*, **295**, L1–L4.

Barnes, J. W. and Fortney, J. J. (2003). Measuring the oblateness and rotation of transiting extrasolar giant planets. *Astrophys. J.*, **588**, 545–56.

(2004). Transit detectability of ring systems around extrasolar giant planets. *Astrophys. J.*, **616**, 1193–203.

Basri, G. and Martin, E. L. (1999). PPL 15: The first brown dwarf spectroscopic binary. *Astron. J.*, **118**, 2460–5.

Bate, M. R. (1998). Collapse of a molecular cloud core to stellar densities: the first three-dimensional calculations. *Astrophys. J.*, **508**, L95–8.

(2000). Predicting the properties of binary stellar systems: the evolution of accreting protobinary systems. *Mon. Not. Roy. Astron. Soc.*, **314**, 33–53.

Bate, M. R. and Bonnell, I. A. (2005). The origin of the initial mass function and its dependence on the mean Jeans mass in molecular clouds. *Mon. Not. Roy. Astron. Soc.*, **356**, 1201–21.

Bate, M. R., Bonnell, I. A., and Bromm, V. (2002a). The formation mechanism of brown dwarfs. *Mon. Not. Roy. Astron. Soc.*, **332**, L65–8.

(2002b). The formation of close binary systems by dynamical interactions and orbital decay. *Mon. Not. Roy. Astron. Soc.*, **336**, 705–13.

(2003a). The formation of a star cluster: predicting the properties of stars and brown dwarfs. *Mon. Not. Roy. Astron. Soc.*, **339**, 577–99.

Bate, M. R., Lubow, S. H., Ogilvie, G. I., and Miller, K. A. (2003b). Three-dimensional calculations of high- and low-mass planets embedded in protoplanetary discs. *Mon. Not. Roy. Astron. Soc.*, **341**, 213–29.

Beckwith, S. V. W., Henning, T., and Nakagawa, Y. (2000). Dust properties and assembly of large particles in protoplanetary disks. In *Protostars and Planets IV*, ed. V. Mannings, A. P. Boss, and S. S. Russell. Tucson: University of Arizona Press, 533–58.

Beckwith, S. V. W. and Sargent, A. I. (1991). Particle emissivity in circumstellar disks. *Astrophys. J.*, **381**, 250–8.

Beckwith, S. V. W., Sargent, A. I., Chini, R. S., and Guesten, R. (1990). A survey for circumstellar disks around young stellar objects. *Astron. J.*, **99**, 924–45.

Beichman, C. A., Bryden, G., Rieke, G. H. *et al.* (2005). Planets and infrared excesses: preliminary results from a Spitzer MIPS survey of solar-type stars. *Astrophys. J.*, **622**, 1160–70.

Bell, K. R., Cassen, P., Klahr, H., and Henning, Th. (1997). The structure and appearance of protostellar accretion disks: limits on disk flaring. *Astrophys. J.*, **486**, 372–87.

Benz, W., Slattery, W. L., and Cameron, A. G. W. (1987). The origin of the Moon and the single impact hypothesis II. *Icarus*, **71**, 30–45.

Berlin, J. (2003). Mineralogisch-petrographische Untersuchungen an der chondritischen Breccie Rumuruti, Dipl. thesis, FU Berlin, pp. 115.

Beust, H. and Morbidelli, A. (1996). Mean-motion resonances as a source for infalling comets toward Beta Pictoris. *Icarus*, **120**, 358–70.

(2000). Falling evaporating bodies as a clue to outline the structure of the $\beta$ Pictoris young planetary system. *Icarus*, **143**, 170–88.

Beust, H., Vidal-Madjar, A., Ferlet, R., and Lagrange-Henri, A. M. (1990). The Beta Pictoris circumstellar disk. X – Numerical simulations of infalling evaporating bodies. *Astron. Astrophys.*, **236**, 202–16.

Biller, B., Close, L., Lenzen, R. *et al.* (2004). Suppressing speckle noise for simultaneous differential extrasolar planet imaging (SDI) at the VLT and MMT. *SPIE*, **5490**, 389–97.

Binney, J. and Tremaine, S. (1987). *Galactic Dynamics*. Princeton, NJ: Princeton University Press.

Birck, J.-L. and Allègre, C. J. (1985). Evidence for the presence of $^{53}$Mn in the early Solar System. *Geophys. Res. Lett.*, **12**, 745–8.

Bizzarro, M., Baker, J. A., and Haack, H. (2004). Mg isotope evidence for contemporaneous formation of chondrules and refractory inclusions. *Nature*, **431**, 275–8.

  (2005). Timing of crust formation on differentiated asteroids inferred from Al–Mg chronometry. *Geophys. Res. Abstr.*, **7**, abstr. no. 08410.

Black, D. C. (1991). It's all in the timing. *Nature*, **352**, 278–9.

Blum, J. (2004). Grain growth and coagulation. In *Astrophysics of Dust*, ed. A. N. Witt, G. C. Clayton, and B. T. Draine. ASP Conference Proceedings, Vol. 309, San Francisco: Astronomical Society of the Pacific, 369–91.

Blum, J. and Münch, M. (1993). Experimental investigations on aggregate–aggregate collisions in the early Solar Nebula. *Icarus*, **106**, 151–67.

Blum, J. and Schräpler, R. (2004). Structure and mechanical properties of high-porosity macroscopic agglomerates formed by random ballistic deposition. *Phys. Rev. Lett.*, **93**, 115503.

Blum, J. and Wurm, G. (2000). Experiments on sticking, restructuring, and fragmentation of preplanetary dust aggregates. *Icarus*, **143**, 138–46.

Blum, J., Wurm, G., Kempf, S., and Henning, T. (1996). The Brownian motion of dust particles in the Solar Nebula: an experimental approach to the problem of pre-planetary dust aggregation. *Icarus*, **124**, 441–51.

Blum, J., Wurm, G., Kempf, S. *et al.* (2000). Growth and form of planetary seedlings: results from a microgravity aggregation experiment. *Phy. Rev. Lett.*, **85**, 2426–9.

Blum, J., Wurm, G., and Poppe, T. (1999b). The CODAG sounding rocket experiment to study aggregation of thermally diffusing dust particles. *Adv. Space Res.*, **23**, 1267–70.

Blum, J., Wurm, G., Poppe, T. *et al.* (1999a). The cosmic dust aggregation experiment CODAG. *Measure. Sci. Tech.*, **10**, 836–44.

Bockelée-Morvan, D., Gautier, D., Hersant, F., Huré, J.-M., and Robert, F. (2002). Turbulent radial mixing in the Solar nebula as the source of crystalline silicates in comets. *Astron. Astrophys.*, **384**, 1107–18.

Bodenheimer, P. (1974). Calculations of the early evolution of Jupiter. *Icarus*, **23**, 319–25.

Bodenheimer, P., Grossman, A. S., DeCampli, W. M., Marcy, G., and Pollack, J. B. (1980). Calculations of the evolution of the giant planets. *Icarus*, **41**, 293–308.

Bodenheimer, P., Hubickyj, O., and Lissauer, J. J. (2000). Models of the *in situ* formation of detected extrasolar giant planets. *Icarus*, **143**, 2–14.

Bodenheimer, P., Laughlin, G., and Lin, D. N. C. (2003). On the radii of extrasolar giant planets. *Astrophys. J.*, **592**, 555–63.

Bodenheimer, P. and Lin, D. N. C. (2002). Implications of extrasolar planets for understanding planet formation. *Annu. Rev. Earth Planet. Sci.*, **30**, 113–48.

Bodenheimer, P., Lin, D. N. C., and Mardling, R. A. (2001). On the tidal inflation of short-period extrasolar planets. *Astrophys. J.*, **548**, 466–72.

Bodenheimer, P. and Pollack, J. B. (1986). Calculations of the accretion and evolution of giant planets: the effects of solid cores. *Icarus*, **67**, 391–408.

Bond, I. A., Udalski, A., Jaroszynski, M. *et al.* (2004). OGLE 2003-BLG-235/MOA 2003-BLG-53: A planetary microlensing event. *Astrophys. J.*, **606**, L155–8.

Bonnell, I. A., Bate, M. R., Clarke, C. J., and Pringle, J. E. (1997). Accretion and the stellar mass spectrum in small clusters. *Mon. Not. Roy. Astron. Soc.*, **285**, 201–8.

Bordé, P., Rouan, D., and Léger, A. (2003). Exoplanet detection capability of the CoRoT space mission. *Astron. Astrophys.*, **405**, 1137–44.

Boss, A. P. (1988). Protostellar formation in rotating interstellar clouds. VII – Opacity and fragmentation. *Astrophys. J.*, **331**, 370–6.

(1989). Protostellar formation in rotating interstellar clouds. VIII – Inner core formation. *Astrophys. J.*, **346**, 336–49.

(1996). Evolution of the Solar Nebula. III. Protoplanetary disks undergoing mass accretion. *Astrophys. J.*, **469**, 906–20.

(1997). Giant planet formation by gravitational instability. *Science*, **276**, 1836–9.

(1998a). *Looking for Earths*. New York: Wiley.

(1998b). Evolution of the Solar Nebula. IV. Giant gaseous protoplanet formation. *Astrophys. J.*, **503**, 923–37.

(1998c). Astrometric signatures of giant-planet formation. *Nature*, **395**, 141–143.

(2000). Possible rapid gas giant planet formation in the Solar Nebula and other protoplanetary disks. *Astrophys. J.*, **536**, L101–4.

(2001a). Formation of planetary-mass objects by protostellar collapse and fragmentation. *Astrophys. J.*, **551**, L167–70.

(2001b). Gas giant protoplanet formation: disk instability models with thermodynamics and radiative transfer. *Astrophys. J.* **563**, 367–373.

(2002a). Stellar metallicity and the formation of extrasolar gas giant planets. *Astrophys. J.*, **567**, L149–53.

(2002b). Evolution of the Solar Nebula. V. Disk instabilities with varied thermodynamics. *Astrophys. J.*, **576**, 462–72.

(2003). Rapid formation of outer giant planets by disk instability. *Astrophys. J.*, **599**, 577–81.

(2004). Convective cooling of protoplanetary disks and rapid giant planet formation. *Astrophys. J.*, **610**, 456–63.

Boss, A. P., Basri, G., Kumar, S. S. *et al.* (2003). Nomenclature: brown dwarfs, gas giant planets, and ? In *Brown Dwarfs, Proceedings of IAU Symposium 211*, ed. Martin, E., San Francisco: ASP, 529.

Boss A. P. and Vanhala H. A. T. (2001). Injection of newly synthesized elements into the protosolar cloud. *Phil. Trans. R. Soc. Lond.*, **A 359**, 2005–17.

Boss, A. P., Wetherill, G. W., and Haghighipour, N. (2002). Rapid formation of ice giant planets. *Icarus*, **156**, 291–5.

Bouchy, F., Bazot, M., Santos, N. C. *et al.* (2005). The acoustic spectrum of the planet-hosting star $\mu$ Arae. *Astron. Astrophys.*, **440**, 609–14.

Bouchy, F., Pont, F., Melo, C. *et al.* (2005). Doppler follow-up of OGLE transiting companions in the Galactic bulge. *Astron. Astrophys.*, **431**, 1105–21.

Bouchy, F., Pont, F., Santos, N. C. *et al.* (2004). Two new "very hot Jupiters" among the OGLE transiting candidates. *Astron. Astrophys.*, **421**, L13–16.

Boudet, N., Mutschke, H., Nayral, C. *et al.* (2005), Temperature-dependence of the submillimeter absorption coefficient of amorphous silicate grains. *Astrophys. J.*, **633**, 272–81.

Bouwman, J., de Koter, A., Dominik, C., and Waters, L. B. F. M. (2003). The origin of crystalline silicates in the Herbig Be star HD 100546 and in comet Hale-Bopp. *Astron. Astrophys.*, **401**, 577–92.

Bouwman, J., de Koter, A., van den Ancker, M. E., and Waters, L. B. F. M. (2000). The composition of the circumstellar dust around the Herbig Ae stars AB Aur and HD 163296. *Astron. Astrophys.*, **360**, 213–26.

Bouwman, J., Meeus, G., de Koter, A. *et al.* (2001). Processing of silicate dust grains in Herbig Ae/Be systems. *Astron. Astrophys.*, **375**, 950–62.

Bouy, H., Brandner, W., Martín, E. L. *et al.* (2003). Multiplicity of nearby free-floating ultracool dwarfs: a Hubble Space Telescope WFPC2 search for companions. *Astron. J.*, **126**, 1526–54.

Bradley J. P., Keller L. P., Snow, T. P. *et al.* (1999). An infrared spectral match between GEMS and interstellar grains. *Science*, **285**, 1716–18.

Brandenburg, A., Nordlund, Å., Stein, R. F., and Torkelsson, U. (1995). Dynamo-generated turbulence and large-scale magnetic fields in a Keplerian shear flow. *Astrophys. J.*, **446**, 741–54.

Brazzle R. H., Pravdivtseva O. V., Meshik A. P., and Hohenberg C. M. (1999). Verification and interpretation of the I-Xe chronometer. *Geochim. cosmochim. acta*, **63**, 739–60.

Briceño, C., Luhman, K. L., Hartmann, L., Stauffer, J. R., and Kirkpatrick, J. D. (2002). The initial mass function in the Taurus star-forming region. *Astrophys. J.*, **580**, 317–35.

Briceño, C., Vivas, A. K., Calvet, N. *et al.* (2001). The CIDA-QUEST large-scale survey of Orion OB1: evidence for rapid disk dissipation in a dispersed stellar population. *Science*, **291**, 93–6.

Bridges, F. G., Supulver, K. D., Lin, D. N. C., Knight, R., and Zafra, M. (1996). Energy loss and sticking mechanisms in particle aggregation in planetesimal formation. *Icarus*, **123**, 422–35.

Brown, T. M. (2001). Transmission spectra as diagnostics of extrasolar giant planet atmospheres. *Astrophys. J.*, **553**, 1006–26.

(2003). Expected detection and false alarm rates for transiting Jovian planets. *Astrophys. J.*, **593**, L125–L128.

Brown, T. M., Charbonneau, D., Gilliland, R. L., Noyes, R. W., and Burrows, A. (2001). Hubble Space Telescope time-series photometry of the transiting planet of HD 209458. *Astrophys. J.*, **552**, 699–709.

Brown, T. M., Libbrecht, K. G., and Charbonneau, D. (2002). A search for CO absorption in the transmission spectrum of HD 209458b. *Pub. Ast. Soc. Pac.*, **114**, 826–32.

Brownlee, D. E., Burnett, D., Clark, B. *et al.* (1996). Stardust: comet and interstellar dust sample return mission. In *Physics, Chemistry, and Dynamics of Interplanetary Dust*, ed. B. A. S. Gustafson and M. S. Hanner (ASP Conf. Series 104). San Fransisco: Astronomical Society of the Pacific.

Brush, S. G. (1990). Theories of the origin of the solar system 1956–1985. *Rev. Mod. Phys.*, **62**, 43–112.

Bryden, G., Chen, X., Lin, D. N. C., Nelson, R. P., and Papaloizou, J. C. B. (1999). Tidally induced gap formation in protostellar disks: gap clearing and suppression of protoplanetary growth. *Astrophys. J.*, **514**, 344–67.

Bryden, G., Różyczka, M., Lin, D. N. C., and Bodenheimer, P. (2000). On the interaction between protoplanets and protostellar disks. *Astrophys. J.*, **540**, 1091–101.

Buffon, G.-L. Leclerc, Comte de (1749). *Histoire naturelle, générale et particulière, avec la description du cabinet du roi.* Paris: Imprimerie Royal.

Burgasser, A. J., Kirkpatrick, J. D., Reid, I. N. *et al.* (2003). Binarity in brown dwarfs: T dwarf binaries discovered with the Hubble Space Telescope wide field planetary camera 2. *Astrophys. J.*, **586**, 512–26.

Burke, C. J., Gaudi, B. S., DePoy, D. L., Pogge, R. W., and Pinsonneault, M. H. (2004). Survey for transiting extrasolar planets in stellar systems. I. Fundamental parameters of the open cluster NGC 1245. *Astron. J.*, **127**, 2382–97.

Burrows, A., Hubbard, W. B., Lunine, J. I., and Liebert, J. (2001). The theory of brown dwarfs and extrasolar giant planets. *Rev. Mod. Phys.*, **73**, 719–65.

Burrows, A., Hubeny, I., Hubbard, W. B., Sudarsky, D., and Fortney, J. J. (2004). Theoretical radii of transiting giant planets: The case of OGLE-TR-56b. *Astrophys. J.*, **610**, L53–6.

Burrows, A., Marley, M. S., Hubbard, W. B. *et al.* (1997). A nongray theory of extrasolar giant planets and brown dwarfs. *Astrophys. J.*, **491**, 856.

Burrows, A., Sudarsky, D., and Hubbard, W. B. (2003). A theory for the radius of the transiting giant planet HD 209458b. *Astrophys. J.*, **594**, 545–51.

Butler, R. P., Marcy, G. W., Williams, E. *et al.* (1996). Attaining Doppler precision of 3 m s⁻¹. *PASP*, **108**, 500–9.

Butler, R. P., Tinney, C. G., Marcy, G. W. *et al.* (2001). Two new planets from the Anglo-Australian Planet Search. *Astrophys. J.*, **555**, 410–17.

Butler, P., Vogt, S. S., Marcy, G. W. *et al.* (2004). A Neptune-mass planet orbiting the nearby M dwarf GJ 436. *Astrophys. J.*, **617**, 580–8.

Cabot, W. (1984). The nonaxisymmetric baroclinic instability in thin accretion disks. *Astrophys. J.*, **277**, 806–12.

Cai, K., Durisen, R. H., and Mejía, A. C. (2004). Boundary conditions of radiative cooling in gravitationally unstable protoplanetary disks. *AAS Meeting #204*, abstract #62.19.

Calvet, N., D'Alessio, P., Hartmann, L. *et al.* (2002). Evidence for a developing gap in a 10 Myr old protoplanetary disk. *Astrophys. J.*, **568**, 1008–16.

Calvet, N., Muzerolle, J., Briceño, C., *et al.* (2004). The mass accretion rates of intermediate-mass T Tauri stars. *Astron. J.*, **128**, 1294–318.

Cameron, A. G. W. (1969). The pre-Hayashi phase of stellar evolution. In *Low-luminosity Stars*, ed. S. S. Kumar. New York: Gordon and Breach.

(1973). Accumulation processes in the primitive solar nebula. *Icarus*, **18**, 407–50.

(1978). Physics of the primitive solar accretion disk. *Moon and Planets*, **18**, 5–40.

Carpenter, J. M., Wolf, S., Schreyer, K., Launhardt, R., and Henning, Th. (2005). Evolution of cold circumstellar dust around Solar-type stars. *Astron. J.*, **129**, 1049–62.

Carter, B. D., Butler, R. P., Tinney, C. G. *et al.* (2003). A planet in a circular orbit with a 6 year period. *Astrophys. J.*, **593**, L43–6.

Casanova, S., Montmerle, T., Feigelson, E. D., and André, P. (1995). ROSAT X-ray sources embedded in the rho Ophiuchi cloud core. *Astrophys. J.*, **439**, 752–70.

Cassen P. and Woolum D. (1999). In *Encyclopedia of the Solar System*, ed. Weissman, P. R., McFadden, L. A., and Johnson, T. V. Academic Press, 35.

Castellano, T., Jenkins, J., Trilling, D. E., Doyle, L., and Koch, D. (2000). Detection of planetary transits of the star HD 209458 in the Hipparcos data set. *Astrophys. J.*, **532**, L51–3.

Chabrier, G. (2003). Galactic stellar and substellar initial mass function. *Pub. Astron. Soc. Pac.*, **115**, 763–95.

Chabrier, G., Barman, T., Baraffe, I., Allard, F., and Hauschildt, P. H. (2004). The evolution of irradiated planets: application to transits. *Astrophys. J.*, **603**, L53–6.

Chalonge, D., Barbier, D., and Vigroux, E. (1942). Étude spectrophotométrique de l'Éclipse de Lune des 2 et 3 mars 1942. *Annales d'astrophysique*, **5**, 1–57.

Chamberlin, T. C. (1899). On Lord Kelvin's address on the age of the Earth as an abode to life. *Science*, **9**, 889; **10**, 11.

(1903). The origin of ocean basins on the planetesimal hypothesis. *Bull. Geolog. Soc. Am.*, **14**, 548 (abstract).

Chambers, J. E. (2001). Making more terrestrial planets. *Icarus*, **152**, 205–24.

Chambers, J. E. and Wetherill, G. W. (1998). Making the terrestrial planets: N-body integrations of planetary embryos in three dimensions. *Icarus*, **136**, 304–27.

Chandrasekhar, S. (1960). The hydrodynamic stability of viscid flow between coaxial cylinders. *Proc. Nat. Acad. Sci. USA*, **46**, 137, 141.

(1961). *Hydrodynamic and Hydromagnetic Stability* (International Series of Monographs on Physics). Oxford: Clarendon.

Charbonneau, D., Allen, L. E., Megeath, S. T. *et al.* (2005). Detection of thermal emission from an extrasolar planet. *Astrophys. J.*, **626**, 523–9.

Charbonneau, D., Brown, T. M., Latham, D. W., and Mayor, M. (2000). Detection of planetary transits across a Sun-like star. *Astrophys. J.*, **529**, L45–8.

Charbonneau, D., Brown, T. M., Noyes, R. W., and Gilliland, R. L. (2002). Detection of an extrasolar planet atmosphere. *Astrophys. J.*, **568**, 377–84.

Charnoz, S. and Morbidelli, A. (2003). Coupling dynamical and collisional evolution of small bodies: an application to the early ejection of planetesimals from the Jupiter–Saturn region. *Icarus*, **166**, 141–56.

Chauvin, G., Lagrange, A.-M., Dumas, C. *et al.* (2004). A giant planet candidate near a young brown dwarf. Direct VLT/NACO observations using IR wavefront sensing. *Astron. Astrophys.*, **425**, L29–32.

Chavanis, P. H. (2000). Trapping of dust by coherent vortices in the Solar nebula. *Astron. Astrophys.*, **356**, 1089–111.

Chen, C. H. and Kamp, I. (2004). Are giant planets forming around HR 4796A? *Astrophys. J.*, **602**, 985–92.

Chen J. H. and Wasserburg G. J. (1981). The isotopic composition of uranium and lead in Allende inclusions and meteoritic phosphates. *Earth Planet. Sci. Lett.*, **5**, 15–21.

Cherniak D., Lanford W. A., and Ryerson F. J. (1991). Lead diffusion in apatite and zircon using ion implantation and Rutherford backscattering techniques. *Geochim. cosmochim. acta*, **55**, 1663–73.

Chiang, E. I. and Murray, N. (2002). Eccentricity excitation and apsidal resonance capture in the planetary system $\upsilon$ Andromedae. *Astrophys. J.*, **576**, 473–7.

Chiang, E. I. and Murray-Clay, R. A. (2004). The circumbinary ring of KH 15D. *Astrophys. J.*, **607**, 913–20.

Chokshi, A., Tielens, A. G. G. M., and Hollenbach, D. (1993). Dust coagulation. *Astrophys. J.*, **407**, 806–19.

Clarke, C. J., Bonnell, I. A., and Hillenbrand, L. A. (2000). The Formation of Stellar Clusters, in *Protostars and Planets IV*, ed. V. Mannings, A. Boss, and S. Russell (University of Arizona Press, Tucson), p. 151.

Clarke, C. J., Gendrin, A., and Sotomayor, M. (2001), The dispersal of circumstellar discs: the role of the ultraviolet switch. *Mon. Not. R. Astron. Soc.*, **328**, 485–91.

Clayton, R. N. (1993). Oxygen isotopes in meteorites. *Annu. Rev. Earth Planet. Sci.*, **21**, 115–49.

Clayton, R. N. and Mayeda, T. K. (1984). The oxygen isotope record in Morchison and other carbonaceous chondrites. *Earth Planet. Sci. Lett.*, **67**, 151–61.

Close, L., Lenzen, R., Guirado, J. C. *et al.* (2005). A dynamical calibration of the mass–luminosity relation at very low stellar masses and young ages. *Nature*, **433**, 286–9.

Close, L. M., Siegler, N., Freed, M., and Biller, B. (2003). Detection of nine M8.0-L0.5 binaries: the very low mass binary population and its implications for brown dwarf and very low mass star formation. *Astrophys. J.*, **587**, 407–22.

Close, L. M., Siegler, N., Potter, D., Brandner, W., and Liebert J. (2002). An adaptive optics survey of M8-M9 stars: discovery of four very low mass binaries with at least one system containing a brown dwarf companion. *Astrophys. J.*, **567**, L53–7.

Cody, A. M., and Sasselov, D. D. (2002). HD 209458: physical parameters of the parent star and the transiting planet. *Astrophys. J.*, **569**, 451–8.

Colwell, J. E. (2003). Low velocity impacts into dust: results from the COLLIDE-2 microgravity experiment. *Icarus*, **164**, 188–96.

Colwell, J. E. and Taylor, M. (1999). Low-velocity microgravity impact experiments into simulated regolith. *Icarus*, **138**, 241–8.

Cuzzi, J. N., Davis, S. S., and Dobrovolskis, A. R. (2003). Blowing in the wind. II. Creation and redistribution of refractory inclusions in a turbulent protoplanetary nebula. *Icarus*, **166**, 385–402.

Cuzzi, J. N., Dobrovolskis, A. R., and Champney, J. M. (1993). Particle-gas dynamics in the midplane of a protoplanetary nebula. *Icarus*, **106**, 102–34.

Cuzzi, J. N., Dobrovolskis, A. R., and Hogan, R. C. (1996). Turbulence, chondrules and planetesimals. In *Chondrules and the Protoplanetary Disk*, ed. Hewins, R., Jones, R., and Scott, E. R. D. Cambridge: Cambridge University Press, 35.

Cuzzi, J. N., Hogan, R. C., Paque, J. M., and Dobrovolskis, A. R. (2001). Size-selective concentration of chondrules and other small particles in protoplanetary nebula turbulence. *Astrophys. J.*, **546**, 496–508.

Cuzzi, J. and Zahnle, K. (2004). Material enhancement in protoplanetary nebulae by particle drift through evaporation fronts. *Astrophys. J.*, **614**, 490–6.

D'Alessio, P., Calvet, N., and Hartmann, L. (2001). Accretion disks around young objects. III. Grain growth. *Astrophys. J.*, **553**, 321–34.

D'Angelo, G., Henning, T., and Kley, W. (2002). Nested-grid calculations of disk–planet interaction. *Astron. Astrophys.*, **385**, 647–70.
(2003b). Thermohydrodynamics of circumstellar disks with high-mass planets. *Astrophys. J.*, **599**, 548–576.

D'Angelo, G., Kley, W., and Henning, T. (2003a). Orbital migration and mass accretion of protoplanets in three-dimensional global computations with nested grids. *Astrophys. J.*, **586**, 540–61.

Davidsson, B. J. R. and Gutierrez, P. J. (2004). Estimating the nucleus density of comet 19P/Borelly. *Icarus*, **168**, 392–408.

de la Fuente Marcos, C. and Barge, P. (2001). The effect of long-lived vortical circulation on the dynamics of dust particles in the mid-plane of a protoplanetary disc. *Mon. Not. R. Astron. Soc.*, **323**, 601–14.

DeCampli, W. M. and Cameron, A. G. W. (1979). Structure and evolution of isolated giant gaseous protoplanets. *Icarus*, **38**, 367–91.

Decin, G., Dominik, C., Waters, L. B. F. M., and Waelkens, C. (2003). Age dependence of the Vega phenomenon: observations. *Astrophys. J.*, **598**, 636–44.

Delgado-Donate, E. J., Clarke, C. J., Bate, M. R., and Hodgkin, S. T. (2004). On the properties of young multiple stars, *Mon. Not. Roy. Astron. Soc.*, **351**, 617–29.

Deming, D., Brown, T. M., Charbonneau, D., Harrington, J., and Richardson, L. J. (2005). A new search for carbon monoxide absorption in the transmission spectrum of the extrasolar planet HD 209458b. *Astrophys. J.*, **622**, 1149–59.

Deming, D., Seager, S., Richardson, L. J., Harrington, J. (2005). Infrared radiation from an extrasolar planet. *Nature*, **434**, 740–3.

Dent, W. R. F. (1988), Infrared radiative transfer in dense disks around young stars. *Astrophys. J.*, **325**, 252–65.

Descartes, R. (1644). *Principia Philosophiae*. Amsterdam: Elsevier, in Latin.

Desch, S. J. and Connolly Jr., H. C. (2002). A model of the thermal processing of particles in Solar nebula shocks: application to the cooling rates of chondrules. *Met. Plan. Sci.*, **37**, 183.

Dodd R. T. (1969). Metamorphism of the ordinary chondrites: a review. *Geochim. cosmochim. acta*, **33**, 161–203.

Dominik, C. and Decin, G. (2003). Age dependence of the Vega phenomenon: theory. *Astrophys. J.*, **598**, 626–35.

Dominik, C. and Nübold, H. (2002). Magnetic aggregation: dynamics and numerical modeling. *Icarus*, **157**, 173–86.

Dominik, C. and Tielens, A. (1995). Resistance to rolling in the adhesive contact of two spheres. *Phil. Mag. A*, **72**, 783–803.

  (1997). Coagulation of dust grains and the structure of dust aggregates in space. *Astrophys. J.*, **480**, 647–73.

Dorschner, J., Begemann, B., Henning, Th., Jäger, C., and Mutschlee, H. (1995). Steps toward interstellar silicate mineralogy. II. Study of Mg–Fe–silicate glasses of variable composition. *Astron. Astrophys.*, **300**, 503.

Dotter, A. L. and Chaboyer, B. C. (2002). The impact of pollution on stellar evolution models. *Bull. Am. Astronomi. Soc.*, **34**, 1127.

Draine, B. T. and Lee, H. M. (1984). Optical properties of interstellar graphite and silicate grains. *Astrophys. J.*, **285**, 89–108.

Dubrulle, B., Morfill, G., and Sterzik, M. (1995). The dust subdisk in the protoplanetary nebula. *Icarus*, **114**, 237–46.

Dullemond, C. P. (2002). The 2-D structure of dusty disks around Herbig Ae/Be stars. I. Models with grey opacities. *Astron. Astrophys.*, **395**, 853–62.

Dullemond, C. P. and Dominik, C. (2004). The effect of dust settling on the appearance of protoplanetary disks. *Astron. Astrophys.*, **421**, 1075–86.

  (2005). Dust coagulation in protoplanetary disks: a rapid depletion of small grains. *Astron. Astrophys.*, **434**, 971–86.

Duquennoy, A. and Mayor, M. (1991). Multiplicity among Solar-type stars in the Solar Neighbourhood. II - Distribution of the orbital elements in an unbiased sample. *Astron. Astrophys.*, **248**, 485–524.

Duren, R. M., Dragon, K., Gunter, S. Z. et al. (2004). Systems engineering for the Kepler mission: a search for terrestrial planets. *Proc. SPIE*, **5497**, 16–27.

Dutrey, A., Guilloteau, S., Duvert, G. et al. (1996). Dust and gas distribution around T Tauri stars in Taurus-Auriga. I. Interferometric 2.7 mm continuum and $^{13}$CO J = 1-0 observations. *Astron. Astrophys.*, **309**, 493–504.

Dutrey, A., Lecavelier des Etangs, A., and Augereau, J.-C. (2004). *The observation of circumstellar disks: dust and gas components*. In *Comets II*, ed. C. Festou, H. V. Keller, and H. A. Weaver, pp. 81–95.

Duvert, G., Guilloteau, S., Ménard, F., Simon, M., and Dutrey, A. (2000). A search for extended disks around weak-lined T Tauri stars. *Astron. Astrophys.*, **355**, 165–70.

Dyson, F. (1999). *The Sun, the Genome, and the Internet*. Oxford: Oxford University Press.

ESA (1997). The Hipparcos and Tycho catalogues. *VizieR Online Data Catalog*, **1239**, 0.

Ehrenfreund, P., Fraser, H. J. et al. (2003). Physics and chemistry of icy particles in the universe: answers from microgravity. *Planet. Space Sci.*, **51**, 473–94.

Eisner, J. A. and Carpenter, J. M. (2003). Distribution of circumstellar disk masses in the young cluster NGC 2024. *Astrophys. J.*, **598**, 1341–9.

Elmegreen, B. (1979). On the disruption of a protoplanetary disk by a T Tauri-like solar wind. *Astron. Astrophys.*, **80**, 77–8.

Elmegreen, B. G. (1999). A prediction of brown dwarfs in ultracold molecular gas. *Astrophys. J.*, **522**, 915–20.

Evans, N. J. (1999). Physical conditions in regions of star formation. *Annu. Rev. Astron. Astrophys.*, **37**, 311–62.

Fabian, A. C. and Podsiadlowski, Ph. (1991). Binary precursor for planet? *Nature*, **353**, 801.

Fabian, D., Jäger, C., Henning, T., Dorschner, J., and Mutschke, H. (2000). Steps toward interstellar silicate mineralogy. V. Thermal evolution of amorphous magnesium silicates and silica. *Astron. Astrophys.*, **364**, 282–92.

Fazio, G. G., Hora, J. L., Allen, L. E. *et al.* (2004). The Infrared Array Camera (IRAC) for the Spitzer Space Telescope. *Astrophys. J. Suppl.*, **154**, 10–17.

Feigelson, E. D., Casanova, S., Montmerle, T., and Guibert, J. (1993). ROSAT X-ray study of the Chamaeleon I dark cloud. I. The stellar population. *Astrophys. J.*, **416**, 623–46.

Ferlet, R., Vidal-Madjar, A., and Hobbs, L. M. (1987). The Beta Pictoris circumstellar disk. V-time variations of the CA II-K line. *Astron. Astrophys.*, **185**, 267–70.

Fischer, D. A. and Valenti, J. (2005). The planet-metallicity correlation. *Ap. J.*, **622**, 1102.

Fischer, D. A. and Valenti, J. A. (2003). Metallicities of stars with extrasolar planets. In *Scientific Frontiers in Research on Extrasolar Planets*, ASP Conference Series, Vol. 294, ed. Deming, D. and Seager, S. San Francisco: Astronomical Society of the Pacific, 117–28.

Fischer, D., Valenti, J. A., and Marcy, G. (2004). Spectral analysis of stars on planet-search surveys. In *IAU Symposium #219: Stars as Suns: Activity, Evolution, and Planets*, ed. Dupree, A. K. San Francisco: Astronomical Society of the Pacific, 29.

Fischer, O., Henning, T., and Yorke, H. W. (1996). Simulation of polarization maps. II. The circumstellar environment of pre-main sequence objects. *Astron. Astrophys.*, **308**, 863–85.

Fleming, T., Stone, J., and Hawley, J. (2000). The effect of resistivity on the nonlinear stage of the magnetorotational instability in accretion disks. *Astrophys. J.*, **530**, 464–77.

Floss C. and Stadermann F. J. (2004). Isotopically primitive interplanetary dust particles of cometary origin: evidence from nitrogen isotopic compositions. *Lunar Planet. Sci.*, **XXXV**, A1281–2.

Font, A. S., McCarthy, I. G., Johnstone, D., and Ballantyne, D. R. (2004). Photoevaporation of circumstellar disks around young stars. *Astrophys. J.*, **607**, 890–903.

Ford, E. B. (2004). Choice of observing schedules for astrometric planet searches. *Pub. Ast. Soc. Pac.*, **116**, 1083–92.

(2005). Quantifying the uncertainty in the orbits of extrasolar planets. *Astron. J.*, **129**, 1706–17.

Ford, E. B., Havlickova, M., and Rasio, F. A. (2001). Dynamical instabilities in extrasolar planetary systems containing two giant planets. *Icarus*, **150**, 303–13.

Ford, E. B., Rasio, F. A., and Yu, K. (2003). Dynamical instabilities in extrasolar planetary systems. *ASP Conf. Ser. 294: Scientific Frontiers in Research on Extrasolar Planets*, 181–8.

Ford, E. B. and Tremaine, S. (2003). Planet-finding prospects for the Space Interferometry Mission. *Pub. Ast. Soc. Pac.*, **115**, 1171–86.

Forrest, W. J., Sargent, B., Furlan, E. *et al.* (2004). Mid-infrared spectroscopy of disks around classical T Tauri stars. *Astrophys. J. Suppl.*, **154**, 443–7.

Franchi, I. A., Baker, L., Bridges, J. C., Wright, I. P., and Pillinget, C. T. (2001). Oxygen isotopes and the early Solar System. *Phil. Trans. R. Soc. Lond.*, **A 359**, 2019–35.

Fridlund, C. V. M. (2003). DARWIN - the scientific constraints. In *Towards Other Earths: DARWIN/TPF and the Search for Extrasolar Terrestrial Planets*, ed. Fridlund, M. and Henning, Th. *ESA*, **SP-539**, 293–8.

Frink, S., Quirrenbach, A., Fischer, D., Röser, S., and Schilbach, E. (2001). A strategy for identifying the grid stars for the Space Interferometry Mission. *Pub. Ast. Soc. Pac.*, **113**, 173–87.

Fromang, S., Terquem, C., and Balbus, S. A. (2002). The ionization fraction in models of protoplanetary discs. *Mon. Not. R. Astron. Soc.*, **329**, 18–28.

Gail, H.-P. (2001). Radial mixing in protoplanetary accretion disks. I. Stationary disk models with annealing and carbon combustion. *Astron. Astrophys.*, **378**, 192–213.

(2002). Radial mixing in protoplanetary accretion disks. III. Carbon dust oxidation and abundance of hydrocarbons in comets. *Astron. Astrophys.*, **390**, 253–65.

(2003). Formation and evolution of minerals in accretion disks and stellar outflows. In *Astromineralogy*, ed. T. K. Henning. Heidelberg: Springer.

(2004). Radial mixing in protoplanetary accretion disks. IV. Metamorphosis of the silicate dust complex. *Astron. Astrophys.*, **413**, 571–91.

Gammie, C. F. (1996). Layered accretion in T Tauri disks. *Astrophys. J.*, **457**, 355–62.

(2001). Nonlinear outcome of gravitational instability in cooling, gaseous disks. *Astrophys. J.*, **533**, 174–83.

Gatewood, G. and Eichhorn, H. (1973). An unsuccessful search for a planetary companion of Barnard's star. *Astron. J.*, **78**, 769–76.

Gaudi, B. S., Seager, S., and Mallen-Ornelas, G. (2005). On the period distribution of close-in extrasolar giant planets. *Astrophys. J.*, **623**, 472–81.

Gehrels, N. (1986). Confidence limits for small numbers of events in astrophysical data. *Astrophys. J.*, **303**, 336–46.

Gilliland, R. L., Brown, T. M., Guhathakurta, P. *et al.* (2000). A lack of planets in 47 Tucanae from a Hubble Space Telescope search. *Astrophys. J.*, **545**, L47–51.

Gilmour, J. D., Pravdivtseva, O. V., Busfield, A., and Hohenberg, C. M. (2004). I–Xe and the chronology of the early Solar System. *Workshop on Chondrites and the Protoplanetary Disk*, abstr. no. 9054.

Gilmour, J. D. and Saxton, J. M. (2001). A timescale of formation of the first solids. *Phil. Trans. R. Soc. Lond.*, **A 359**, 2037–48.

Gizis, J. E., Reid, I. N., Knapp, G. R. *et al.* (2003). Hubble Space Telescope observations of binary very low mass stars and brown dwarfs. *Astron. J.*, **125**, 3302–10.

Glassgold, A. E., Feigelson, E. D., and Montmerle, T. (2000). Effects of energetic radiation in young Stellar objects. In *Protostars & Planets IV*, ed. V. G. Mannings, A. P. Boss, and S. Russell. Tucson: University of Arizona Press.

Glassgold, A. E., Lucas, R., and Omont, A. (1986). Molecular ions in the circumstellar envelope of IRC + 10216. *Astron. Astrophys.* **157**, 35–48.

Glassgold, A. E., Najita, J., and Igea, J. (1997). X-ray ionization of protoplanetary disks. *Astrophys. J.*, **480**, 344–50.

Godon, P. and Livio, M. (2000). The formation and role of vortices in protoplanetary disks. *Astrophys. J.*, **537**, 396–404.

Goldreich, P., Lithwick, Y. and Sari, R. (2004). Planet formation by coagulation: a focus on Uranus and Neptune. *Annu. Rev. Astron. Astrophys.*, **42**, 549–601.

Goldreich, P. and Sari, R. (2003). Eccentricity evolution for planets in gaseous disks. *Astrophys. J.*, **585**, 1024–37.

Goldreich, P. and Tremaine, S. (1979). The excitation of density waves at the Lindblad and corotation resonances by an external potential. *Astrophys. J.*, **233**, 857–71.

(1980). Disk–satellite interactions. *Astrophys. J.*, **241**, 425–41.

Goldreich, P. and Ward, W. R. (1973). The formation of planetesimals. *Astrophys. J.*, **183**, 1051–61.

Gonzalez, G. (1997). The stellar metallicity-giant planet connection. *Mon. Not. R. Astron. Soc.*, **285**, 403–12.

Goodman, J. and Rafikov, R. R. (2001). Planetary torques as the viscosity of protoplanetary disks. *Astrophys. J.*, **552**, 793–802.

Göpel C., Manhès G., and Allègre C. J. (1994). U–Pb systematics of phosphates from equilibrated ordinary chondrites. *Earth Planet. Sci. Lett.*, **121**, 153–71.

Goswami J. N. and Lal, D. (1979). Formation of the parent bodies of the carbonaceous chondrites. *Icarus*, **10**, 510.

Goswami J. N. and MacDougall, J. D. (1983). Nuclear track and compositional studies of olivines in CI and CM chondrites. *J. Geophys. Res. Suppl.*, **88**, A755–64.

Gould, A., Pepper, J., and DePoy, D. L. (2003). Sensitivity of transit searches to habitable-zone planets. *Astrophys. J.*, **594**, 533–7.

Goździewski, K. (2003). A dynamical analysis of the HD 37124 planetary system. *Astron. Astrophys.*, **398**, 315–25.

Grady, C. A., Woodgate, B., Torres, C. A. O. *et al.* (2004). The environment of the optically brightest Herbig Ae star, HD 104237. *Astrophys. J.*, **608**, 809–30.

Gratton, R., Feldt, M., Schmid, H. M. *et al.* (2004). The science case of the CHEOPS planet finder for VLT. *SPIE*, **5492**, 1010–21.

Greenzweig, Y. and Lissauer, J. J. (1992). Accretion rates of protoplanets. I – Gaussian distributions of planetesimal velocities. *Icarus*, **100**, 440–63.

Grossman, L. and Larimer, J. W. (1974). Early chemical history of the Solar System. *Rev. Geophys. Space Phys.*, **12**, 71–101.

Gu, P., Bodenheimer, P. H., and Lin, D. N. C. (2004). The internal structural adjustment due to tidal heating of short-period inflated giant planets. *Astrophys. J.*, **608**, 1076–94.

Gu, P., Lin, D. N. C., and Bodenheimer, P. H. (2003). The effect of tidal inflation instability on the mass and dynamical evolution of extrasolar planets with ultrashort periods. *Astrophys. J.*, **588**, 509–34.

Guillot, T. (1999). A comparison of the interiors of Jupiter and Saturn. *Planet. Space Science*, **47**, 1183–200.

(2005). The interiors of giant planets: models and outstanding questions. *Ann. Rev. Earth Planet. Sci.*, **33**, 493–530.

Guillot, T., Gautier, D. and Hubbard, W. B. (1997). New constraints on the composition of Jupiter from Galileo measurements and interior models. *Icarus*, **130**, 534–9.

Guillot, T., Stevenson, D. J., Hubbard, W. B. and Saumon, D. (2004). The interior of Jupiter. In *Jupiter: The Planet, Satellites and Magnetosphere*, ed. Bagenal, F., Dowling, T. E., and McKinnon, W. B. Cambridge: Cambridge University Press, 35–57.

Gullbring, E., Hartmann, L., Briceno, C., and Calvet, N. (1998). Disk accretion rates for T Tauri stars. *Astrophys. J.*, **492**, 323.

Habing, H. J., Dominik, C., Jourdain de Muizon, M. *et al.* (2001). Incidence and survival of remnant disks around main-sequence stars. *Astron. Astrophys.*, **365**, 545–61.

Haghighipour, N. and Boss, A. P. (2003). On pressure gradients and rapid migration of solids in a nonuniform solar nebula. *Astrophys. J.*, **583**, 996–1003.

Haisch, K. E., Lada, E. A., and Lada, C. J. (2001). Disk frequencies and lifetimes in young clusters. *Astrophys. J.*, **553**, L153–56.

Halbwachs, J. L., Arenou, F., Mayor, M., Udry, S., and Queloz, D. (2000). Exploring the brown dwarf desert with Hipparcos. *Astron. Astrophys.*, **355**, 581–94.

Halbwachs, J.-L., Mayor, M., and Udry, S. (2005). Statistical properties of exoplanets. IV. The period-eccentricity relations of exoplanets and of binary stars. *Astron. Astrophys.*, **431**, 1129.

Halliday A. N., Lee D.-C., Forcelli, D. *et al.* (2001). The rates of accretion, core formation and volatile loss in the early Solar System. *Phil. Trans. R. Soc. Lond.*, **A 359**, 2095.

Hamilton, C. M., Herbst, W., Shih, C., and Ferro, A. J. (2001). Eclipses by a circumstellar dust feature in the pre-main-sequence star KH 15D. *Astrophys. J.*, **554**, L201–4.

Hanner, M. S. (1999). The silicate material in comets. *Space Science Reviews*, **90**, 99–108.

Hanner, M. S., Gehrz, R. D., Harker, D. E. *et al.* (1999). Thermal emission from the dust coma of comet Hale-Bopp and the composition of the silicate grains. *Earth Moon and Planets*, **79**, 247–64.

Hartman, J. D., Bakos, G., Stanek, K. Z., and Noyes, R. W. (2004). HATnet variability survey in the high stellar density "Kepler field" with millimagnitude image subtraction photometry. *Astron. J.*, **128**, 1761–83.

Hartmann, L. (2001). Physical conditions of protosolar matter. *Phil. Trans. R. Soc. Lond.*, **A 359**, 2049–60.

Hartmann, L., Calvet, N., Gullbring, E., and D'Alessio, P. (1998). Accretion and the evolution of T Tauri disks. *Astrophys. J.*, **495**, 385–400.

Hartmann, L. and Kenyon, S. J. (1996). The FU Orionis phenomenon. *Annu. Rev. Astron. Astrophys.*, **34**, 207–40.

Hartmann, W. K. (1978). Planet formation: mechanism of early growth. *Icarus*, **33**, 50–61.

Hawley, J. F., Gammie, C. F., and Balbus, S. A. (1995). Local three-dimensional magnetohydrodynamic simulations of accretion disks. *Astrophys. J.*, **440**, 742–63.

Hayashi, C. (1981). Structure of the Solar Nebula, growth and decay of magnetic fields and effects of magnetic and turbulent viscosities on the nebula. *Prog. Theor. Phys.*, **70**, 35–53.

Hayashi, C., and Nakano, T. (1963). Evolution of stars of small masses in the pre-Main Sequence stages. *Prog. Theor. Phys.*, **30**, 460.

Heim, L., Blum, J., Preuss, M., and Butt, H. (1999). Adhesion and friction forces between spherical micrometer-sized particles. *Phys. Rev. Let.*, **83**, 3328–31.

Henney, W. J. and O'Dell, C. R. (1999). A Keck high-resolution spectroscopic study of the Orion Nebula proplyds. *Astron. J.*, **118**, 2350–68.

Henning, T. (2001). Frontiers of radiative transfer. *IAU Symposium No. 200, ASP Conf. Ser.*, 567–72.

Henning, T., Michel, B., and Stognienko, R. (1995). Dust opacities in dense regions. *Planet. Space Sci.*, **43**, 1333–43.

Henry, G. W., Marcy, G. W., Butler, R. P., and Vogt, S. S. (2000). A transiting "51 Peg-like" planet. *Astrophys. J.*, **529**, L41–4.

Hersant, F., Dubrulle, B., and Huré, J.-M. (2005). Turbulence in circumstellar disks. *Astron. Astrophys.*, **429**, 531–42.

Herschel, W. (1811). Astronomical observations relating to the construction of the heavens, arranged for the purpose of a critical examination, the result of which appears to throw some new light upon the organization of the celestial bodies. *Phil. Trans. R. Soc. London*, **101**, 269.

Hester, J. J., Scowen, P. A., Sankrit, R. *et al.* (1996). Hubble Space Telescope WFPC2 imaging of M16: photoevaporation and emerging young stellar objects. *Astron. J.*, **111**, 2349.

Higa, M., Arakawa, M., and Maeno, N. (1998). Size dependence of restitution coefficients of ice in relation to collision strength. *Icarus*, **133**, 310–20.

Hines, D., Backman, D. E., Bouwman, J. *et al.* (2005). The Formation and Evolution of Planetary systems (FEPS): Discovery of an unusual debris system associated with HD 12039. *Astrophys. J.*, **638**, 1070–9.

Hobbs, L. M., Vidal-Madjar, A., Ferlet, R., Albert, C. E., and Gry, C. (1985). The gaseous component of the disk around Beta Pictoris. *Astrophys. J.*, **293**, L29–33.

Hodgson, L. S. and Brandenburg, A. (1998). Turbulence effects in planetesimal formation. *Astron. Astrophys.*, **330**, 1169–74.

Hogerheijde, M. R. (2004). From infall to rotation around young stars: The origin of protoplanetary disks. In *Star Formation at High Angular Resolution (ASP Conference Series, Vol. S-221)*, ed. Jayawardhana, R., Burton, M. G., and Bourke, T. L. San Fransisco: ASP.

Hollenbach, D., Johnstone, D., Lizano, S., and Shu, F. (1994). Photoevaporation of disks around massive stars and application to ultracompact H II regions. *Astrophys. J.*, **428**, 654–69.

Hollenbach, D. J., Yorke, H. W., and Johnstone, D. (2000). Disk dispersal around young stars. In *Protostars and Planets IV*, ed. V. Mannings, A. P. Boss, and S. S. Russell. Tucson: University of Arizona Press, 401.

Hoppe, P. and Zinner, E. (2000). Presolar dust grains from meteorites and their stellar sources. *J. Geophys. Res.*, **105**, 10371.

Houck, J. R., Roellig, T. L., van Cleve, J. *et al.* (2004). The Infrared Spectrograph (IRS) on the Spitzer Space Telescope. *Astrophys. J. Suppl.*, **154**, 18–24.

Hoyle, F. (1960). On the origin of the solar nebula. *Q. J. R. A. S.*, **1**, 28–55.

Hubbard, W. B., Guillot, T., Marley, M. S. *et al.* (1999). Comparative evolution of Jupiter and Saturn. *Plan. Space Sci.*, **47**, 1175–82.

Hubickyj, O., Bodenheimer, P., and Lissauer, J. J. (2004). Evolution of gas giant planets using the core accretion model. *RevMexAA (Serie de Conferencias)*, **22**, 83–6.

　(2005). Accretion of the gaseous envelope of Jupiter around a 5–10 Earth-mass core. *Icarus*, **179**, 415–31.

Humayun, M. and Clayton, R. N. (1995). Precise determination of the isotopic composition of potassium: application to terrestrial rocks and lunar soils. *Geochim. cosmochim. acta*, **59**, 2115.

Hure, J.-M. (2000). On the transition to self-gravity in low mass AGN and YSO accretion discs. *Astrophys. J.*, **358**, 378–94.

Huss, G. R. and Lewis, R. S. (1995). Presolar diamond, SiC, and graphite in primitive chondrites – abundances as a function of meteorite class and petrologic type. *Geochim. cosmochim. acta*, **59**, 115–60.

Ida, S. and Lin, D. N. C. (2004). Toward a deterministic model of planetary formation. I. A desert in the mass and semimajor axis distributions of extrasolar planets. *Astrophys. J.*, **604**, 388–413.

Ida, S. and Makino, J. (1993). Scattering of planetesimals by a protoplanet – slowing down of runaway growth. *Icarus*, **106**, 210.

Ida, S. and Nakazawa, K. (1989). Collisional probability of planetesimals revolving in the solar gravitational field. III. *Astron. Astrophys.*, **224**, 303–15.

Igea, J. and Glassgold, A. E. (1999). X-ray ionization of the disks of young stellar objects. *Astrophys. J.*, **518**, 848–58.

Ikoma, M., Emori, H., and Nakazawa, K. (2001). Formation of giant planets in dense nebulae: critical core accretion mass revisited. *Astrophys. J.*, **553**, 999–1005.

Ikoma, M., Nakazawa, K., and Emori, H. (2000). Formation of giant planets: dependences on core accretion rate and grain opacity. *Astrophys. J.*, **537**, 1013–25.

Ilgner, M., Henning, T., Markwick, A. J., and Millar, T. J. (2004). Transport processes and chemical evolution in steady accretion disk flows. *Astron. Astrophys.*, **415**, 643–59.

Inaba, S. and Ikoma, M. (2003). Enhanced collisional growth of a protoplanet that has an atmosphere. *Astron. Astrophys.*, **410**, 711–23.

Inaba, S. and Wetherill, G. W. (2001). Formation of Jupiter: core accretion model with fragmentation. In *Lunar and Planetary Institute Conference Abstracts*, no. 1384.

Inaba, S., Wetherill, G. W., and Ikoma, M. (2003). Formation of gas giant planets: core accretion models with fragmentation and planetary envelope. *Icarus*, **166**, 46–62.

Ivlev, A. V., Morfill, G. E., and Konopka, U. (2002). Coagulation of charged microparticles in neutral gas and charge-induced gel transitions. *Phys. Rev. Lett.*, **89**, 195502–4.

Janiczek, P. M. (1983). Remarks on the transit of Venus expedition of 1874. In *Sky with Ocean Joined, Proceedings of the Sesquicentennial Symposia of the US Naval Observatory, December 5–8, 1980*. ed. Dick, S. J. and Doggett, L. E. Washington: US Naval Observatory, 53.

Jang, H. D. and Friedlander, S. K. (1998). Restructuring of chain aggregates of titania nanoparticles in the gas phase. *Aerosol Sci. Tech.*, **29**, 81–91.

Jang-Condell, H. and Sasselov, D. D. (2005). Type I migration in a nonisothermal protoplanetary disk. *Astrophys. J.*, **619**, 1123–31.

Jeans, J. H. (1919). *Problems of cosmogony and stellar dynamics*, Adams Prize Essay for 1917. Cambridge: Cambridge University Press.

(1929). *Astronomy and Cosmogony*, 2nd edn. Cambridge: Cambridge University Press.

Jeffery P. M. and Reynolds J. H. (1961). Origin of excess $^{129}$Xe in stone meteorites. *J. Geophys. Res.*, **66**, 3582–3.

Jefferys, H. (1917). The motion of tidally-distorted masses, with special reference to theories of cosmogony. *Mem. R. Astron. Soc.*, **62**, 1.

(1918). On the early history of the solar system. *MNRS*, **78**, 424.

Jenkins, J. M. and Doyle, L. R. (2003). Detecting reflected light from close-in extrasolar giant planets with the Kepler photometer. *Astrophys. J.*, **595**, 429–45.

Jessberger, E. K., Stephan, T., Rost, D. *et al.* (2001). Properties of interplanetary dust: information from collected samples. in *Interplanetary Dust*, ed. S. F. Dermott, H. Fechtig, E. Grün, and B. A. S. Gustafson. Heidelberg: Springer, p. 253.

Jewitt, D. C. and Luu, J. X. (1993). Discovery of the candidate Kuiper Belt object 1992 QB1. *Nature*, **362**, 730–732

Johansen, A., Andersen, A. C., and Brandenburg, A. (2004). Simulations of dust-trapping vortices in protoplanetary discs. *Astron. Astrophys.*, **417**, 361–74.

Johansen, A. and Klahr, H. H. (2005). *Dust diffusion in protoplanetary disks by magnetorotational turbulence. ApJ*, **634**, 1353–71.

Johansen, A., Klahr, H., and Henning, Th. (2006). *Gravoturbulent formation of planetesimals. Astrophys. J.*, **636**, 1121–34.

Johnson, B. M. and Gammie, C. F. (2005). *Linear theory of thin, radially-stratified disks. ApJ*, **626**, 978–90.

Johnstone, D., Fich, M., Mitchell, G. F., and Moriarty-Schieven, G. (2001). Large area mapping at 850 microns. III. Analysis of the clump distribution in the Orion B molecular cloud. *Astrophys. J.*, **559**, 307–17.

Johnstone, D., Hollenbach, D., and Bally, J. (1998). Photoevaporation of disks and clumps by nearby massive stars: application to disk destruction in the Orion Nebula. *Astrophys. J.*, **499**, 758–76.

Johnstone, D., Wilson, C. D., Moriarty-Schieven, *G. et al.* (2000). Large-area mapping at 850 microns. II. Analysis of the clump distribution in the $\rho$ Ophiuchi molecular cloud. *Astrophys. J.*, **545**, 327–39.

Jones A. P. (2001). Interstellar and circumstellar grain formation and survival. *Phil. Trans. R. Soc. Lond.*, **A 359**, 1961.

Jones, H. (2005). Exoplanet metallicities and periods. In *IAU Symposium #219: Stars as Suns: Activity, Evolution, and Planets*, ed. Dupree, A. K. San Francisco: Astronomical Society of the Pacific, in press.

Jones, H. R. A. (2004). What is known about the statistics of extrasolar planetary systems? *The Search for Other Worlds: Fourteenth Astrophysics Conference, AIP Conference Proceedings*, **713**, 17–26.

Jones, H. R. A., Butler, R. P., Tinney, C. G. *et al.* (2004). The distribution of exo-planet properties with semimajor axis. In *ASP Conf. Ser.* 321 *Extrasolar Planets, Today and Tomorrow*, ed. Beaulieu, J. P., Lecavelier, A., and Terquem, C., 298–304.

Jorissen, A., Mayor, M., and Udry, S. (2001). The distribution of exoplanet masses. *Astron. Astrophys.*, **379**, 992–8.

Kalas, P. and Jewitt, D. (1995). Asymmetries in the Beta Pictoris dust disk, *Astron. J.*, **110**, 794.

Kant, I. (1755). *Allgemeine Naturgeschichte und Theorie des Himmels*. Koenigsberg and Leipzig: Johann Friederich Petersen.

Kary, D. M. and Lissauer, J. J. (1994). Numerical simulations of planetary growth. In *Numerical Simulations in Astrophysics*, ed. Franco J., Lizano, S., Aguilar, S., and Daltabuit, E. Cambridge: Cambridge University Press, 364–73.

Kearns, K. E. and Herbst, W. (1998). Additional periodic variables in NGC 2264. *Astron. J.*, **116**, 261–5.

Kempf, S., Pfalzner, S., and Henning, T. (1999). N-particle simulations of dust growth. I. Growth driven by Brownian motion. *Icarus*, **141**, 388–98.

Kenyon, S. J. and Bromley, B. C. (2004a). Detecting the dusty debris of terrestrial planet formation. *Astrophys. J.*, **602**, L133–6.

  (2004b). Collisional cascades in planetesimal disks. II. Embedded planets. *Astron. J.*, **127**, 513–30.

Kessel, O., Yorke, H. W., and Richling, S. (1998). Photoevaporation of protostellar disks. III. The appearance of photoevaporating disks around young intermediate mass stars. *Astron. Astrophys.*, **337**, 832–46.

Kim, J. S., Hines, D. C., Backman, D. E. *et al.* (2005). The Formation and Evolution of Planetary systems: cold outer disks associated with Sun-like stars. *Astrophys. J.*, **632**, 659–69.

Kita N. T., Huss G. R., Tachibana, S. *et al.* (2004). Constraints on the origin of chondrules and CAIs from short-lived and long-lived radionuclides. *Workshop on Chondrites and the Protoplanetary Disk*, abstr. no. #9064.

Kita N. T., Nagahara H., Togashi S., and Morishita Y. (2000). A short duration of chondrule formation in the Solar nebula: evidence from Al-26 in Semarkona ferromagnesian chondrules. *Geochim. cosmochim. acta*, **64**, 3913–22.

Klahr, H. (2004). The global baroclinic instability in accretion disks. II. Local linear analysis. *Astrophys. J.*, **606**, 1070–82.

Klahr, H. and Bodenheimer, P. (2006). Formation of giant planets by concurrent accretion of solids and gas inside an anti-cyclonic vortex, *Astrophys. J.*, submitted.

Klahr, H. H. and Bodenheimer, P. (2003). Turbulence in accretion disks: vorticity generation and angular momentum transport via the global baroclinic instability. *Astrophys. J.*, **582**, 869–92.

Klahr, H. and Henning, T. (1997). Particle-trapping eddies in protoplanetary accretion disks. *Icarus*, **128**, 213–29.

Klahr, H., Henning, T., and Kley, W. (1999). On the azimuthal structure of thermal convection in circumstellar disks. *Astrophys. J.*, **514**, 325–43.

Klahr, H. and Lin, D. N. C. (2001). Dust distribution in gas disks: a model for the ring around HR 4796A. *Astrophys. J.*, **554**, 1095–109.

Kleine, T., Mezger, K., and Münker, C. (2003). Constraints on the age of the Moon from 182W. *Met. Planet. Sci. Suppl.*, **38**, Abstract no. 5212.

Kleine, T., Mezger, K., Palme, H., Scherer, E., and Münker, C. (2005). Early core formation in asteroids and late accretion of chondrite parent bodies: evidence from 182Hf-182W in CAIs, metal-rich chondrites and iron meteorites. *Geochim. Cosmochim. Acta*, **69**, 5805–8.

Kleine, T., Münker, C., Mezger, K., and Palme, H. (2002). Rapid accretion and early core formation on asteroids and the terrestrial planets from Hf-W chronometry. *Nature*, **418**, 952–5.

Klerner S. (2001). Materie im frühen Sonnensystem: die Entstehung von Chondren, Matrix und refraktären Forsteriten. Phd thesis, University of Cologne, pp. 125.

Klerner S. and Palme H. (2000). Formation of chondrules and matrix in carbonaceous chondrites. *Met. Plan. Sci.*, **35**, A89.

Klessen, R. S., Burkert, A., and Bate, M. R. (1998). Fragmentation of molecular clouds: the initial phase of a stellar cluster. *Astrophys. J.*, **501**, L205–8.

Kley, W. (1998). On the treatment of the Coriolis force in computational astrophysics. *Astron. Astrophys.*, **338**, L37–L41.

   (1999). Mass flow and accretion through gaps in accretion discs. *Mon. Not. R. Astron. Soc.*, **303**, 696–710.

Kley, W., D'Angelo, G., and Henning, T. (2001). Three-dimensional simulations of a planet embedded in a protoplanetary disk. *Astrophys. J.*, **547**, 457–64.

Kley, W., Lee, M.-H., Murray, N., and Peale, S. (2005). Modeling the resonant planetary system GJ 876. *Astron. Astrophys.*, **437**, 727–42.

Knödlseder J., Cervino M., Le Duigou J. M. *et al.* (2002). Gamma-ray line emission from OB associations and young open clusters. II. The Cygnus region. *Astron. Astrophys.*, **390**, 945–60.

Koerner, D. W., Chandler, C. J., and Sargent, A. I. (1995). Aperture synthesis imaging of the circumstellar dust disk around DO Tauri. *Astrophys. J.*, **452**, L69–72.

Kokubo, E. and Ida, S. (1996). On runaway growth of planetesimals. *Icarus*, **123**, 180–91.

   (1998). Oligarchic growth of protoplanets. *Icarus*, **131**, 171–8.

   (2000). Formation of protoplanets from planetesimals in the Solar nebula. *Icarus*, **143**, 15–27.

   (2002). Formation of protoplanet systems and diversity of planetary systems. *Astrophys. J.*, **581**, 666-80.

Koller, J., Li, H., and Lin, D. N. C. (2003). Vortices in the co-orbital region of an embedded protoplanet. *Astrophys. J.*, **596**, L91–4.

Kominami, J. and Ida, S. (2002). The effect of tidal interaction with a gas disk on formation of terrestrial planets. *Icarus*, **157**, 43–56.

Konacki, M., Torres, G., Jha, S. *et al.* (2003b). An extrasolar planet that transits the disk of its parent star. *Nature*, **421**, 507.

Konacki, M., Torres, G., Sasselov, D. D. *et al.* (2004). The transiting extrasolar giant planet around the star OGLE-TR-113. *Astrophys. J.*, **609**, L37–40.

(2003a). High-resolution spectroscopic follow-up of OGLE planetary transit candidates in the galactic bulge: two possible Jupiter-mass planets and two blends. *Astrophys. J.*, **597**, 1076–91.

(2005). A transiting extrasolar planet around the star OGLE-TR-10. *Astrophys. J.*, **624**, 372–7.

Königl, A. (1991). Disk accretion onto magnetic T Tauri stars. *Astrophys. J.*, **370**, L39–L43.

Kornet, K., Bodenheimer, P., and Różyczka, M. (2002). Models of the formation of the planets in the 47 UMa system. *Astron. Astrophys.*, **396**, 977–86.

Kornet, K., Bodenheimer, P., Różyczka, M., and Stepinski, T. F. (2005). Formation of giant planets in disks with different metallicities. *Astron. Astrophys.*, **430**, 1133–1138.

Kortenkamp, S. J., Wetherill, G. W., and Inaba, S. (2001). Runaway growth of planetary embryos facilitated by massive bodies in a protoplanetary disk. *Science*, **293**, 1127–9.

Korycansky, D. G. and Pollack, J. B. (1993). Numerical calculations of the linear response of a gaseous disk to a protoplanet. *Icarus*, **102**, 150–65.

Kouchi, A., Kudo, T., Nakano, H. *et al.* (2002). Rapid growth of asteroids owing to very sticky interstellar organic grains. *Astrophys. J.*, **566**, L121–4.

Kozasa, T., Blum, J., and Mukai, T. (1992). Optical properties of dust aggregates. I – Wavelength dependence. *Astron. Astrophys.* **263**, 423–32.

Krause, M. and Blum, J. (2004). Growth and form of planetary seedlings: results from a sounding rocket microgravity aggregation experiment. *Phys. Rev. Lett.*, **93**, 021103.

Krot, A., Scott, E., and Reipurth, B. (2005) *Proceedings of Chondrites and the Protoplanetary Disk*, ASP Conference Series. San Francisco: Astronomical Society of the Pacific.

Krügel, E. and Siebenmorgen, R. (1994). Dust in protostellar cores and stellar disks. *Astron. Astrophys.*, **288**, 929–41.

Kuiper, G. P. (1951). On the origin of the solar system. In *Astrophysics – A Topical Symposium*, ed. J. A. Hynek. New York: McGraw-Hill.

Kumar, S. (1963). The structure of stars of very low mass. *Astrophys. J.*, **137**, 1121.

Kunihiro T., Rubin A. E., McKeegan K. D., and Wasson J. T. (2004). Initial $^{26}$Al/$^{27}$Al in carbonaceous–chondrite chondrules: too little $^{26}$Al to melt asteroids. *Geochim. cosmochim. acta*, **68**, 2947–57.

Kürster, M., Endl, M., Rouesnel, F. *et al.* (2003). The low-level radial velocity variability in Barnard's star (=GJ 699). *Astron. Astrophys.*, **403**, 1077–87.

Kusaka, T., Nakano, T., and Hayashi, C. (1970). Growth of solid particles in the primordial Solar Nebula. *Prog. Theor. Phys. Suppl.*, **44**, 1580–95.

Lada, C. J. and Lada, E. A. (2003). Embedded clusters in molecular clouds. *Ann. Rev. Astron. Astrophys.*, **41**, 57–115.

Lada, C. J. and Wilking, B. A. (1984). The nature of the embedded population in the Rho Ophiuchi dark cloud – mid-infrared observations. *Astrophys. J.*, **287**, 610–21.

Lada, E. A. (2003). Evolution of circumstellar disks in young stellar clusters. *BAAS*, **35**, #24.06, 730.

Lamers, H. J. G. L. M., Lecavelier des Etangs, A., and Vidal-Madjar, A. (1997). $\beta$ Pictoris light variations: II. Scattering by a dust cloud. *Astron. Astrophys.*, **328**, 321–30.

Laplace, P. S. (1796). *Exposition de système du monde*. Paris: Circle-Sociale. English translation: H. H. Harte. (1830). *The System of the World*. Dublin: University Press.

Larimer J. W. (1979). The condensation and fractionation of refractory lithophile elements. *Icarus*, **40**, 446–54.

Larson, R. B. (1969). Numerical calculations of the dynamics of collapsing proto-star. *Mon. Not. Roy. Astron. Soc.*, **145**, 271.

LaTourrette, T. and Wasserburg, G. J. (1998). Mg diffusion in anorthite: implications for the formation of early Solar System planetesimals. *Earth Planet. Sci. Lett.*, **158**, 91–108.

Lattanzi, M. G., Casertano, S., Sozzetti, A., and Spagna, A. (2002). The GAIA astrometric survey of extra-solar planets. In *GAIA: A European Space Project*, ed. Bienaymé, O. and Turon, C. *EAS Publication Series*, **2**, 207–14.

Laughlin, G. and Adams, F. C. (1997). Possible stellar metallicity enhancements from the accretion of planets. *Astrophys. J.*, **491**, L51.

Laughlin, G., Bodenheimer, P., and Adams, F. C. (2004). The core accretion model predicts few Jovian-mass planets orbiting red dwarfs. *Astrophys. J.*, **612**, L73–6.

Laughlin, G. and Chambers, J. E. (2001). Short-term dynamical interactions among extrasolar planets. *Astrophys. J.*, **551**, L109–13.

Laughlin, G and Różyczka, M. (1996). The effect of gravitational instabilities on protostellar disks. *Astrophys. J.*, **456**, 279–91.

Laughlin, G., Steinacker, A., and Adams, F. C. (2004). Type I planetary migration with MHD turbulence. *Astrophys. J.*, **608**, 489–96.

Lauretta, D., Leshin, L. A., and McSween, Jr. H. Y. (2005). *Meteorites and the Early Solar System II*. Tucson: University of Arizona Press.

Laws, C., Gonzalez, G., Walker, K. M. *et al.* (2003). Parent stars of extrasolar planets. VII. New abundance analyses of 30 systems. *Astron. J.*, **125**, 2664–77.

Lecavelier des Etangs, A. (1999). A library of stellar light variations due to extra-solar comets. *A&AS*, **140**, 15–20.

Lecavelier des Etangs, A., Delevil, M., Vidal-Madjar, A. *et al.* (1994). Pictoris: evidence of light variations. In *Circumstellar Dust Disks and Planet Formation*, Editions Frontières, 93.

Lecavelier des Etangs, A., Deleuil, M., Vidal-Madjar, A. *et al.*, (1995). β Pictoris: evidence of light variations. *Astron. Astrophys.*, **299**, 557.

Lecavelier des Etangs, A., Nitschelm, C., Olsen, E. H. *et al.* (2005). A photometric survey of stars with circumstellar material. *A&A*, **439**, 571–4.

Lecavelier des Etangs, A., Vidal-Madjar, A., Burki, G. *et al.* (1997). β Pictoris light variations: I. The planetary hypothesis. *Astron. Astrophys.*, **328**, 311–20.

Lecavelier des Etangs, A., Vidal-Madjar, A., and Ferlet, R. (1999). Photometric stellar variation due to extra-solar comets. *Astron. Astrophys.*, **343**, 916–22.

Lecavelier des Etangs, A., Vidal-Madjar, A., McConnell, J. C., and Hébrard, G. (2004). Atmospheric escape from hot Jupiters. *Astron. Astrophys.*, **418**, L1–4.

Lee, M. H. and Peale, S. J. (2002). Dynamics and origin of the 2:1 orbital resonances of the GJ 876 planets. *Astrophys. J.*, **567**, 596–609.

Lee T., Papanastassiou D. A., and Wasserburg G. J. (1976). Demonstration of $^{26}$Mg excess in Allende and evidence for $^{26}$Al. *Geophys. Res. Lett.*, **3**, 109–12.

Lee, T., Shu, F. H., Shang, H., Glassgold, A. E., and Rehm, K. E. (1998). Protostellar cosmic rays and extinct radioactivities in meteorites. *Astrophys. J.*, **506**, 898.

Levison, H. F., Lissauer, J. J., and Duncan, M. J. (1998). Modeling the diversity of outer planetary systems. *Astron. J.*, **116**, 1998–2014.

Levison, H. F. and Stewart, G. R. (2001). Remarks on modeling the formation of Uranus and Neptune. *Icarus*, **153**, 224–8.

Lewis, M. C. and Stewart, G. R. (2000). Collisional dynamics of perturbed planetary rings. I. *Astronom. J.*, **120**, 3295–310.

Li, H., Colgate, S. A., Wendroff, B., and Liska, R. (2001). Rossby wave instability of thin accretion disks. III. Nonlinear simulations. *Astrophys. J.*, **551**, 874–96.

Li, H., Finn, J. M., Lovelace, R. V. E., and Colgate, S. A. (2000). Rossby wave instability of thin accretion disks. II. Detailed linear theory. *Astrophys. J.*, **533**, 1023–34.

Lin, D. N. C. and Papaloizou, J. (1979). On the structure of circumbinary accretion disks and the tidal evolution of commensurable satellites. *Mon. Not. R. Astron. Soc.*, **188**, 191–201.

(1980). On the structure and evolution of the primordial solar nebula, *Mon. Not. R. Astron. Soc.*, **191**, 37–48.

Lin, D. N. C. and Papaloizou, J. C. B. (1993). On the tidal interaction between protostellar disks and companions. *Protostars and Planets III*, ed. Levy, E. and Lumine, J. I. Arizona: University of Arizona Press, 749–835.

Lin, D. N. C., Papaloizou, J., and Kley, W. (1993). On the nonaxisymmetric convective instabilities in accretion disks. *Astrophys. J.*, **416**, 689–99.

Lin, D. N. C., Woosley, S. E., and Bodenheimer, P. (1991). Formation of a planet orbiting pulsar 1829–10 from the debris of a supernova explosion. *Nature*, **353**, 827–9.

Lissauer, J. J. (1987). Timescales for planetary accretion and the structure of the protoplanetary disk. *Icarus*, **69**, 249–65.

(1995). Urey prize lecture: On the diversity of plausible planetary systems. *Icarus*, **114**, 217–36.

Livio, M. and Pringle, J. E. (2003). Metallicity, planetary formation and migration. *Mon. Not. Roy. Astr. Soc.*, **346**, L42–4.

Love, S. G. and Pettit, D. R. (2004). Fast, repeatable clumping of solid particles in microgravity. *Lunar Planet. Sci. Conf.*, **35**, 1119.

Lovelace, R. V. E., Li, H., Colgate, S. A., and Nelson, A. F. (1999). Rossby wave instability of Keplerian accretion disks. *Astrophys. J.*, **513**, 805–10.

Lovis, C., Mayor, M., Bouchy, F. *et al.* (2005). The HARPS search for southern extra-solar planets. III. Three Saturn-mass planets around HD 93083, HD 101930 and HD 102117. *Astron. Astrophys.*, **437**, 1121–6.

Low, C. and Lynden-Bell, D. (1976). The minimum Jeans mass or when fragmentation must stop. *Mon. Not. Roy. Astron. Soc.*, **176**, 367–90.

Lubow, S. H., Seibert, M., and Artymowicz, P. (1999). Disk accretion onto high-mass planets. *Astrophys. J.*, **526**, 1001–12.

Lugmair, G. W. and Galer, S. J. G. (1992). Age and isotopic relationships among the angrites Lewis Cliff 86010 and Angra dos Reis. *Geochim. cosmochim. acta*, **56**, 1673–94.

Lugmair, G. W. and Shukolyukov, A. (1998). Early Solar System timescales according to 53 Mn–53 Cr systematics. *Geochim. cosmochim. acta*, **62**, 2863–86.

(2001). Early Solar System events and timescales. *Met. Planet. Sci.*, **36**, 1017.

Lugo, J., Lizano, S., and Garay, G. (2004). Photoevaporated disks around massive young stars. *Astrophys. J.*, **614**, 807–17.

Luhman K. L. (2004a). The first discovery of a wide binary brown dwarf. *Astrophys. J.*, **614**, 398–403.

(2004b). New brown dwarfs and an updated initial mass function in Taurus. *Astrophys. J.*, **617**, 1216–32.

Luhman, K. L., Briceño, C., Stauffer, J. R. *et al.* (2003a). New low-mass members of the Taurus star-forming region. *Astrophys. J.*, **590**, 348–56.

Luhman, K. L., Stauffer, J. R., Muench, A. A. *et al.* (2003b). A census of the young cluster IC 348. *Astrophys. J.*, **593**, 1093–115.

Macintosh, B. A., Bauman, B., Wilhelmsen Evans, J. *et al.* (2004). Extreme adaptive optics planet imager: overview and status. *SPIE*, **5490**, 359–69.

Mamajek, E. E., Meyer, M. R., Hinz, P. M. *et al.* (2004). Constraining the lifetime of circumstellar disks in the terrestrial planet zone: a mid-infrared survey of the 30 Myr old Tucana-Horologium Association. *Astrophys. J.*, **612**, 496–510.

Mamajek, E. E., Meyer, M. R., and Liebert, J. (2002), Post-T Tauri Stars in the nearest OB association. *Astron. J.*, **124**, 1670–94.

Mannings, V. and Emerson, J. P. (1994). Dust in discs around T Tauri stars: grain growth? *Mon. Not. R. Astron. Soc.*, **267**, 361–78.

Marcy, G. W. and Butler, R. P. (1995). 51 Pegasi. *IAU Circ.*, **6251**, 1.

(1998). Detection of Extrasolar Giant Planets. *Ann. Rev. Astr. Astrophys.*, **36**, 57–97.

(2000). Planets orbiting other Suns. *Pub. Ast. Soc. Pac*, **112**, 137–40.

Marcy, G. W., Butler, R. P., Fischer, D. A. *et al.* (2002). A planet at 5 AU around 55 Cancri. *Astrophys. J.*, **581**, 1375–88.

Marcy, G. W., Butler, R. P., Fischer, D. *et al.* (2004). A Doppler planet survey of 1330 FGKM stars. *ASP Conference Proceedings*, **321**, 3.

Marcy, G. W., Butler, R. P., Fischer, D. A., and Vogt, S. S. (2003). Properties of extrasolar planets. In *ASP Conf. Ser. 294: Scientific Frontiers in Research on Extrasolar Planets*, 1–16.

Marcy, G. W., Cochran, W. D., and Mayor, M. (2000). Extrasolar planets around main-sequence stars. In *Protostars and Planets IV*, ed. V. Mannings, A. P. Boss, and S. S. Russell. Tucson: University of Arizona Press, 1285.

Marley, M. S., (1999). Interiors of the giant planets. In *Encyclopedia of the Solar System*, ed. P. Weissman, L. A. McFadden, and T. V. Johnson. Academic Press, 339–55.

Marois, C., Doyon, R., Nadeau, D. *et al.* (2005). TRIDENT: an infrared differential imaging camera optimized for the detection of methanated substellar companions *PASP, Astron. Astrophys.*, **117**, 745–56.

Marshall, J. and Cuzzi, J. N. (2001). Electrostatic enhancement of coagulation in protoplanetary nebulae. *LPI*, **32**, 1262.

Martín, E. L., Barrado y Navascués, D., Baraffe, I., Bouy, H., and Dahm, S. (2003). A Hubble Space Telescope wide field planetary camera 2 survey for brown dwarf binaries in the α Persei and Pleiades open clusters. *Astrophys. J.*, **594**, 525–32.

Martín, E. L., Guenther, E. W., Caballero, J. A. *et al.* (2004). NAHUAL: A high-resolution IR spectrograph for the GTC. *Astronomische Nachrichten*, **325**, 132.

Marzari, F. and Weidenschilling, S. J. (2002). Eccentric extrasolar planets: the jumping Jupiter model. *Icarus*, **156**, 570–9.

Masset, F. S. (2001). On the co-orbital corotation torque in a viscous disk and its impact on planetary migration. *Astrophys. J.*, **558**, 453–62.

Masset, F. S. (2002). The co-orbital corotation torque in a viscous disk: numerical simulations. *Astron. Astrophys.*, **387**, 605–23.

Masset, F. S. and Ogilvie, G. I. (2004). On the saturation of corotation resonances: a numerical study. *Astrophys. J.*, **615**, 1000–10.

Masset, F. S. and Papaloizou, J. C. B. (2003). Runaway migration and the formation of hot Jupiters. *Astrophys. J.*, **588**, 494–508.

Masunaga, H. and Inutsuka, S. (1999). Does "tau ~ 1" terminate the isothermal evolution of collapsing clouds? *Astrophys. J.*, **510**, 822–7.

Matsuyama, I., Johnstone, D., and Hartmann, L. (2003). Viscous diffusion and photoevaporation of stellar disks. *Astrophys. J.*, **582**, 893–904.

Mayer, L., Quinn, T., Wadsley, J., and Stadel, J. (2002). Formation of giant planets by fragmentation of protoplanetary disks. *Science*, **298**, 1756–59.

(2004). The evolution of gravitationally unstable protoplanetary disks: fragmentation and possible giant planet formation. *Astrophys. J.*, **609**, 1045–64.

Mayor, M., Pepe, F., Queloz, D. *et al.* (2003). Setting new standards with HARPS. *The Messenger*, **114**, 20.

Mayor, M. and Queloz, D. (1995). A Jupiter-mass companion to a solar-type star. *Nature* **378**, 355.

Mayor, M., Udry, S., Naef, D. *et al.* (2004). The CORALIE survey for southern extra-solar planets. XII. Orbital solutions for 16 extra-solar planets discovered with CORALIE. *Astron. Astrophys.*, **415**, 391–402.

Mazeh, T., Naef, D., Torres, G. *et al.* (2000). The spectroscopic orbit of the planetary companion transiting HD 209458. *Astrophys. J.*, **532**, L55–8.

McArthur, B. E., Endl, M., Cochran, W. D. *et al.* (2004). Detection of a Neptune-mass planet in the $\rho^1$ Cancri system using the Hobby–Eberly telescope. *Astrophys. J.*, **614**, L81–4.

McCarthy, C., Butler, R. P., Tinney, C. G. *et al.* (2004). Multiple companions to HD 154857 and HD 160691. *ApJ*, **617**, 575.

McCarthy, C. and Zuckerman, B. (2004). The brown dwarf desert at 75–1200 AU. *Astron. J.*, **127**, 2871–84.

McCaughrean, M. J. and O'Dell, C. R. (1996). Direct imaging of circumstellar disks in the Orion Nebula. *Astron. J.*, **111**, 1977-86.

McKeegan, K. D., Chaussidon, M., and Robert, F. (2000). Incorporation of short-lived $^{10}$Be in a calcium-aluminum-rich inclusion from the Allende meteorite. *Science*, **289**, 1334–7.

Meakin, P. (1991). Fractal aggregates in geophysics. *Revi. Geophys.*, **29**, 317–54.

Meeus, G., Waters, L. B. F. M., Bouwman, J. *et al.* (2001). ISO spectroscopy of circumstellar dust in 14 Herbig Ae/Be systems: towards an understanding of dust processing. *Astron. Astrophys.*, **365**, 476–90.

Mejía, A. C., Durisen, R. H., Pickett, M. K., and Cai, K. (2005). The thermal regulation of gravitational instabilities in protoplanetary disks. II. Extended simulations with varied cooling rates. *Astrophys. J.*, **619**, 1098–113.

Men'shchikov, A. B. and Henning, T. (1997). Radiation transfer in circumstellar disks. *Astron. Astrophys.*, **318**, 879–907.

Men'shchikov, A. B., Henning, T., and Fischer, O. (1999), Self-consistent model of the dusty torus around HL Tauri. *Astrophys. J.*, **519**, 257–78.

Menou, K. and Goodman, J. (2004). Low-mass protoplanet migration in T Tauri $\alpha$-disks. *Astrophys. J.*, **606**, 520–31.

Messenger, S., Keller, L. P., Stodermann, F. J., Walker, R. M., and Zinner, E. (2003). Samples of stars beyond the Solar System: silicate grains in interplanetary dust. *Science*, **300**, 105.

Metchev, S. A., Hillenbrand, L., and Meyer, M. (2004). Ten-micron observations of nearby young stars. *Astrophys. J.*, **600**, 435–50.

Meyer, M. R., Hillenbrand, L. A., Backman, D. E. *et al.* (2004). The formation and evolution of planetary systems: first results from a Spitzer Legacy Science Program. *Astrophys. J. Suppl.*, **154**, 422–427.

Meyer, M. R., Hillenbrand, L. A., Backman, D. E. *et al.* (2005). The formation and evolution of planetary systems: placing our Solar System in context with Spitzer. *Pub. Ast. Soc. Pac.*, submitted.

Meyer-Vernet, N. and Sicardy, B. (1987). On the physics of resonant disk–satellite interaction. *Icarus*, **69**, 157–75.

Michel, P., Benz, W., Tanga, P., and Richardson, D. C. (2001). Collisions and gravitational reaccumulation: forming asteroid families and satellites. *Science*, **294**, 1695.

Miyake, K. and Nakagawa, Y. (1993). Effects of particle size distribution on opacity curves of protoplanetary disks around T Tauri stars. *Icarus*, **106**, 20–41.

Miyake, K. and Nakagawa, Y. (1995). Dust particle settling in passive disks around T Tauri stars: models and IRAS observations. *Astrophys. J.*, **441**, 361–84.

Miyamoto M., Fujii N., and Takeda H. (1981). Ordinary chondrite parent body: an internal heating model. *Proc. Lunar Sci. Conf.*, **12B**, 1145–52.

Mizuno, H. (1980). Formation of the giant planets. *Prog. Theor. Phys.*, **64**, 544–57.

(1989). Grain growth in the turbulent accretion disk Solar Nebula. *Icarus*, **80**, 189–201.

Mizuno, H., Markiewicz, W. J., and Völk, H. J. (1988). Grain growth in turbulent protoplanetary accretion disks. *Astron. Astrophys.*, **195**, 183–92.

Mizuno, H., Nakazawa, K., and Hayashi, C. (1978). Instability of a gaseous envelope surrounding a planetary core and formation of giant planets. *Prog. Theor. Phys.*, **60**, 699–710.

Morfill, G. E. and Völk, H. J. (1984). Transport of dust and vapor and chemical fractionation in the early protosolar cloud. *Astrophys. J.*, **287**, 371–95.

Motte, F., André, P., and Neri, R. (1998). The initial conditions of star formation in the rho Ophiuchi main cloud: wide-field millimeter continuum mapping. *Astron. Astrophys.*, **336**, 150–72.

Mouillet, D., Larwood, J. D., Papaloizou, J. C. B., and Lagrange, A. M. (1997). A planet on an inclined orbit as an explanation of the warp in the Beta Pictoris disc. *Mon. Not. R. Astron. Soc.*, **292**, 896.

Moulton, F. R. (1905). On the evolution of the Solar System. *Astrophys. J.*, **22**, 165.

Moutou, C., Pont, F., Bouchy, F., and Mayor, M. (2004). Accurate radius and mass of the transiting exoplanet OGLE-TR-132b. *Astron. Astrophys.*, **424**, L31–4.

Murray, N., Chaboyer, B., Arras, P., Hansen, B., and Noyes, R. W. (2001). Stellar pollution in the Solar neighborhood. *Astrophys. J.*, **555**, 801–15.

Naef, D., Mayor, M., Beuzit, J. L. *et al.* (2004). The ELODIE survey for northern extra-solar planets. III. Three planetary candidates detected with ELODIE. *Astron. Astrophys.*, **414**, 351–359.

Naef, D., Mayor, M., Beuzit, J.-L. *et al.* (2005). The ELODIE planet search: synthetic view of the survey audits global detection threshold. *Proceedings of the 13th Cambridge Workshop on Cool Stars, Stellar Systems and the Sun*, ESA SP series, **560**, 833.

Nagahara, H. and Ozawa, K. (1996). Evaporation of forsterite in $H_2$ gas. *Geochim. cosmochim. acta*, **60**, 1445.

Najita, J., Carr, J. S., and Mathieu, R. D. (2003). Gas in the terrestrial planet region of disks: CO fundamental emission from T Tauri stars. *Astrophys. J.*, **589**, 931–52.

Nakagawa, Y., Nakazawa, K., and Hayashi, C. (1981). Growth and sedimentation of dust grains in the primordial Solar Nebula. *Icarus*, **45**, 517–28.

Nakagawa, Y., Sekiya, M., and Hayashi, C. (1986). Settling and growth of dust particles in a laminar phase of a low-mass Solar Nebula. *Icarus*, **67**, 375–90.

Nakajima, T., Oppenheimer, B. R., Kulkarni, S. R. *et al.* (1995). Discovery of a cool brown dwarf. *Nature*, **378**, 463.

Nakamura T., Nagao K., Metzler K., and Takaoka N. (1999). Heterogeneous distribution of Solar and cosmogenic noble gases in CM chondrites and implications for the formation of CM parent bodies. *Geochim. cosmochim. acta*, **63**, 257–73.

Natta, A. (1999). Star Formation. In *Infrared Space Astronomy, Today and Tomorrow*, ed. F. Casoli, F. David, and J. Lequeux, EDP-Sciences/Springer-Verlag.

Natta, A., Testi, L., Neri, R., Shepherd, D. S., and Wilner, D. J. (2004). A search for evolved dust in Herbig Ae stars. *Astron. Astrophys.*, **416**, 179–86.

Nelson, A. F. (2000). Planet formation is unlikely in equal-mass binary systems with $a \sim 50$ AU. *Astrophys. J.*, **537**, L65–8.

Nelson, R. P. (2005). On the orbital evolution of low mass protoplanets in turbulent, magnetised disks. *Astron. Astrophys.*, **443**, 1067–85.

Nelson, A. F. and Benz, W. (2003a). On the early evolution of forming Jovian planets. I. Initial conditions, systematics, and qualitative comparisons to theory. *Astrophys. J.*, **589**, 556–77.

Nelson, A. F. and Benz, W. (2003b). On the early evolution of forming Jovian planets. II. Analysis of accretion and gravitational torques. *Astrophys. J.*, **589**, 578–604.

Nelson, A. F., Benz, W., Adams, F. C., and Arnett, D. (1998). Dynamics of circumstellar disks. *Astrophys. J.*, **502**, 342–71.

Nelson, A. F., Benz, W., and Ruzmaikina, T. V. (2000). Dynamics of circumstellar disks. II. Heating and cooling. *Astrophys. J.*, **529**, 357–390.

Nelson, R. P. and Papaloizou, J. C. B. (2002). Possible commensurabilities among pairs of extrasolar planets. *Mon. Not. R. Astron. Soc.*, **333**, L26–30.

    (2004). The interaction of giant planets with a disc with MHD turbulence – IV. Migration rates of embedded protoplanets. *Mon. Not. R. Astron. Soc.*, **350**, 849–64.

Nelson, R. P., Papaloizou, J. C. B., Masset, F.S., and Kley, W. (2000). The migration and growth of protoplanets in protostellar discs. *Mon. Not. R. Astron. Soc.*, **318**, 18–36.

Nguyen, A. N. and Zinner, E. (2004). Discovery of ancient silicate stardust in a meteorite. *Science*, **303**, 1496–9.

Nitschelm, C., Lecavelier des Etangs, A., Vidal-Madjar, A. *et al.* (2000). A three-year Strömgren photometric survey of suspected beta Pictoris-like stars. *A&AS*, **145**, 275-81.

Nittler, L. R. (2003). Presolar stardust in meteorites: recent advances and scientific frontiers. *Earth Planet. Sci. Lett.*, **209**, 259.

Norton, O. R. (2002). *The Cambridge Encyclopedia of Meteorites*. Cambridge: Cambridge University Press.

Nübold, H., Poppe, T., Rost, M., Dominik, C., and Glassmeier, K.-H. (2003). Magnetic aggregation II. Laboratory and microgravity experiments. *Icarus*, **165**, 195–214.

Nuth, J. A. III, Berg, O., Faris, J., and Wasilewski, P. (1994). Magnetically enhanced coagulation of very small iron grains. *Icarus*, **107**, 155–63.

Ogilvie, G. I. and Lubow, S. H. (2003). Saturation of the corotation resonance in a gaseous disk. *Astrophys. J.*, **587**, 398–406.

O'Neill, H. St. C. and Palme, H. (1998). Composition of the silicate Earth: implications for accretion and core formation. In *The Earth's Mantle: Structure, Composition and Evolution - the Ringwood Volume*, ed. I. Jackson. Cambridge: Cambridge University Press, 3–126.

Onishi, T., Mizuno, A., Kawamura, A., Tachihara, K., and Fukui, Y. (2002). A complete search for dense cloud cores in Taurus. *Astrophys. J.*, **575**, 950–73.

Ossenkopf, V. (1993). Dust coagulation in dense molecular clouds: the formation of fluffy aggregates. *Astron. Astrophys.*, **280**, 617–46.

Ott, U. (2001). Presolar grains in meteorites: an overview and some implications. *Planet. Space Sci.*, **49**, 763.

Owen, T., Mahaffy, P., Nieman, H. B. *et al.* (1999). A low-temperature origin for the planetesimals that formed Jupiter. *Nature*, **402**, 269–70.

Paardekooper, S.-J. and Mellema, G. (2004). Planets opening dust gaps in gas disks. *Astron. Astrophys.*, **425**, L9–12.

Pack, A., Yurimoto, H., and Palme, H. (2004). Petrographic and oxygen-isotopic study of refractory forsterites from R-chondrite Dar al Gani 013 (R3.5-6), unequilibrated ordinary and carbonaceous chondrites. *Geochim. cosmochim. acta*, **68**, 1135–57.

Padoan, P. and Nordlund, Å. (2002). The stellar initial mass function from turbulent fragmentation. *Astrophys. J.*, **576**, 870–9.

Padoan, P., Zweibel, E., and Nordlund, Å. (2000). Ambipolar drift heating in turbulent molecular clouds. *Astrophys. J.*, **540**, 332–41.

Palme, H. (2001). Chemical and isotopic heterogeneity in protosolar matter. *Phil. Trans. R. Soc. Lond.*, **A 359**, 2061.

Palme H. and Jones A. (2004). Solar System abundances of the elements. In *Treatise on Geochemistry 1: Meteorites, comets and planets*, ed. A. M. Davis, Amsterdam: Elsevier. 41–61.

Pantin, E., Els, S., Marchis, F. *et al.* (2000). First detection of a dust disk around Iota Horologii, a southern star orbitted by an extrasolar giant planet. *AAS/Division of Planetary Sciences Meeting*, **32**, 6533.

Papaloizou, J. C. B. and Larwood, J. D. (2000). On the orbital evolution and growth of protoplanets embedded in a gaseous disc. *Mon. Not. R. Astron. Soc.*, **315**, 823–33.

Papaloizou, J. C. B. and Nelson, R. P. (2003). The interaction of a giant planet with a disc with MHD turbulence – I. The initial turbulent disc models. *Mon. Not. R. Astron. Soc.*, **339**, 983–2.

Papaloizou, J. C. B., Nelson, R. P., and Masset, F. (2001). Orbital eccentricity growth through disc-companion tidal interaction. *Astron. Astrophys.*, **366**, 263–75.

Papaloizou, J. C. B. and Terquem, C. (1999). Critical planetary core masses in protoplanetary disks and the formation of short-period giant planets. *Astrophys. J.*, **521**, 823–38.

Papike, J. J. (1998). Planetary materials. *Rev. Mineral.*, **36**.

Paulson, D. B., Saar, S. H., Cochran, W. D., and Henry, G. W. (2004). Searching for planets in the Hyades. III. The quest for short-period planets. *Astron. J.*, **127**, 1644–52.

Pellas P., Fieni C., Trieloff M., and Jessberger, E. K. (1997). The cooling history of the Acapulco meteorite as recorded by the $^{244}$Pu and $^{40}$Ar–$^{39}$Ar chronometers. *Geochim. cosmochim. acta*, **61**, 3477–501.

Perri, F. and Cameron, A. G. W. (1974). Hydrodynamic instability of the solar nebula in the presence of a planetary core. *Icarus*, **22**, 416–25.

Pickett, B. K., Cassen, P., Durisen, R. H., and Link, R. (1998). The effects of thermal energetics on three-dimensional hydrodynamic instabilities in massive protostellar disks. *Astrophys. J.*, **504**, 468–91.

(2000b). The effects of thermal energetics on three-dimensional hydrodynamic instabilities in massive protostellar disks. II. High-resolution and adiabatic evolutions. *Astrophys. J.*, **529**, 1034–53.

Pickett, B. K., Durisen, R. H., Cassen, P., and Mejía, A. C. (2000a). Protostellar disk instabilities and the formation of substellar companions. *Astrophys. J.*, **540**, L95–8.

Pickett, B. K., Mejía, A. C., and Durisen, R. H. (2003). The thermal regulation of gravitational instabilities in protoplanetary disks. *Astrophys. J.*, **590**, 1060–80.

Podolak, M. (2003). The contribution of small grains to the opacity of protoplanetary atmospheres. *Icarus*, **165**, 428–37.

Podolak, M., Pollack, J. B., and Reynolds, R. T. (1988). Interactions of planetesimals with protoplanetary atmospheres. *Icarus*, **73**, 163–79.

Podsiadlowski, Ph. (1993). Planet formation scenarios. In *Planets around Pulsars*, ASP Conference Series, Vol. 36, ed. J. A. Phillips, S. E. Thorsett, and S. R. Kulkarni. San Francisco: ASP.

Podsiadlowski, Ph., Pringle, J. E., and Rees, M. J. (1991). The origin of the planet orbiting PSR1829–10. *Nature*, **352**, 783–4.

Pollack, J. B. and Bodenheimer, P. (1989). Theories of the origin and evolution of the giant planets. In *Origin and Evolution of Planetary and Satellite Atmospheres*, ed. S. K. Atreya, J. B. Pollack, and M. S. Matthews. Tucson: University of Arizona Press. 564–604.

Pollack, J. B., Hubickyj, O., Bodenheimer, P. *et al.* (1996). Formation of the giant planets by concurrent accretion of solids and gas. *Icarus*, **124**, 62–85.

Polnau, E. and Lugmair, G. W. (2001). Mn–Cr isotope systematics in the two ordinary chondrites Richardton (H5) and Ste. Marguerite (H4). *Lun. Planet. Sci.*, XXXII, abstr. no. #1527.

Pont, F., Bouchy, F., Queloz, D. *et al.* (2004). The "missing link": A 4-day period transiting exoplanet around OGLE-TR-111. *Astron. Astrophys.*, **426**, L15–L18.

Poppe, T. (2003). Sintering of highly porous silica-particle samples: analogues of early Solar System aggregates. *Icarus*, **164**, 139–48.

Poppe, T., Blum, J., and Henning, T. (1997). Generating a jet of deagglomerated small particles in vacuum. *Rev. Sci. Instr.*, **68**, 2529–33.

(1999). New experiments on the stickiness of micron-sized preplanetary dust aggregation. *Adv. Space Res.*, **23**, 1197–200.

(2000a). Analogous experiments on the stickiness of micron-sized preplanetary dust. *Astrophys. J.*, **533**, 454–71.

(2000b). Experiments on collisional grain charging of micron-sized preplanetary dust. *Astrophys. J.*, **533**, 472–80.

Poppe, T. and Schräpler, R. (2005). Further experiments on collisional tribocharging of cosmic grains. *Astron. Astrophys.*, **438**, 1–9.

Pringle, J. E. (1981). Accretion discs in astrophysics. *Ann. Rev. Astron. Astrophys.*, **19**, 137–162.

Prinz, D. K. (1974). The spatial distribution of Lyman alpha on the Sun. *Astrophys. J.*, **187**, 369–75.

Przygodda, F., van Boekel, R., Àbrahàm *et al.* (2003). Evidence for grain growth in T Tauri disks. *Astron. Astrophys.*, **412**, L43–L6.

Queloz, D., Eggenberger, A., Mayor, M. *et al.*, (2000). Detection of a spectroscopic transit by the planet orbiting the star HD209458. *Astron. Astrophys.*, **359**, L13–17.

Quillen, A. C. (2002). Using a Hipparcos-derived Hertzsprung–Russell diagram to limit the metallicity scatter of stars in the Hyades: are stars polluted? *Astron. J.*, **124**, 400–3.

Rafikov, R. R. (2004). Fast accretion of small planetesimals by protoplanetary cores. *Astron. J.*, **128**, 1348–63.

(2005). Can giant planets form by direct gravitational instability? *Astrophys. J.*, **621**, L69–72.

Rebolo, R., Zapatero-Osorio, M. R., and Martin, E. L. (1995). Discovery of a brown dwarf in the Pleiades star cluster. *Nature*, **377**, 129.

Rees, M. J. (1976). Opacity-limited hierarchical fragmentation and the masses of protostars. *Mon. Not. Roy. Astron. Soc.*, **176**, 483–6.

Reid, I. N. (2002). On the nature of stars with planets. *Pub. Ast. Soc. Pac.*, **114**, 306–29.

Reid, I. N., Gizis, J. E., Kirkpatrick, J. D., and Koerner, D. W. (2001). A search for l dwarf binary systems. *Astron. J.*, **121**, 489–502.

Reid, I. N., Kirkpatrick, J. D., Liebert, J. *et al.* (1999). L dwarfs and the substellar mass function. *Astrophys. J.*, **521**, 613–29.

Reipurth, B. and Clarke, C. (2001). The formation of brown dwarfs as ejected stellar embryos. *Astron. J.*, **122**, 432–9.

Rice, W. K. M. and Armitage, P. J. (2003). On the formation timescale and core masses of gas giant planets. *Astrophys. J.*, **598**, L55–8.

Rice, W. K. M., Armitage, P. J., Bate, M. R., and Bonnell, I. A. (2003a). The effect of cooling on the global stability of self-gravitating protoplanetary discs. *Mon. Not. R. Astron. Soc.*, **339**, 1025–30.

(2003b). Astrometric signatures of self-gravitating protoplanetary discs. *Mon. Not. R. Astron. Soc.*, **338**, 227–32.

Rice, W. K. M., Lodato, G., Pringle, J. E., Armitage, P. J., and Bonnell, I. A. (2004). Accelerated plantesimal growth in self-gravitating protoplanetary discs. *Mon. Not. R. Astron. Soc.*, **355**, 543–52.

Richard, D. (2003). On non-linear hydrodynamic instability and enhanced transport in differentially rotating flows. *Astron. Astrophys.*, **408**, 409–14.

Richard, D. and Zahn, J. (1999). Turbulence in differentially rotating flows. What can be learned from the Couette-Taylor experiment? *Astron. Astrophys.*, **347**, 734–8.

Richling, S. and Yorke, H. W. (1997). Photoevaporation of protostellar disks. II. The importance of UV dust properties and ionizing flux. *Astron. Astrophys.*, **327**, 317–24.

(2000). Photoevaporation of protostellar disks. V. Circumstellar disks under the influence of both extreme-ultraviolet and far-ultraviolet radiation. *Astrophys. J.*, **539**, 258–72.

Rieke, G. H., Young, E. T., Engelbracht, C. W. *et al.* (2004). The Multiband Imaging Photometer for Spitzer (MIPS). *Astrophys. J. Suppl.*, **154**, 25–9.

Rivera, E. J. and Lissauer, J. J. (2001). Dynamical models of the resonant pair of planets orbiting the star GJ 876. *Astrophys. J.*, **558**, 392–402.

Robichon, N. (2002). Detection of Transits of Extrasolar Planets with GAIA. In *GAIA: A European Space Project*, ed. O. Bienaymé and C. Turon. *EAS Publications Series*, **2**, 215–21.

Robichon, N. and Arenou, F. (2000). HD 209458 planetary transits from Hipparcos photometry. *Astron. Astrophys.*, **355**, 295–8.

Rodmann, J., Henning, Th., Chandler, C. J., Mundy, L. G., and Wilner, D. J. (2006). *Astron. Astrophys.*, **446**, 211–21.

Rubin, A. E. and Wasson, J. T. (1995). Variations of chondrite properties with heliocentric distance. *Meteoritics*, **30**, 569.

Ruden, S. P. (2004). Evolution of photoevaporating protoplanetary disks. *Astrophys. J.* **605**, 880–91.

Ruden, S. P. and Pollack, J. B. (1991). The dynamical evolution of the proto Solar Nebula. *Astrophys. J.*, **375**, 740–60.

Rüdiger, G. and Hollerbach, R. (2004). *Geophysical and Astrophysical Dynamo Theory* Chichester: Wiley.

Russell, H. N. (1935). *The Solar System and Its Origin*. New York: Macmillan.

Russell, S. S., Gounelle, M., and Hutchison, R. (2001). Origin of short-lived radionuclides. *Phil. Trans. R. Soc. Lond.*, **A 359**, 1991.

Russell, S. S., Srinivasan, G., Huss, G. R., Wasserburg, G. J. and MacPherson, G. J. (1996). Evidence for widespread $^{26}$Al in the solar nebula and constraints for nebula timescales. *Science*, **273**, 757–62.

Ryu, D. and Goodman, J. (1992). Convective instability in differentially rotating disks. *Science*, **388**, 438–450.

Sackett, P. D. (1999). Searching for unseen planets via occultation and microlensing. *NATO Sci. Ser.*, **532**, 189–227.

Safronov, V. S. (1960). On the gravitational instability in flattened systems with axial symmetry and non-uniform rotation. *Ann. Astrophys.*, **23**, 979–82.

(1969). *Evoliutsiia doplanetnogo oblaka* (English translation: Evolution of the protoplanetary cloud and formation of Earth and the planets, NASA Tech. Transl. F-677, Jerusalem: Israel Sci. Transl. 1972).

Salpeter, E. E. (1955). The luminosity function and stellar evolution. *Astrophys. J.*, **121**, 161.

Sano, T., Miyama, S. M., Umebayashi, T., and Nakano, T. (2000). Magnetorotational instability in protoplanetary disks. II. Ionization state and unstable regions. *Astrophys. J.*, **543**, 486–501.

Sano, T. and Stone, J. (2002a). The effect of the Hall term on the nonlinear evolution of the magnetorotational instability. I. Local axisymmetric simulations. *Astrophys. J.*, **570**, 314–28.

(2002b). The effect of the Hall term on the nonlinear evolution of the magnetorotational instability. II. Saturation level and critical magnetic Reynolds number. *Astrophys. J.*, **577**, 534–53.

Santos, N. C., Bouchy, F., Mayor, M. *et al.* (2004a). The HARPS survey for southern extrasolar planets. II. A 14 Earth-masses exoplanet around $\mu$ Arae. *Astron. Astrophys.*, **426**, L19.

Santos, N. C., Israelian, G., and Mayor, M. (2000). Chemical analysis of eight recently discovered extrasolar planet host stars. *Astron. Astrophys.*, **363**, 228–38.

(2001). The metal-rich nature of stars with planets. *Astron. Astrophys.*, **373**, 1019.

(2004b). Spectroscopic [Fe/H] for 98 extrasolar planet-host stars. Exploring the probability of planet formation. *Astron. Astrophys.*, **415**, 1153.

Saumon, D., Chabrier, G., and van Horn, H. M. (1995). An equation of state for low-mass stars and giant planets. *Astrophys. J. Suppl.*, **99**, 713–41.

Saumon, D. and Guillot, T. (2004). Shock compression of deuterium and the interiors of Jupiter and Saturn. *Astrophys. J.*, **609**, 1170–80.

Scally, A. and Clarke, C. J. (2001). Destruction of protoplanetary discs in the Orion Nebula Cluster. *Mon. Not. R. Astron. Soc.*, **325**, 449–56.

Schmitt, W., Henning, T., and Mucha, R. (1997). Dust evolution in protoplanetary accretion disks. *Astron. Astrophys.*, **325**, 569–84.

Schneider, J. (1994). On the occultations of a binary star by a circum-orbiting dark companion. *Planet. Space Sci.*, **42**, 539–44.

Schneider, J. and Chevreton, M. (1990). The photometric search for Earth-sized extrasolar planets by occultation in binary systems. *Astron. Astrophys.*, **232**, 251–7.

Schöller, M. and Glindemann, A. (2003). The VLT Interferometer - hunting for planets. In *Towards Other Earths: DARWIN/TPF and the Search for Extrasolar Terrestrial Planets*, ed. Fridlund M. and Henning, Th., *ESA*, **SP-539**, 109–20.

Schräpler, R. and Henning, T. (2004). Dust diffusion, sedimentation, and gravitational instabilities in protoplanetary disks. *Astrophys. J.*, **614**, 960–78.

Schultz-Grunow, F. (1959). Zur Stabilitt der Couette-Strmung. *Z. Angewandte Mathematik und Mechanik*, **39**, 101–10.

Schumann, T. E. W. (1940), Theoretical aspects of the size distribution of fog particles. *Quart. J. Roy. Meteorol. Soc.*, **66**, 195.

Scott E. R. D., Love S. G., and Krot A. N. (1996). Formation of chondrules and chondrites in the protoplanetary nebula. In *Chondrules and the Protoplanetary*

*Disk*, ed. R. H. Hewins, R. H. Jones, and E. R. D. Scott. Cambridge: Cambridge University Press.

Seager, S. and Hui, L. (2002). Constraining the rotation rate of transiting extrasolar planets by oblateness measurements. *Astrophys. J.*, **574**, 1004–10.

Sears, D. W. G. (2004). *The Origin of Chondrules and Chondrites*. Cambridge: Cambridge University Press.

Sears D. W. G. and Dodd R. T. (1988). Overview and classification of meteorites. In *Meteorites and the Early Solar System*, ed. J. F. Kerridge and M. S. Matthews. Tucson: University of Arizona Press.

Sekiya, M. and Takeda, H. (2003). Were planetesimals formed by dust accretion in the Solar Nebula? *Earth, Planets, and Space*, **55**, 263–9.

Semenov, D., Henning, T., Helling, C., Ilgner, M., and Sedlmayr, E. (2003). Rosseland and Planck mean opacities for protoplanetary discs. *Astron. Astrophys.*, **410**, 611–21.

Semenov, D., Wiebe, D., and Henning, Th. (2004). Reduction of chemical networks. II. Analysis of the fractional ionisation in protoplanetary discs. *Astron. Astrophys.*, **417**, 93–106.

Shakura, N. I. and Sunayev, R. A. (1973). Black holes in binary systems. Observational appearance. *Astron. Astrophys.*, **24**, 337–55.

Shao, M., ed. (2003). Interferometry in Space. *Proceedings of the SPIE*, **4852**, 2003.

Shao, M. (2004). Science overview and status of the SIM project. *SPIE*, **5491**, 328.

Sheehan, D. P., Davis, S. S., Cuzzi, J. N., and Estberg, G. N. (1999). Rossby wave propagation and generation in the protoplanetary nebula. *Icarus*, **142**, 238–48.

Shmidt, O. (1944). A meteoric theory of the origin of the Earth and planets. *C. R. Dokl. Acad. Sci. URSS*, **45**, 229.

Shu, F. H., Adams, F. C., and Lizano, S. (1987). Star formation in molecular clouds – observation and theory. *Ann. Rev. Astron. Astrophys.*, **25**, 23–81.

Shu, F., Johnstone, D., and Hollenbach, D. (1993). Photoevaporation of the solar nebula and the formation of the giant planets. *Icarus*, **106**, 92–102.

Shu, F. H., Najita, J., Ostriker, E. *et al.* (1994). Magnetocentrifugally driven flows from young stars and disks. I. A generalized model. *Astrophys. J.*, **429**, 781–96.

Shu F. H., Shang H., and Lee T. (1996). Toward an astrophysical theory of chondrites. *Science*, **271**, 1545–52.

Shu, F. H., Tremaine, S., Adams, F. C., and Ruden, S. P. (1990). SLING amplification and eccentric gravitational instabilities in gaseous disks. *Astrophys. J.*, **358**, 495–514.

Shukolyukov, A. and Lugmair, G. W. (1993) Life iron-60 in the early Solar System. *Science*, **259**, 1138–42.

Shuping, R., Bally, J., Morris, M., and Throop, H. (2003). Evidence for grain growth in the protostellar disks of Orion. *Astrophys. J.*, **587**, L109–12.

Siegler, N., Close, L. M., Cruz, K. L., Martin, E. L., and Reid, N. (2005). Discovery of two very low mass binaries: final results of an adaptive optics survey of nearby M6.0-M7.5 stars. *Astrophys. J.*, **621**, 1023–32.

Sigurdsson, S., Richer, H. B., Hansen, B. M., Stairs, I. H., and Thorsett, S. E. (2003). A young white dwarf companion to pulsar B1620-26: evidence for early planet formation. *Science*, **301**, 193–6.

Silk, J. (1977a). On the fragmentation of cosmic gas clouds. II – Opacity-limited star formation. *Astrophys. J.*, **214**, 152–60.

   (1977b). On the fragmentation of cosmic gas clouds. III – The initial stellar mass function. *Astrophys. J.*, **214**, 718–24.

Silverstone, M. D., Meyer, M. R., Mamajek, E. E. *et al.* (2006). The Formation and Evolution of Planetary Systems (FEPS): primordial warm dust evolution from 3–30 Myr around Sun-like stars. *Astrophys. J.*, **639**, 1138–46.

Simon, M., Dutrey, A., and Guilloteau, S. (2000). Dynamical masses of T Tauri stars and calibration of pre-main-sequence evolution. *Astrophys. J.*, **545**, 1034–43.

Sivaramakrishnan, A., Koresko, C. D., Makidon, R. B. et al. (2001). Ground-based coronagraphy with high-order adaptive optics. *Astrophys. J.*, **552**, 397–408.

Slesnick, C. L., Hillenbrand, L. A., and Carpenter, J. M. (2004). The spectroscopically determined substellar mass function of the Orion Nebula cluster. *Astrophys. J.*, **610**, 1045–63.

Smoluchowski, M. (1916). Drei Vorträge über Diffusion, Brownsche Bewegung und Koagulation von Kolloidteilchen. *Physikalische Zeitschrift*, **17**, 557–85.

Sozzetti, A. (2004). On the possible correlation between the orbital periods of extrasolar planets and the metallicity of the host stars. *Mon. Not. R. Astron. Soc.*, **354**, 1194–200.

Sozzetti, A., Casertano, S., Brown, R. A., and Lattanzi, M. G. (2002). Narrow-angle astrometry with the Space Interferometry Mission: The search for extrasolar planets. I. Detection and characterization of single planets. *Pub. Ast. Soc. Pac.*, **114**, 1173–96.

(2003). Narrow-angle astrometry with the Space Interferometry Mission: The search for extrasolar planets. II. Detection and characterization of planetary systems. *Pub. Ast. Soc. Pac.*, **115**, 1072–104.

Spangler, C., Sargent, A. I., Silverstone, M. D., Becklin, E. E., and Zuckerman, B. (2001). Dusty debris around Solar-type stars: temporal disk evolution. *Astrophys. J.*, **555**, 932–44.

Spitzer, L. and Tomasko L. G. (1968). Heating of HI regions by energetic particles. *Astrophys. J.*, **152**, 971–86.

Spruit, H. C. (2001). Accretion disks. In *The Neutron Star – Black Hole Connection (NATO ASI Elounda 1999)*, ed. C. Kouvelitou and V. Connaughton. Dordrecht: Kluwer Academic Publishers.

Squires, K. D. and Eaton, J. K. (1991). Preferential concentration of particles by turbulence. *Phys. Fluids A*, **3**, 1169–78.

Stepinski, T. F. (1992). Generation of dynamomagnetic fields in the primordial solar nebula. *Icarus*, **97**, 130–41.

Stern, S. A. and Weissman, P. R. (2001). Rapid collisional evolution of comets during the formation of the Oort cloud. *Nature*, **409**, 589–91.

Sternberg, A., Hoffmann T. L., and Pauldrach, A. W. A. (2004). Ionizing photon emission rates from O- and early B-type stars and clusters. *Astrophys. J.*, **599**, 1334–43.

Sterzik, M. F., Durisen, R. H., and Zinnecker, H. (2003). How do binary separations depend on cloud initial conditions? *Astron. Astrophys.*, **411**, 91–7.

Sterzik, M. F. and Morfill, G. E. (1994). Evolution of protoplanetary disks with condensation and coagulation. *Icarus*, **111**, 536–46.

Stevenson, D. J. and Lunine, J. I. (1988). Rapid formation of Jupiter by diffuse redistribution of water vapor in the Solar Nebula. *Icarus*, **75**, 146–55.

Stöffler, D., Keil, K., and Scott, E. R. D. (1991). Shock metamorphism of ordinary chondrites. *Geochim. cosmochim. acta*, **55**, 3845–67.

Störzer, H. and Hollenbach, D. (1999). Photodissociation region models of photoevaporating circumstellar disks and application to the proplyds in Orion. *Astrophys. J.*, **515**, 669–84.

Strom, K. M., Strom, S. E., Edwards, S., Cabrit, S., and Skrutskie, M. F. (1989). Circumstellar material associated with solar-type pre-main-sequence stars – a possible constraint on the timescale for planet building. *Astron. J.*, **97**, 1451–70.

Supulver, K. D., Bridges, F. G., and Lin, D. N. C. (1995). The coefficient of restitution of ice particles in glancing collisions: experimental results for unfrosted surfaces. *Icarus*, **113**, 188–99.

Supulver, K. D., Bridges, F. G., Tiscareno, S., Lievore, J., and Lin, D. N. C. (1997). The sticking properties of water frost produced under various ambient conditions. *Icarus*, **129**, 539–54.

Suttner, G. and Yorke, H. W. (1999). Dust coagulation in infalling protostellar envelopes. I. Compact grains. *Astrophys. J.*, **524**, 857–66.

(2001). Early dust evolution in protostellar accretion disks. *Astrophys. J.*, **551**, 461–77.

Tachibana S. and Huss G. R. (2003). The initial abundance of Fe-60 in the Solar System. *Astrophys. J.*, **588**, L41–4.

Tajima, N. and Nakagawa, Y. (1997). Evolution and dynamical stability of the proto-giant-planet envelopes. *Icarus*, **126**, 282–92.

Takeuchi, T. and Lin, D. N. C. (2002), Radial flow of dust particles in accretion disks. *Astrophys. J.*, **581**, 1344–55.

(2003). Surface outflow in optically thick dust disks by radiation pressure. *Astrophys. J.*, **593**, 524–533.

Tanaka, H., Himeno, Y., and Ida, S. (2005). Dust growth and settling in protoplanetary disks and disk spectral energy distributions. I. Laminar disks. *Astrophys. J.*, **625**, 414–26.

Tanaka, H., Takeuchi, T., and Ward, W. R. (2002). Three-dimensional interaction between a planet and an isothermal gaseous disk. I. Corotation and Lindblad torques and planet migration. *Astrophys. J.*, **565**, 1257–74.

Tanga, P., Babiano, A., Dubrulle, B., and Provenzale, A. (1996). Forming planetesimals in vortices. *Icarus*, **121**, 158–70.

Tennekes, H. and Lumley, J. L. (1972). *First Course in Turbulence*. Cambridge: MIT Press.

Testi, L., Natta, A., Shepherd, D. S., and Wilner, D. J. (2003). Large grains in the disk of CQ Tau. *Astron. Astrophys.*, **403**, 323–8.

Testi, L. and Sargent, A. I. (1998). Star formation in clusters: a survey of compact millimeter-wave sources in the Serpens core. *Astrophys. J.*, **508**, L91–4.

Thamm, E., Steinacker, J., and Henning, T. (1994). Ambiguities of parametrized dust disk models for young stellar objects. *Astron. Astrophys.*, **287**, 493–502.

Thi, W. F., Blake, G. A., van Dishoeck, E. F. *et al.* (2001). Substantial reservoirs of molecular hydrogen in the debris disk around young stars. *Nature*, **409**, 60–3.

Thommes, E. W. (2005). A safety net for fast migrators: interactions between gap-opening and sub-gap-opening bodies in a protoplanetary disk. *ApJ*, **626**, 1033–44.

Thommes, E. W., Duncan, M. J., and Levison, H. F. (1999). The formation of Uranus and Neptune in the Jupiter–Saturn region of the Solar System. *Nature*, **402**, 635–8.

(2002). The formation of Uranus and Neptune among Jupiter and Saturn. *Astron. J.*, **123**, 2862–83.

(2003). Oligarchic growth of giant planets. *Icarus*, **161**, 431–55.

Thommes, E. W. and Lissauer, J. J. (2003). Resonant inclination excitation of migrating giant planets. *Astrophys. J.*, **597**, 566–80.

Throop, H. B. and Bally, J. (2005). Can photoevaporation trigger planetesimal formation? *Astrophys. J.*, **623**, L149–52.

Throop, H. B., Bally, J., Esposito, L. W., and McCaughrean, M. J. (2001). Evidence for dust grain growth in young circumstellar disks. *Science*, **292**, 1686–9.

Toomre, A. (1964). On the gravitational stability of a disk of stars. *Astrophys. J.*, **139**, 1217–38.

Torres, G., Konacki, M., Sasselov, D. D., and Jha, S. (2004a). Testing blend scenarios for extrasolar transiting planet candidates. I. OGLE-TR-33: A false positive. *Astrophys. J.*, **614**, 979–89.

(2004b). New data and improved parameters for the extrasolar transiting planet OGLE-TR-56b. *Astrophys. J.*, **609**, 1071–75.

(2005). Testing blend scenarios for extrasolar transiting planet candidates. II. OGLE-TR-56. *Astrophys. J.*, **619**, 558–69.

Trieloff M., Jessberger E. K., Herrwerth I. *et al.* (2003). $^{244}$Pu and $^{40}$Ar-$^{39}$Ar thermochronometries reveal structure and thermal history of the H-chondrite parent asteroid. *Nature*, **422**, 502–6.

Trieloff, M., Kunz, J. and Allègre, C. J. (2002). Noble gas systematics of the Réunion mantle plume source and the origin of primordial noble gases in Earth's mantle. *Earth Planet. Sci. Lett.*, **200**, 297–313.

Trieloff, M., Kunz, J., Claque, D. A. V., Harrison, D., and Allègre, C. J. (2000). The nature of pristine noble gases in mantle plumes. *Science*, **288**, 1036–8.

Trilling, D. E., Lunine, J. I., and Benz, W. (2002). Orbital migration and the frequency of giant planet formation. *Astron. Astrophys.* **394**, 241–251.

Turner G., Enright M. C., and Cadogan P. H. (1978). The early history of chondrite parent bodies inferred from $^{40}$Ar-$^{39}$Ar ages. *Proc. Lunar Planet. Sci. Conf.*, **9th**, 989–1025.

Udalski, A., Paczynski, B., Zebrun, K. *et al.* (2002a). The Optical Gravitational Lensing Experiment. Search for planetary and low-luminosity object transits in the Galactic Disk. Results of 2001 campaign. *Acta astronomica*, **52**, 1–37.

Udalski, A., Pietrzynski, G., and Szymanski, M (2003). The Optical Gravitational Lensing Experiment. Additional planetary and low-luminosity object transits from the OGLE 2001 and 2002 observational campaigns. *Acta astronomica*, **53**, 133–49.

Udalski, A., Szewczyk, O., Zebrun, K. *et al.* (2002). The Optical Gravitational Lensing Experiment. Planetary and low-luminosity object transits in the Carina fields of the Galactic Disk. *Acta astronomica*, **52**, 317–59.

Udalski, A., Szymanski, M. K., Kubiak, M. *et al.* (2004). The Optical Gravitational Lensing Experiment. Planetary and low-luminosity object transits in the fields of the Galactic Disk. Results of the 2003 OGLE observing campaigns. *Acta astronomica*, **54**, 313.

Udalski, A., Zebrun, K., Szymanski, M. *et al.* (2002b). The Optical Gravitational Lensing Experiment. Search for planetary and low-luminosity object transits in the Galactic Disk. Results of 2001 campaign – Supplement. *Acta astronomica*, **52**, 115–28.

Udry, S., Mayor, M. and Santos, N. C. (2003). Statistical properties of exoplanets. I. The period distribution: constraints for the migration scenario. *Astron. Astrophys.*, **407**, 369.

Umebayashi, T. (1983). The densities of charged particles in very dense interstellar clouds. *Prog. Theor. Phys.*, **69**, 480–502.

Umebayashi, T. and Nakano, T. (1981). Fluxes of energetic particles and the ionization rate in very dense interstellar clouds. *PASJ*, **33**, 617–35.

Urey, H. C. (1951). The origin and development of the Earth and other terrestrial planets. *Geochim. cosmochim. acta*, **1**, 209; **2**, 263.

Urpin, V. A. (1984). Hydrodynamic flows in accretion disks. *Sov. Astron.*, **28**, 50.

Urpin, V. A. and Brandenburg, A. (1998). Magnetic and vertical shear instabilities in accretion discs, *Mon. Not. R. Astron. Soc.*, **294**, 399–406.

Völk, H. J., Jones, F. C., Morfill, G. E., and Röser, S. (1980), Collisions between grains in a turbulent gas. *Astron. Astrophys.*, **85**, 316–25.

Valenti, J. A. and Fischer, D. A. (2005). Spectroscopic properties of cool stars. I1040 F, G, and K dwarfs. *Astrophys. J., Suppl. Ser.* **159**, 141–66.

van Albada, T. S. (1968). Numerical integrations of the N-body problem. *Bull. Astron. Inst. Neth.*, **19**, 479.

van Boekel, R., Min, M., Leinert, C. *et al.* (2004). The building blocks of planets within the 'terrestrial' region of protoplanetary disks. *Nature*, **432**, 479–82.

van Boekel, R., Waters, L. B. F. M., Dominik, C. *et al.* (2003). Grain growth in the inner regions of Herbig Ae/Be star disks. *Astron. Astrophys.*, **400**, L21–4.

van de Kamp, P. (1963). Astrometric study of Barnard's star from plates taken with the 24-inch Sproul refractor. *Astron. J.*, **68**, 515–21.

(1975). Astrometric study of Barnard's star from plates taken with the Sproul 61-cm refractor. *Astron. J.*, **80**, 658–61.

van Schmus W. R. and Wood J. A. (1967). A chemical-petrologic classification for the chondritic meteorites. *Geochim. cosmochim. acta*, **31**, 747–65.

Vance, S. (2003). Methanol sticking and the accretion of centimeter-sized particles in the Solar Nebula. *Met. and Planet. Sci.*, **38**, 5273.

Velikhov, E. (1959). Stability of an ideally conducting liquid flowing between cylinders rotating in a magnetic field. *Sov. Phys. Jetp*, **9**, 995-998.

Vidal-Madjar, A., Désert, J.-M., Lecavelier des Étangs, A. *et al.* (2004). Detection of oxygen and carbon in the hydrodynamically escaping atmosphere of the extrasolar planet HD209458b. *Astron. Astrophys.*, **604**, L69–72.

Vidal-Madjar, A., Ferlet, R., Hobbs, L. M., Gry, C., and Albert, C. E. (1986). The circumstellar gas cloud around Beta Pictoris. II. *Astron. Astrophys.*, **167**, 325–32.

Vidal-Madjar, A. and Lecavelier des Etangs, A. (2004). "Osiris" (HD 209458b), an evaporating planet, *APS Conf. Ser.*, **321**, 152–9.

Vidal-Madjar, A., Lecavelier des Étangs, A., Désert, J.-M. *et al.* (2003). An extended upper atmosphere around the extrasolar planet HD209458b. *Nature*, **422**, 143–6.

Vidal-Madjar, A., Lecavelier des Etangs, A., and Ferlet, R. (1998). $\beta$ Pictoris, a young planetary system? A review. *Planet. Space Sci.*, **46**, 629–48.

Vogt, S. S., Butler, R. P., Marcy, G. W. *et al.* (2005). Five new multi-component planetary systems. *Astrophys. J.*, **632**, 638–58.

Vogt, S. S., Marcy, G. W., Butler, R. P., and Apps, K. (2000). Six new planets from the Keck Precision Velocity Survey. *Astrophys. J.*, **536**, 902–14.

Ward, W. R. (1986). Density waves in the Solar Nebula – differential Lindblad torque. *Icarus*, **67**, 164–80.

(1991). Horseshoe orbit drag. *Lunar and Planetary Institute Conference Abstracts*, 1463.

(1992). Coorbital corotation torque. *Lunar and Planetary Institute Conference Abstracts*, 1491.

(1997). Protoplanet migration by nebula tides. *Icarus*, **126**, 261–81.

Wardle, M. (1999). The Balbus–Hawley instability in weakly ionized discs. *Mon. Not. R. Astron. Soc.*, **307**, 849–56.

Wasserburg, G. J., Gallino, R., and Busso, M. (1998). A test of the Supernova Trigger Hypothesis with $^{60}$Fe and $^{26}$Al. *Astrophys. J.*, **500**, L189.

Watson, P. K., Mizes, H., Castellanos, A., and Perez, A. (1997). The packing of fine, cohesive powders. In *Powders and Grains 97*, ed. R. Behringer and J. T. Jenkins. Rotterdam: Balkema, 109–112.

Wehrstedt, M. and Gail, H.-P. (2002). Radial mixing in protoplanetary accretion disks. II. Time dependent disk models with annealing and carbon combustion. *Astron. Astrophys.*, **385**, 181–204.

Weidenschilling, S. J. (1977). Aerodynamics of solid bodies in the Solar Nebula. *Mon. Not. R. Astron. Soc.*, **180**, 57–70.

(1980). Dust to planetesimals – settling and coagulation in the Solar Nebula. *Icarus*, **44**, 172–89.

(1984). Evolution of grains in a turbulent Solar Nebula. *Icarus*, **60**, 553–67.

(1997a). Planetesimals from stardust. *ASP Conf. Ser. 122: From Stardust to Planetesimals*, 281.

(1997b). The origin of comets in the Solar Nebula: A unified model. *Icarus*, **126**, 290.

Weidenschilling, S. J. and Cuzzi, J. N. (1993). Formation of planetesimals in the Solar Nebula. In *Protostars and Planets III*, ed. E. H. Levy and J. I. Lunine, Tucson: University of Arizona Press, 1031–60.

Weidenschilling, S. and Ruzmaikina, T. (1994). Coagulation of grains in static and collapsing protostellar clouds. *Astrophys. J.*, **430**, 713–26.

Weinberger, A. J., Becklin, E. E., Schneider, G. *et al.* (2002). Infrared views of the TW Hydra disk. *Astrophys. J.*, **566**, 409–18.

Weissman, P. R. (1984). The VEGA particulate shell – comets or asteroids? *Science*, **224**, 987–9.

Weizsäcker, C. F. von (1944). Über die Entstehung des Planetensystems. *Z. Astrophys.*, **22**, 319. Translation, "On the origin of the planetary system" Report No. RSIC-138 [=AD-4432290] Huntsville AL: Redstone Scientific Information Center.

Wendt, F. (1933). Turbulente Stroemungen zwischen zwei rotierenden konaxialen Zylindern. *Ingenieur-Archiv*, **4**, 577.

Werner, M. W., Roellig, T. L., Low, F. J. *et al.* (2004). The Spitzer Space Telescope mission. *Astrophys. J. Suppl.*, **154**, 1–9.

Wetherill, G. W. (1996). The formation and habitability of extra-solar planets. *Icarus*, **119**, 219-38.

(1980). Formation of the terrestrial planets. *Ann. Rev. Astron. Astrophys.*, **18**, 77–113.

Wetherill, G. W. and Stewart, G. R. (1989). Accumulation of a swarm of small planetesimals. *Icarus*, **77**, 330–57.

(1993). Formation of planetary embryos – effects of fragmentation, low relative velocity, and independent variation of eccentricity and inclination. *Icarus*, **106**, 190.

Whipple, F. L. (1972). On certain aerodynamic processes for asteroids and comets. In *From Plasma to Planet*, ed. Elvius, A. New York: Wiley Interscience Division, 211.

Whitworth, A. P. and Zinnecker, H. (2004). The formation of free-floating brown dwarves and planetary-mass objects by photo-erosion of prestellar cores. *Astron. Astrophys.*, **427**, 299–306.

Wilden, B. S., Jones, B. F., Lin, D. N. C., and Soderblom, D. R. (2002). Evolution of the lithium abundance of Solar-type stars. X. Does accretion affect the lithium dispersion in the Pleiades? *Astron. J.*, **124**, 2799–812.

Wilde, S. A., Valley, J. W., Peck, W. H., and Graham C. M. (2001). Evidence from detrital zircons for the existence of continental crust and oceans on the Earth 4.4 Gyr ago. *Nature*, **409**, 175–8.

Winn, J. N., Holman, M. J., Johnson, J. A., Stanek, K. Z., and Garnavich, P. M. (2004). KH 15D: gradual occultation of a pre-main-sequence binary. *Astrophys. J.*, **603**, L45–8.

Wittenmyer, R. A., Welsh, W. F., Orosz, J. A. *et al.* (2004). The orbital ephemeris of HD 209458b. *APS Conf. Ser.*, **321**, 215–16.

Wolf, S. and D'Angelo, G. (2005). On the observability of giant protoplanets in circumstellar disks. *Astrophys. J.*, **619**, 1114–22.

Wolf, S., Gueth, F., Henning, T., and Kley, W. (2002). Detecting planets in protoplanetary disks: a prospective study. *Astrophys. J.*, **566**, L97–9.

Wolf, S., Padgett, D. L., and Stapelfeldt, K. R. (2003). The circumstellar disk of the Butterfly star in Taurus. *Astrophys. J.*, **588**, 373–86.

Wolszczan, A. (1993). PSR1257 + 12 and its planetary companion. In *Planets around Pulsars*, ASP Conference Series, Vol. 36, ed. J. A. Phillips, S. E. Thorsett, and S. R. Kulkarni. San Francisco: ASP.

Wolszczan, A. and Frail, D. A. (1992). A planetary system around the millisecond pulsar PSR1257 + 12. *Nature*, **355**, 145–147.

Wood J. A. and Pellas P. (1991). What heated the parent meteorite planets? In *The Sun in Time*, ed. C. P. Sonett, M. S. Giampapa, and M. S. Matthews, Tucson: University of Arizona Press, 741–760.

Wood, K., Wolff, M. J., Bjorkman, J. E., and Whitney, B. (2002). The spectral energy distribution of HH 30 IRS: constraining the circumstellar dust size distribution. *Astrophys. J.*, **564**, 887–95.

Wooden D. H., Butner, H. M., Harker, D. E., and Woodward, C. E. (2000). Mg-rich silicate crystals in comet Hale-Bopp: ISM relics or Solar Nebula condensates? *Icarus*, **43**, 126.

Wooden, D. H., Harker, D. E., Woodward, C. E. *et al.* (1999). Silicate mineralogy of the dust in the inner coma of Comet C/1995 01 (Hale-Bopp) pre- and postperihelion. *Astrophys. J.*, **517**, 1034–58.

Wright, J. T., Marcy, G. W., Butler, R. P., and Vogt, S. S. (2004). Chromospheric Ca II emission in nearby F, G, K, and M stars. *Astrophys. J. Suppl.*, **152**, 261–95.

Wuchterl, G. (1991a). Hydrodynamics of giant planet formation II: Model equations and critical mass. *Icarus*, **91**, 39–52.

(1991b). Hydrodynamics of giant planet formation III: Jupiter's nucleated instability. *Icarus*, **91**, 53–64.

(1993). The critical mass for protoplanets revisited: massive envelopes through convection. *Icarus*, **106**, 323–34.

(1995). Giant planet formation: a comparative view of gas accretion. *Earth, Moon and Planets*, **67**, 51–65.

(1999). Convection and giant planet formation. In *Theory and Tests of Convection in Stellar Structure*, ASP Conference Series, Vol. 173, ed. Gimènez, A., Guinan, E. F., and Montesinos, B. San Francisco: Astronomical Society of the Pacific. 185–8.

Wuchterl, G., Guillot, T., and Lissauer, J. J. (2000). Giant planet formation. In *Protostars and Planets IV*, ed. Mannings, V., Boss, A. P., and Russell, S. Tucson: University of Arizona Press, 1081–109.

Wünsch, R., Klahr, H. H. and Różyczka, M. N. (2005). 2-D models of layered protoplanetary disks: I. The ring instability. *Mon. Not. R. Astron. Soc.*, **362**, 361–8.

Wurm, G. (2003). The formation of terrestrial planets. In *Towards Other Earths: DARWIN/TPF and the Search for Extrasolar Terrestrial Planets*, ed. Fridlund, M. and Henning, Th. ESA Special Publication SP-539, Heidelberg: ESA, 151–61.

Wurm, G. and Blum, J. (1998). Experiments in preplanetary dust aggregation. *Icarus*, **132**, 125–36.

Wurm, G. and Blum, J. (2000). An experimental study on the structure of cosmic dust aggregates and their alignment by motion relative to gas. *Astrophys. J.*, **529**, L57–60.

Wurm, G., Blum, J., and Colwell, J. E. (2001a). A new mechanism relevant to the formation of planetesimals in the Solar Nebula. *Icarus*, **151**, 318–21.

(2001b). Aerodynamical sticking of dust aggregates. *Phys. Rev. E*, **64**, 046301–9.

Wurm, G., Paraskov, G., and Krauß, O. (2004a). On the importance of gas flow through porous bodies for the formation of planetesimals. *Astrophys. J.*, **606**, 983–7.

(2005a). Ejection of dust by elastic waves in collisions between millimeter- and centimeter-sized dust aggregates at 16.5 to 37.5 m/s impact velocities. *Phys. Rev. E*, **71**, 021304 1–10.

(2005b). Growth of planetesimals by impacts at 25m/s. *Icarus*, **178**, 253–63.

Wurm, G., Relke, H., and Dorschner, J. (2003). Experimental study of light scattering by large dust aggregates consisting of micron-sized $SiO_2$ monospheres. *Astrophys. J.*, **595**, 891–9.

Wurm, G., Relke, H., Dorschner, J., and Krauß, O. (2004b). Light scattering experiments with micron-sized dust aggregates: results on ensembles of $SiO_2$ monospheres and of irregularly shaped graphite particles. *J. Quant. Spectros. Rad. Trans.*, **89**, 371–384.

Wurm, G. and Schnaiter, M. (2002). Coagulation as unifying element for interstellar polarization. *Astrophys. J.*, **567**, 370–5.

Yin, Q., Jacobsen, S. B., Yamashita, K. *et al.* (2002). A short timescale for terrestrial planet formation from Hf-W chronometry of meteorites. *Nature*, **418**, 949–52.

Yorke, H. and Bodenheimer, P. (1999). The formation of protostellar disks. III. The influence of gravitationally induced angular momentum transport on disk structure and appearance. *Astrophys. J.*, **525**, 330–42.

Yorke, H.W, Bodenheimer P., and Laughlin G. (1995). The formation of protostellar disks II. Disks around intermediate-mass stars. *Astron. Astrophys.*, **443**, 199–208.

Yorke, H. W. and Welz, A. (1996). Photoevaporation of protostellar disks. I. The evolution of disks around early B stars. *Astron. Astrophys.*, **315**, 555–64.

Youdin, A. and Chiang, E. (2004). Particle pileups and planetesimal formation. *Astrophys. J.*, **601**, 1109–19.

Youdin, A. N. and Shu, F. H. (2002). Planetesimal formation by gravitational instability. *Astrophys. J.*, **580**, 494–505.

Young, E. D. (2001). The hydrology of carbonaceous chondrite parent bodies and the evolution of planet progenitors. *Phil. Trans. R. Soc. Lond.*, **A 359**, 2095.

Zinner, E. and Göpel, C. (2002). Aluminum-26 in H4 chondrites: implications for its production and its usefulness as a fine-scale chronometer for early Solar System events. *Met. Planet. Sci.*, **37**, 1001–13.

Zucker, S. and Mazeh, T. (2001). Derivation of the mass distribution of extrasolar planets with MAXLIMA, a maximum likelihood algorithm. *Astrophys. J.*, **562**, 1038–44.

# Index

abundances, element 64, 214, 257
Alfven, Hannes 3
ambipolar diffusion 42, 50
angular momentum transport 2, 46, 122, 258
asteroids 66, 78, 166, 257
asteroid belt 25, 28
astrometric detection 11, 160, 185, 196, 201, 250, 252
Atacama Large Millimeter Array (ALMA) 201, 253

Barnard's star 10
Beta Pictoris 153, 157
Biomarker 161, 255
brown dwarfs
    accretion-ejection scenario for 243
    evaporation scenario for 246
    metallicity of 237
    minimum mass of 238
    formation of 36, 241
    binary 240
Brownian motion 95, 119

calcium–aluminum-rich inclusions (CAIs) 67, 257
cosmochemistry 64, 131, 258
Cameron, Andrew G.W. 4
chondrules 61, 67, 122, 257
Chthonian 160
circumstellar disks (*see also* protoplanetary disk)
cluster–cluster aggregation (CCA) 94
coagulation 39, 112
    equation of 94, 125
CORALIE instrument 203
core accretion, gas capture model
    codes for 169
    critical core mass in 8, 138, 166
    crossover mass in 172
    history of 165
    sequence for 164
    theory of 6, 130, 163, 184, 192
core formation
    gas giants 129
    ice giants 141
    masses in 198
    terrestrial planets 84

CoRoT mission 152, 206, 253
comets 15, 20, 58, 66
    extrasolar 157

Darwin Mission 190, 201, 254
Dead Zone 51
debris disks 20, 25, 27
Descartes, Rene 1, 165
deuterium burning 237, 249
disk instability model (*see* gravitational instability
    model of planet formation) 3, 163, 192
disk time scales
    observed 22, 26
    viscous 46
    photoevaporation 36
    thermal 47
    dynamical 47
Doppler method (*see* radial velocity technique) 11,
    179, 252, 203
dust aggregation experiment (CODAG) 95
dust–gas interaction 107
dust grain collisions 91
    compaction by 101
    cross sections for 94, 118
    grain charging by 96, 106, 109
    restructuring by 101
    velocity of 101, 118
dust particles
    adhesion of 91, 94
    alignment of 96
    depletion of 127
    dynamics of 58
    growth of 120
    polarization by 117
    porosity of 93
    radial drift of 122
    settling of 58, 120, 138, 173
    trapping of 59

ELODIE instrument 203
extrasolar planets 168, 179, 203
    atmospheres of 147
    earth-mass 186, 193, 205, 224